D1000802

Springer Tracts in Modern Physics
Volume 184

Now also Available Online

Starting with Volume 165, Springer Tracts in Modern Physics is part of the [SpringerLink] service. For all customers with standing orders for Springer Tracts in Modern Physics, we offer the full text in electronic form via [SpringerLink] free of charge. Please contact our librarian who can receive a password for free access to the full articles by registration at:

http://link.springer.de/series/stmp/reg_form.htm

If you do not have a standing order, you can nevertheless browse through the table of contents of the volumes and the abstracts of each article at:

http://link.springer.de/series/stmp/

There you will also find more information about the series.

Springer
New York
Berlin
Heidelberg
Hong Kong
London
Milan
Paris
Tokyo

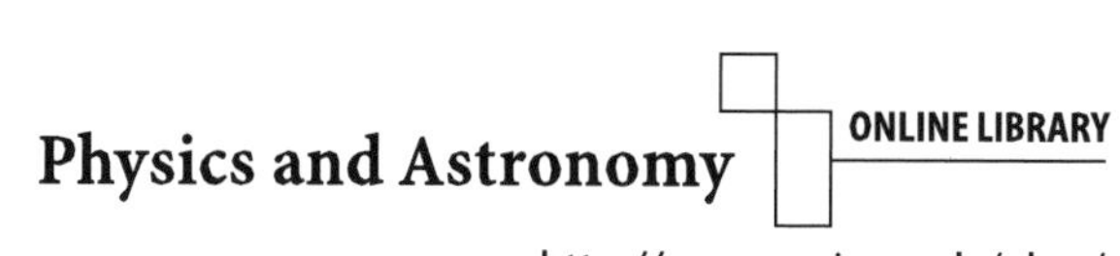

http://www.springer.de/phys/

Springer Tracts in Modern Physics

Springer Tracts in Modern Physics provides comprehensive and critical reviews of topics of current interest in physics. The following fields are emphasized: elementary particle physics, solid-state physics, complex systems, and fundamental astrophysics.
Suitable reviews of other fields can also be accepted. The editors encourage prospective authors to correspond with them in advance of submitting an article. For reviews of topics belonging to the above mentioned fields, they should address the responsible editor, otherwise the managing editor.
See also http://www.springer.de/phys/books/stmp.html

Michael Schulz

Statistical Physics and Economics

Concepts, Tools, and Applications

With 54 Figures

Michael Schulz
Abteilung Theoretische Physik
Universität Ulm
Albert-Einstein-Allee 11
89069 Ulm
Germany
michael.schulz@physik.uni-ulm.de

Physics and Astronomy Classification Scheme (PACS): 89.65.Gh, 89.75.Fb, 89.75.Da

Library of Congress Cataloging-in-Publication Data
Schulz, Michael, 1959 Jan. 8–
Statistical physics and economics : concepts, tools, and applications / Michael Schulz.
p. cm. — (Springer tracts in modern physics ; 184)
Includes bibliographical references and index.
ISBN 0-387-00282-0 (alk. paper)
1. Economics—Statistical methods. 2. Statistical physics. I. Title. II. Series.
QC1.S797 vol. 184
[HB137]
539 s—dc21
[330'.01'519] 2002044504

ISBN 0-387-00282-0 Printed on acid-free paper.

Printed in the United States of America.

9 8 7 6 5 4 3 2 1 SPIN 10905258

Typesetting: Pages created by the author using a Springer LaTeX macro package.

www.springer-ny.com

Springer-Verlag New York Berlin Heidelberg
A member of BertelsmannSpringer Science+Business Media GmbH

Preface

Econophysics describes phenomena in the development and dynamics of economic systems by using of a physically motivated methodology. First of all, Mandelbrot had analyzed economic and social relations in terms of modern statistical physics. Since then, the number of publications related to this topic has increased irresistible greatly. To be fair to this historical evolution, I point out, however, that physical and economic concepts had already been connected long ago. Terms such as work, power, and efficiency factor have similar physical and economic meanings. Many physical discoveries – for instance in thermodynamics, optics, solid state physics, or chemical physics – correspond to a parallel evolution in the fields of technology and economics.

The term econophysics, or social physics, also is not a recent idea. For example, in the small book *Sozialphysik* published in 1925 [221], R. Lämmel demonstrates how social and economic problems can be understood by applying simple physical relations. Of course, the content of early social physics and the topics of modern econophysics are widely different. Nevertheless, the basic idea (i.e., the description and the explanation of economic phenomena in terms of a physical theory) did not change over the whole time. At this point, an important warning should be pronounced. Econophysics is no substitute for economics. An economic theory differs essentially from what we understand as econophysics. Of course, a short definition of economics is not very simple, even for seasoned economists. A possible working definition may be: Economics is the study of how people choose to use scarce or limited productive resources to produce various commodities and distribute them to various members of society for their consumption. This definition suggests the large variety of disciplines combined under the general term economics: microeconomics, controlling, macroeconomics, finance, environmental economics, and many other scientific branches are usually considered a part of economics.

From this short characterization of economics, it is obvious that the aims of economic investigations and physical research are strongly different. Therefore, the question remains how physical knowledge may contribute to progress in the understanding of the dynamics of economic systems. As mentioned above, it is not the aim of econophysics to replace some or all of the traditional and modern economic sciences by new, physically motivated theories and methods. The key to answering the question is given by two essential terms: the methodology of physics and the statistical physics of complex systems.

The successful evolution of physics during the last three centuries rests on its methodology, which can certainly be described as being analytical. This means that by decomposing a system into its parts, a physicist may try to understand the properties of the whole system.

In particular, the physical experiment plays a central role during the formation of new physical knowledge. Especially, the reproduction of the results in the course of a well-defined experiment backs up physical theories. A well-established theory then allows predictions about very complicated systems that were never analyzed before by physical methods or that cannot be investigated by physical experiments. A traditional example is astronomy. The motion of planets may be observed with sufficiently complicated instruments, but these observations are not reproducible experiments in a physical sense. On the other hand, gravitational law can be checked by various lab experiments. With knowledge of gravitation theory and starting from well-defined initial conditions, we are able to calculate the motion of planets over a sufficiently long period in agreement with astronomical observations.

A similar situation occurs also for complex systems. General evolution laws, limit probability distribution functions, and universal properties may be checked experimentally for simple systems and allow us to formulate a general theory. If we have obtained such a suitable theory about the behavior of several complex systems, we may use this knowledge also for the analysis of more complicated systems.

We should be aware that the degree of complexity of the economic world is extremely high, which means that usually it is not possible to make economic observations under the controlled experimental conditions characteristic of scientific laboratories. As a result of this limitation, the quantitative economic knowledge is far from complete. However, econophysics may give a consequent basis for the interpretation of the structure and dynamics of economic systems or subsystems such as financial markets or national economies.

The main goal of the book is to present some of the most useful theoretical concepts and techniques for understanding the physical ideas behind the evolution and the dynamics of economic systems. But it should be remarked that the concepts and tools presented are also relevant to a much larger class of problems in the natural sciences as well as in the social and medical sciences. The only condition is that the underlying systems be classified as sufficiently complex. From this point of view, the mathematical background and the general theoretical concept used for the analysis of economic systems may be helpful also for the description of social systems, biological organisms, populations, communication networks, biological evolution processes, meteorology, turbulence, granular matter, epidemics, the geosciences, and so on.

The central theme of the book is that of collective and cooperative properties in the behavior of economic units, such as firms, markets, and consumers. It is very important to understand these properties as a consequence of the interaction of a large number of degrees of freedom. This fact allows us to describe economic phenomena using modern physical concepts, such as deterministic chaos, self-organization, scaling laws, renormalization group techniques, and complexity, but also traditional ideas of fluctuation theories, response theory, disorder, and non-reproducibility.

Obviously, an applicable description of a complex system requires the definition of a set of relevant degrees of freedom. The price one has to pay is that one gets practically no information about the remaining irrelevant degrees of freedom. As a consequence, the theoretical basis used for the analysis of economic processes can be described as a probabilistic theory. The more or less empirical specification of the relevant and irrelevant degrees of freedom is influenced by the scales in mind. Characteristic physical scales are time and length scales. In economics, an additional scale, the so-called price scale, has often been taken into account. Econophysics focuses its attention on the description of economic problems in terms of various scales. These scales of interest determine the choice of the relevant degrees of freedom and the mathematical method for solving the underlying problem.

The first two chapters cover important notations of complex systems and the statistical physics of out-of-equilibrium systems considering the dominant scales and the relevant degrees of freedom, respectively. The mathematics is presented in a simple and intuitive way whenever possible with respect to the mathematical rigor.

The third chapter deals with problems related to financial markets. Although finance and financial mathematics offer a large number of different concepts and mathematical instruments to solve various practical problems, the physical concept presented provides a way to derive the complicated, partially anomalous fluctuations of stock prices and exchange rates from general, universal laws. Additionally, this chapter extends the mathematical and physical tools, for instance by introducing the concept of the renormalization group approach, the generalized central limit theorem, and the theory of large fluctuations.

The fourth chapter considers economic problems that are not directly connected with the dynamics of markets. Microeconomics, the limitation of thermodynamic concepts in the economy, environmental economics, and macroeconomics are discussed in terms of deterministic chaos, stability theory, scaling laws, field theories, and self-organized criticality.

In the subsequent chapter, several numerical methods used for the solution of economic problems are discussed and compared with similar physical techniques. Especially, various kinds of Monte Carlo simulations (dynamical, reversed, and quasi-Monte Carlo) and cellular automaton theories will be introduced.

Contents

1. Economy and Complex Systems

1.1 What Is a Complex System?

The aim of this book is to develop concepts and methods that allow us to deal with economic systems from a unifying physically motivated point of view. An economy is usually classified as a manifestation of complex social systems. To clarify this statement, we have to discuss what we mean by complex systems. Unfortunately, an exact definition of complex systems is still an open problem. In a heuristic manner, we may describe them as:

Complex systems are composed of many particles, or objects, or elements that may be of the same or different kinds. The elements may interact in a more or less complicated fashion by more or less nonlinear couplings.

In order to give this formal definition a physical context, we should qualitatively discuss some typical systems that may be denoted truly complex.

The various branches of science offer us numerous examples, some of which turn out to be rather simple, whereas others may be called truly complex. Let us start with a simple physical example. Granular matter is composed of many similar granules. Shape, position, and orientation of the components determine the stability of granular systems. The complete set of the particle coordinates and of all shape parameters defines the actual structure.

Furthermore, under the influence of external force fields, the granules move around in quite an irregular fashion, whereby they perform numerous more or less elastic collisions with each other.

A driven granular system is a standard example of a complex system. The permanent change of the structure due to the influence of external fields and the interaction between the components is a characteristic feature of complex systems.

Another standard complex system is Earth's climate, encompassing all components of the atmosphere, biosphere, cryosphere, and oceans and considering the effects of extraterrestrial processes such as solar radiation and tides.

Computers and information networks are interpreted as a new class of complex systems. This is especially so with respect to hardware dealing with artificial intelligence, where knowledge and learning processing will be replacing the standard algebra of logic.

In biology, we are again dealing with complex systems. Each higher animal consists of various strongly interacting organs with an enormous number of complex functions. Each organ contains many partially very strong specialized cells that cooperate in a well-regulated fashion.

Probably the most complex organ is the human brain, composed of 10^{11} nerve cells. Their collective interaction allows us to recognize visual and acoustic patterns, to speak, or to perform other mental functions. Each living cell is composed of a complicated nucleus, ribosomes, mitochondria, membranes, and other constituents, each of which contain many further components. At the lowest level, we observe many simultaneously acting biochemical processes, such as the duplication of DNA sequences or the formation of proteins.

This hierarchy can also be continued in the opposite direction. Animals themselves form different kinds of societies. Probably the most complex system in our world is the global human society, especially the economy, with its numerous participants (such as managers, employers, and consumers) its capital goods (such as machines, factories, and research centers), its natural resources, its traffic, and its financial systems, which provides us with another large class of complex systems. Economic systems are embedded in the more comprehensive human societies, with their various human activities and their political, ideological, ethical, cultural, or communicative habits.

All of these systems are characterized by permanent structural changes on various scales, indicating an evolution far from thermodynamic equilibrium. The basic constituents of each complex system are its elements, which are expected to have only a few degrees of freedom.

Starting from this microscopic level, we wish to explain the evolution of a complex system at larger scales. Probably, we would arrive at a quite different description of the system at the macroscopic level. Definitely, we have to deal with two problems. First, we have to clarify the scales of interest, and then we have to show that the more or less chaotic motion of the elements contributes to pronounced collective phenomena at macroscopic scales.

The definition of correct microscopic scales as well as suitable macroscopic scales may sometimes be an ambiguous problem. For instance, in ecology we deal with a hierarchy of levels that range from the molecular level through that of animals and humans to that of economic systems and societies. Formally, we can start from the quantum-mechanical level or alternatively from a microscopic, classical many-body system. But in order to describe a complex system at this ultimately microscopic level, we need an enormous amount of information, which nobody is able to handle.

A macroscopic description allows a strong compression of data so that we are no longer concerned with the microscopic motion but rather with properties at large scales. The appropriate choice of the macroscopic level is by no means a trivial problem. It depends strongly on the question in mind. In order to deal with complex systems, we quite often still have to find adequate variables or relevant quantities to describe the properties of these systems. Each macroscopic system contains a set of usually collective large-scale quantities that may be of interest for the underlying problem. We will denote such degrees of freedom as relevant quantities. The knowledge of these quantities permits the characterization of a special feature of the complex system at the macroscopic level. For instance, investigations of financial markets are based mainly on the analysis of temporally variable asset prices. Thus, we may classify these prices as relevant quantities. However, the giant system of the global economy, with its very complex evolution, is hidden behind these price fluctuations. All of these microscopic or macroscopic well-founded degrees of freedom form the huge set of irrelevant observables for the relatively small group of relevant quantities.

The second problem in treating complex systems consists in establishing relations that allow some predictions about the future evolution of the relevant quantities. Unfortunately, the motions of the irrelevant and relevant degrees of freedom of a complex system are normally coupled strongly together. Therefore, an accurate prediction of future values of the relevant degrees of freedom includes automatically the determination of the accurate evolution of the irrelevant degrees of freedom.

This strategy is naturally a hopeless attempt, independent from the underlying complex system. But a formal possibility to estimate an upper boundary for the total number of degrees of freedom is nevertheless offered. To this aim, we may isolate at a certain time the complex system, including a sufficiently large part of its environment. The momentary state of the isolated system is now considered as an initial condition for further evolution. If the further development of the relevant degrees of freedom runs as it would in the open system, the isolated system is sufficiently large for a quantitative description of the contained complex system. From a mathematical point of view, the evolution of the complex system is then embedded in a well-defined initial problem.

Before we start with the mathematical treatment of complex (especially economic) systems, let us now try to define them more rigorously. The question of whether a system is complex or simple, depends strongly on the level of scientific knowledge. An arbitrary system of linear coupled oscillators is today an easily solvable problem. In the lifetime of Galileo, without knowledge of the theory of linear differential equations, one surely would have classified this problem as a complex system in the context of our definition specified above.

A modern definition that is independent of the actual mathematical level is based on the concept of algebraic complexity. To this aim, we must introduce a universal computer that can solve any mathematically reasonable problem after a finite time with a program of finite length.

Without going into details, we point out that such a universal computer can be constructed, at least in a thought experiment as was shown by Turing. Of course, there exist different programs that solve the same problem. As a consequence of number theory, the lengths of the programs solving a particular problem have a lower boundary. This minimum length may be used as a universal measure of the algebraic degree of complexity. Unfortunately, this meaningful definition raises another problem. As can be shown by means of a famous theorem by Gödel, the problem of finding a minimum program cannot be solved in a general fashion. In other words, we must estimate the complexity of a system in an intuitive way and led by the level of scientific knowledge.

The concept of complexity is a very versatile one. In this book, we want to show how various economic problems can be described by methods belonging to quite different disciplines of the physics of complex systems. At the same time, we will see that at a sufficiently abstract level there exist profound analogies between the behavior of economic systems and other complex systems.

1.2 Determinism Versus Chaos

We want to come now to the mathematical treatment of an arbitrary complex system. We proceed from the global system discussed as a thought experiment in the previous section. This system includes both the complex system and a sufficiently large part of its environment. The exact determination of all time-dependent relevant quantities implies the solution of the complete set of microscopic equations of motion of the global system.

The formally complete predictability of the future evolution of the embedded complex system is a consequence of the deterministic Newtonian mechanics. In the sense of classical physics, determinism means that the trajectories of all particles can be computed if their momentum and positions are known at an initial time. The deterministic principle has been shaken twice in modern physics.

First, quantum mechanics tells us that we are not able to make accurate predictions of the trajectory of a particle. However, we can argue that the deterministic character is still conserved as a property of the wave functions. Second, the theory of deterministic chaos has shown that even in classical mechanics predictability cannot be guaranteed without absolutely precise knowledge of the microscopic configuration of the complete global system.

Chaos is not observed in linear systems. A mechanical system is linear if the output from the system is proportional to the input. Mathematically, the signature of a linearity is the superposition principle, which states that the sum of two solutions of the mechanical equations describing the system is again a solution. The theory of linear mechanical systems is fully understood except for some technical problems. The breakdown of the linearity, and therefore the breakdown of the superposition principle, is a necessary condition for the behavior of a nonlinear mechanical system to appear chaotic.

However, nonlinearity alone is not sufficient for the formation of a chaotic regime. For instance, the equation of a simple pendulum is a nonlinear one. The solutions are elliptic functions without any kind of apparent randomness or irregularity. Solitons are further examples of a regular collective motion in a system with nonlinear couplings. In particular, solitons are stabilized due to the balance of nonlinearity and various dispersion effects.

Standard problems of classical mechanics, such as falling bodies, the pendulum, or the dynamics of planetary systems considering only a system composed of the sun and one planet require only a few degrees of freedom. These famous examples allowed the quantitative formulation of mechanics by Galileo and Newton. In other words, these famous pioneers of modern physics treated one- or, at most, two-body problems without any kind of chaotic behavior.

The mathematical processing becomes more and more complicated, and the solution variety increases enormously, if we increase the number of degrees of freedom and therefore the degree of complexity. Obviously, the tendency to be chaotic increases if the number of degrees of freedom increases.

The apparent unpredictability of a deterministic, mechanical many-body system arises from the sensitive dependence on the initial conditions and from the fact that the initial conditions can be measured only approximately in practice due to the finite resolution of any measuring instrument.

In order to understand this statement, we analyze a microscopic, mechanical system with $2N$ degrees of freedom. The dynamics can be rewritten in terms of deterministic Hamilton's equations as

$$\frac{dq_i}{\partial t} = \frac{\partial H}{\partial p_i} \quad \text{and} \quad \frac{dp_i}{\partial t} = -\frac{\partial H}{\partial q_i} , \tag{1.1}$$

where the q_i's ($i = 1, ..., N$) are the generalized coordinates, the p_i's are the generalized momenta conjugate to the q_i, and H is the Hamiltonian of the system. Coordinates and momenta form the set of degrees of freedom. Formally, these microscopic degrees of freedom can be combined into a supervector $\Gamma = \{q_1, ..., q_N, p_1, ..., p_N\}$. Each vector Γ describes a certain microscopic state of the underlying system. Thus, the whole system under consideration can be represented by a point in a $2N$-dimensional space, spanned by a reference frame of $2N$ axes, corresponding to the degrees of freedom $\{q_1, ..., p_N\}$. This space is called the phase space. It plays a fundamental role, being the natural framework of the dynamics of many-body systems.

Practically all trajectories of the system through the $2N$-dimensional phase space are unstable against small perturbations. The stability of an arbitrary trajectory to an infinitesimally small perturbation is studied by an analysis of the so-called Lyapunov exponents. This concept is very

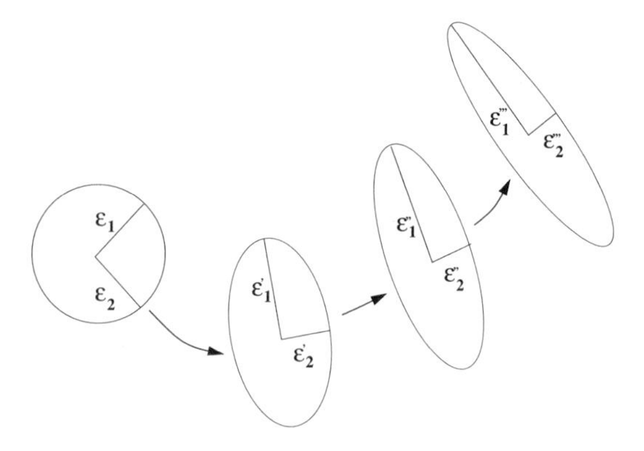

Fig. 1.1. The time evolution of an infinitesimally small ball of initial radius $\varepsilon_1 = \varepsilon_2$ in a schematic phase space. The ball is deformed into an ellipsoid with increasing time.

geometrical. Imagine an infinitesimally small ball of radius $\varepsilon(0)$ containing the initial position of neighboring trajectories. Under the action of the dynamics, the center of the ball may move through the phase space, and the ball will be distorted. Because the ball is infinitesimal, this distortion is governed by a linearized theory. Thus, the ball remains an ellipsoid with the $2N$ principal axes $\varepsilon_\alpha(t)$ (Figure 1.1). Then, the Lyapunov exponents can be defined as

$$\lambda_\alpha = \lim_{t\to\infty} \lim_{\varepsilon(0)\to 0} \frac{1}{t} \frac{\varepsilon_\alpha(t)}{\varepsilon(0)} . \tag{1.2}$$

The limit $\varepsilon(0) \to 0$ is necessary because, for finite radius $\varepsilon(0)$, as t increases, the ball can no longer be adequately represented by an ellipsoid, due to the increase of nonlinear effects. On the other hand, the long time limit $t \to \infty$ is important for gathering enough information to represent the entire trajectory. Obviously, the distance between infinitesimal neighboring trajectories diverges if the real part of at least one Lyapunov exponent is positive.

If the diameter of the initial ball has a finite value, then the initial shape is very violently distorted (Figure 1.2). The ball transforms into an amoebalike body that eventually grows out into extremely fine filaments that spread out over the whole accessible phase space. Such a mixing flow is a characteristic property of systems with a sufficiently high degree of complexity.

There remains the question of whether Lyapunov exponents with positive real part occur in microscopic mechanical systems. We obtain as a direct consequence of the time-reversal symmetry that, for every Lyapunov exponent, another Lyapunov exponent exists with opposite sign. In other words, we should expect a regular behavior only then if the real parts of all Lyapunov exponents vanish.

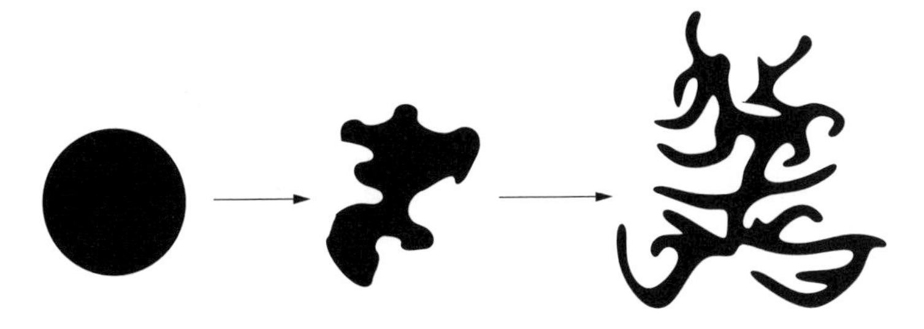

Fig. 1.2. The deformation of a finite ball of the phase space in the course of its time evolution.

However, this situation is practically excluded for complicated many-body systems. Computer simulations have demonstrated also that relatively simple mechanical systems with a few degrees of freedom already show chaotic behavior. All the more, economic systems, such as stock markets and travel networks, or human societies, offer a pronounced chaotic behavior on microscopic scales due to the enormous number of degrees of freedom.

1.3 The Probability Distribution

Although most many-body systems exceed the technical possibilities of the mathematical calculus of mechanics, we are able to calculate the properties of large systems by applying of methods belonging to statistical physics. In order to do this, we have to ask for a general concept describing complex systems at microscopic scales. This description should fulfill two conditions. On the one hand, the approach should consider the apparent unpredictability of the chaotic motion of many-body systems, and on the other hand, it should be a starting point for establishing relations between various relevant quantities at the macroscopic level.

As already mentioned in the previous section, a concrete prediction about the microscopic movement of all particles is impossible if the initial condition cannot be determined exactly. We may, however, give the probability for the realization of a certain microscopic state either on the basis of the preparation of the complex system or due to a suitable empirical estimation.

The intuitive notation of the probability is clear in dice games or coin tossing. The probability is empirically defined by the relative frequency of a given realization repeating the game an infinite number of times. Probability reflects our partial ignorance, as in the outcome of a dice game. This frequency concept of probability, introduced at the beginning of the last century, is not very suitable for the characterization of complex systems such as financial markets, traffic networks, or human societies.

In fact, we are neither able to determine frequencies by a successive repetition in the sense of a scientific experiment nor have we enough information about possible outcomes. An alternative way to overcome this dilemma is the interpretation of the probability as a degree of belief that an event will occur. This concept was the original point of view of Bayes, and it combines a priori judgments and scientific information in a natural way.

Bayesian statistics is very general and can be applied to any thinkable process, independent of the feasibility of repeating the process under the conserved conditions. This definition is particularly suitable for economic or social systems because a repetition in the sense of a scientific experiment is almost always impossible for such systems.

In order to formulate the probability concept more precisely, we use the language of set theory. The elements of such a theory are denoted with respect to the probabilistic properties as events. In particular, each microscopic state represented by a vector Γ of the phase space is denoted as a certain event of the underlying microscopic system.

Events form various sets, including the set of all events Ω and the set of no events. For instance, an arbitrary region of the phase space corresponds to such a set of events, and the whole phase space has to be interpreted as the set Ω. All possible sets of events form a closed system under the operations of union and intersection.

From this abstract point of view, the probability is now defined as a measure $P(A)$ of our degree of belief in the appearance of an arbitrary event contained in the set A. The measure $P(A)$ is always nonnegative, $P(A) \geq 0$, and normalized, $P(\Omega) = 1$. Furthermore, if A and B are two nonoverlapping sets, $A \cap B = \emptyset$, the probability that an event is contained in $A \cup B$ is the probability that the event is an element either of A or of B. This leads to the main axiom of probability theory

$$P(A \cup B) = P(A) + P(B) \qquad \text{for} \qquad A \cap B = \emptyset. \tag{1.3}$$

We generalize this relation to a countable collection of nonoverlapping sets A_i $(i = 1, 2, ...)$ such that $A_i \cap A_j = \emptyset$ for all $i \neq j$ and obtain

$$P\left(\bigcup_i A_i\right) = \sum_i P(A_i). \tag{1.4}$$

After this excursion into set theory, we can continue our original problem. We consider the set of events $A(\Gamma)$ that the system is within an infinitesimal volume element $d\Gamma = \prod_{i=1}^{N} [dq_i dp_i]$ of the phase space centered on the microscopic state Γ.

The main problem is to assign the probability measure $dP(A(\Gamma), t)$ that at time t a microscopic state is an element of $A(\Gamma)$. This is an a priori probability, which is simply assumed with respect to our experience. In other words, this intuitive step reflects the degree of belief. It is convenient to write

$$dP = \rho(q_1, ..., q_s, p_1, ..., p_s, t) dq_1 ... dq_s dp_1 ... dp_s = \rho(\Gamma, t) d\Gamma, \tag{1.5}$$

where the function $\rho(\Gamma, t)$ is denoted as the probability density for the outcome of the state Γ at time t. By definition, the density is a nonnegative quantity. Each finite region R of the phase space is a union of infinitesimal volume elements. Due to (1.4), the probability of finding a microscopic state in this region at time t is

$$P(R,t) = \int_R d\Gamma \rho(\Gamma,t). \tag{1.6}$$

If we expand the region R over the whole phase space, we receive the normalization condition

$$\int \rho(\Gamma,t) d\Gamma = 1. \tag{1.7}$$

This equation corresponds to $P(\Omega) = 1$ in the language of set theory, reflecting our knowledge that the certainty of finding the system somewhere in the phase space is always true.

Finally, it should be remarked that the probabilistic description based on Bayesian statistics is the only concept when we have not enough information about a realistic system, and even if we have this information, it is still a convenient representation for complex systems. Of course, the concrete mathematical structure of a microscopic founded probability distribution may be too complicated for a further treatment. But the probabilistic concept itself permits the theoretical overcoming of the general problem that initial conditions are measurable only with a finite accuracy.

1.4 The Liouville Equation

We want to address the task of whether we can determine the evolution of the probability distribution for a given microscopic system. To this aim, we assume that an initial state Γ_0 was realized with a certain probability $\rho(\Gamma_0, t_0) d\Gamma$.

In the course of the deterministic microscopic motion, the initial state is shifted into another microscopic state along the trajectory $\Gamma(t) = \{q_i(t), p_i(t)\}$ with the boundary condition $\Gamma(0) = \Gamma_0$. In other words, the probability density ρ is conserved along each trajectory of the complex system. This circumstance requires

$$\frac{d\rho}{dt} = \frac{\partial \rho}{\partial t} + \sum_{i=1}^{N} \left[\frac{\partial \rho}{\partial q_i} \dot{q}_i + \frac{\partial \rho}{\partial p_i} \dot{p}_i \right] = 0. \tag{1.8}$$

After replacing the velocities $\dot{q}_i$ and forces $\dot{p}_i$ by the equations (1.1), we arrive at

$$\frac{\partial \rho}{\partial t} + \hat{L}\rho = 0, \tag{1.9}$$

where we have introduced the Liouvillian of the system

$$\hat{L} = \sum_{i=1}^{N} \left[\frac{\partial H}{\partial p_i} \frac{\partial}{\partial q_i} - \frac{\partial H}{\partial q_i} \frac{\partial}{\partial p_i} \right] . \tag{1.10}$$

The relation (1.9) is called the Liouville equation and is the most important equation of statistical physics, just as the Schrödinger equation is the main equation of quantum mechanics. The Liouvillian plays the same role that the Hamiltonian plays in Newtonian mechanics. The Hamiltonian fixes the rules of evolution of any microscopic state of the underlying system. In statistical physics, the Liouvillian again defines the equation of motion, which is now represented by the distribution function ρ.

For all microscopic systems, the mechanical and statistical representations of the evolution are equivalent. The difference between the descriptions lies in the definition of what we call the objects of evolution: points in phase space are the objects of classical mechanics, while distribution functions are the objects of statistical physics.

The meaning of the Liouville equation for the evolution of an economic system and also for many other complex systems lies in the combination of the probabilistic and deterministic features of the evolution process. Let us believe in the sense of Bayesian statistics that an economic situation at time t_0 can be described at the microscopic level by a probability distribution $\rho(\Gamma, t_0)$. Then, the Liouville equation represents a deterministic map of $\rho(\Gamma, t_0)$ onto the probability distribution $\rho(\Gamma, t)$ at a later time $t > t_0$. In other words, the Liouville equation conserves our degree of belief.

1.5 Econophysics

Economics is traditionally oriented toward choice and decision problems. The classical description of economics involves two general aspects that are of particular interest for a physical interpretation of economic processes. First, economics is a discipline that deals only with a certain aspect of reality, namely the items of scarcity. In particular, it concentrates on the optimum manner in which man employs scarce resources.

Second, economics focuses its attention on the behavior of various human decision units, such as households, financial markets, and governmental agencies. The outcome of an economic process was generally considered to be a result of intrinsic or endogenous mechanisms of the economic system itself and of various more or less external or exogenous factors from a purely economic point of view.

However, economists are increasingly becoming aware of the fact that economic systems are embedded in their human and natural environments and that many apparently exogenous factors are a part of a larger complex system of physical, biological, and social mechanisms controlling the evolution of our planet.

The question remains how physics can contribute to understanding economic problems. The great success of physics rests on its methodology, which can certainly be described as analytical. By decomposing a system into its parts, a physicist tries to understand the properties of the whole system. From this point of view, phenomena such as migration, commuting, production decisions, financial transactions, traffic, and transportation can be analyzed by applying of physical methods.

In order to deal with economic systems, we quite often still have to find the relevant quantities to describe the essential properties of these systems. This problem is still the task of economic investigations. The physical contribution may consist of establishing more or less general equations that describe the evolution of relevant quantities and the relations between these quantities on the basis of universal laws.

A second task should be the solution of these equations by applying modern techniques of statistical physics. This is, roughly speaking, the field of econophysics.

Econophysics may give new impulses both to economic decision-making and risk management and to a deeper understanding of systems with an enormous degree of complexity. However, we warn that to understand econophysics as an alternative way that replaces traditional and modern economic sciences. These scientific disciplines still require a profound economic knowledge.

2. Evolution and Probabilistic Concepts

2.1 Some Notations of Probability Theory

2.1.1 Probability Distribution of Relevant Quantities

In the microscopic probability distribution $\rho(\Gamma, t)$, all degrees of freedom are contained equally. Such a function, even if we were able to determine it, would of course be impractical and therefore unusable for the analysis of complex systems because of the large number of contained degrees of freedom.

In general, we are interested in the description of complex systems only on the basis of the relatively small number of relevant degrees of freedom. Such an approach may be denoted as a kind of reductionism. Unfortunately, we are not able to give an unambiguous definition of what degree of freedom is relevant for the description of a complex system and what degree of freedom is irrelevant. As we have mentioned in the previous chapter, the relevant quantities are introduced empirically in accordance with the underlying problem.

To proceed, we split the complete phase space into a subspace of the relevant degrees of freedom and the complementary subspace of the irrelevant degrees of freedom. Then, every microscopic state Γ may be represented as a combination of the set $Y = \{Y_1, Y_2, ...Y_{N_{\mathrm{rel}}}\}$ of N_{rel} relevant degrees of freedom and the set Γ_{irr} of the irrelevant degrees of freedom so that

$$\Gamma = \begin{cases} Y & \text{relevant degrees of freedom} \\ \Gamma_{\mathrm{irr}} & \text{irrelevant degrees of freedom} \end{cases} \tag{2.1}$$

The microscopic probability density may be written as $\rho(\Gamma, t) = \rho(Y, \Gamma_{\mathrm{irr}}, t)$. In order to eliminate the irrelevant degrees of freedom, we integrate the probability density over Γ_{irr}

$$p(Y, t) = \int d\Gamma_{\mathrm{irr}} \rho(Y, \Gamma_{\mathrm{irr}}, t) \,. \tag{2.2}$$

The remaining probability density $p(Y, t)$ is more suitable for describing complex systems. The elimination of all more or less microscopic, irrelevant degrees of freedom corresponds to the transition from the microscopic level to a macroscopic representation. By definition, the probability density $p(Y, t)$ is also normalized:

$$\int dY p(Y,t) = \int dY d\Gamma_{\text{irr}} \rho(Y, \Gamma_{\text{irr}}, t) = \int d\Gamma \rho(\Gamma, t) = 1. \tag{2.3}$$

The integration over all irrelevant degrees of freedom means that we suppose a maximal measure of ignorance of these quantities.

We may think about this description in geometrical terms. The system of relevant degrees of freedom can be represented by a point in the corresponding N_{rel}-dimensional subspace of the phase space. Obviously, an observer records an apparently unpredictable behavior of the evolution of the relevant quantities at the macroscopic scale if he considers only the relevant data.

That is because of the fact that the dynamical evolution of the relevant quantities is governed by the hidden irrelevant degrees of freedom on microscopic scales. Thus, different microscopic trajectories in the phase space can lead to the same macroscopic results in the subspace and, vice versa, identical macroscopic initial configurations may develop into different directions.

We are not able to predict the later evolution completely even if we know the initial conditions precisely. In other words, the restriction onto the subspace of relevant quantities leads to a permanent loss of the degree of belief.

The average of an arbitrary function $f(\Gamma)$ is obtained by adding all values of $f(\Gamma)$ considering the statistical weight $\rho(\Gamma, t)\, d\Gamma$. Hence

$$\overline{f}(t) = \int d\Gamma \rho(\Gamma, t) f(\Gamma). \tag{2.4}$$

The mean value may be a time-dependent quantity due to the time dependence of the probability density. If the function f depends only on the relevant degrees of freedom (i.e., $f = f(Y)$), then we get

$$\overline{f}(t) = \int d\Gamma \rho(\Gamma, t) f(Y) = \int dY p(Y, t) f(Y). \tag{2.5}$$

In this expression, the dynamics of the irrelevant degrees of freedom are again hidden in the distribution function $p(Y, t)$. Obviously, the relevant probability density satisfies all conditions necessary for a sufficient description of a complex system on the level of the selected set of relevant degrees of freedom.

2.1.2 Measures of Central Tendency

Suppose that we consider a probability distribution function $p(Y, t)$ with only one relevant degree of freedom. Generally, each multivariable probability density may be reduced to such a single variable function by integration over all degrees of freedom except one.

Let us now answer the following question. What is the typical value of the outcome of a given problem with a sufficiently high degree of complexity if we know the probability distribution function $p(Y, t)$? Unfortunately, there is no unambiguous answer. The quantity used most frequently for the characterization of the central tendency is the mean or average

$$\overline{y}(t) = \int dY p(Y,t) Y. \tag{2.6}$$

There are two other major measures of central tendency. The probability $P_<(y,t)$ gives the time-dependent fraction of events with values less then y,

$$P_<(y,t) = \int_{-\infty}^{y} dY p(Y,t). \tag{2.7}$$

The function $P_<(y,t)$ increases monotonically with y from 0 to 1. Using (2.7), the central tendency may be characterized by the median $y_{1/2}(t)$. The median is the halfway point in a graded array of values,

$$P_<(y_{1/2},t) = \frac{1}{2}. \tag{2.8}$$

Finally, the most probable value $y_{\max}(t)$ is another quantity describing the mean behavior. This quantity maximizes the density function

$$\left.\frac{\partial p(Y,t)}{\partial Y}\right|_{Y=y_{\max}(t)} = 0. \tag{2.9}$$

If this equation has several solutions, the most probable value $y_{\max}(t)$ is the one with the largest p. Apart from unimodal symmetric probability distribution functions, the three quantities differ. These differences are important for the interpretation of empirical averages obtained from a finite number of observations.

For a few trials, the most probable value will be sampled first and the average made on a few such measures will not be far from $y_{\max}(t)$. In contrast, the empirical average determined from a large but finite number of observations approaches progressively the true average $\overline{y}(t)$.

2.1.3 Measure of Fluctuations Around the Central Tendency

We consider again only one degree of freedom as a relevant quantity. When repeating observations of this variable several times, one expects them to be within an interval anchored at the central tendency. The width of this interval is a measure of the deviations from the central tendency. A possible measure of this width is the average of the absolute value of the spread defined by

$$D_{\mathrm{sp}}(t) = \int_{-\infty}^{\infty} dY \left|Y - y_{1/2}(t)\right| p(Y,t). \tag{2.10}$$

The absolute value of the spread does not exist for probability distribution functions decaying as or slower than Y^{-2} for large Y. Another measure is the standard deviation σ. This quantity is the square root of the variance σ^2:

$$\sigma^2 = \int_{-\infty}^{\infty} dY \left[Y - \overline{y}(t)\right]^2 p(Y,t). \tag{2.11}$$

The standard deviation does not always exist, such as for probability densities $p(Y,t)$ with tails decaying as or slower than Y^{-3}.

2.1.4 Moments and Characteristic Functions

Now, we come back to the general case of a multivariable probability distribution function $p(Y,t)$ with $Y = \{Y_1, Y_2, ...Y_{N_{\text{rel}}}\}$. Here, N_{rel} is the number of relevant degrees of freedom and therefore the dimension of the vector Y. The moments of order n are defined by the average

$$m^{(n)}_{\alpha_1\alpha_2...\alpha_n}(t) = \int dY \left[\prod_{k=1}^{n} Y_{\alpha_k}\right] p(Y,t). \tag{2.12}$$

The first moment $m^{(1)}_{\alpha}(t)$ is the mean $\overline{y}_{\alpha}(t)$ of component α. Therefore, the formal vector $\{m^1_1(t), m^1_2(t), ...\}$ defines in generalization of (2.6) the central tendency of the underlying dynamics. The second moment $m^{(2)}_{\alpha\beta}(t)$ corresponds to the average

$$m^{(2)}_{\alpha\beta}(t) = \int dY p(Y,t) Y_{\alpha} Y_{\beta}. \tag{2.13}$$

These quantities are also denoted as components of the correlation matrix. For the definition (2.12) to be meaningful, the integral on the right-hand side must be convergent. That means a necessary condition for the existence of a moment of order n is that the probability density function decays faster than $|Y|^{-n-N_{\text{rel}}}$ for $|Y| \to \infty$. This is trivially obeyed for probability distribution functions that vanish outside a finite region of the space of relevant degrees of freedom.

Statistical problems are often discussed in terms of moments because they avoid the difficult problem of determining the full functional behavior of the probability density. In principle, the knowledge of all moments is in many realistic cases equivalent to that of the probability distribution function. However, the strict equivalence between the knowledge of the moments and the probability density requires further constraints.

The moments are closely related to the characteristic function, which is defined as the Fourier transform of the probability distribution

$$\hat{p}(k,t) = \int dY \exp\{ikY\} p(Y,t) \tag{2.14}$$

with the N_{rel}-dimensional vector $k = \{k_1, k_2, ..., k_{N_{\text{rel}}}\}$. From here, we obtain the inverse relation

$$p(Y,t) = \frac{1}{(2\pi)^{N_{\text{rel}}}} \int dk \exp\{-ikY\} \hat{p}(k,t). \tag{2.15}$$

Thus, the normalization condition (2.3) is equivalent to $\hat{p}(0,t) = 1$, and the moments of the probability density can be obtained from derivatives of the characteristic function at $k = 0$:

$$m^{(n)}_{\alpha_1\alpha_2\ldots\alpha_n}(t) = (-i)^n \prod_{l=1}^{n} \left(\frac{\partial}{\partial k_{\alpha_l}} \right) \hat{p}(k,t) \Bigg|_{k=0} . \tag{2.16}$$

If all moments exist, the characteristic function may also be presented as the series expansion

$$\hat{p}(k,t) = \sum_{n=0}^{\infty} \sum_{\{\alpha_1,\alpha_2,\ldots\alpha_n\}} \frac{i^n}{n!} m^{(n)}_{\alpha_1\alpha_2\ldots\alpha_n}(t) \left(\prod_{l=1}^{n} k_{\alpha_l} \right) . \tag{2.17}$$

The inversion formula shows that different characteristic functions arise from different probability distribution functions (i.e., the characteristic function $\hat{p}(k,t)$ is truly characteristic). Additionally, the straightforward derivation of the moments by (2.16) makes any determination of the characteristic function directly relevant to measurable quantities.

2.1.5 Cumulants

Another important function is the cumulant generating function, which is defined as the logarithm of the characteristic function

$$\Phi(k,t) = \ln \hat{p}(k,t) . \tag{2.18}$$

This leads to the introduction of the cumulants $c_{\alpha_1\alpha_2\ldots\alpha_n}(t)$ as derivatives of the cumulant generating function at $k = 0$,

$$c^{(n)}_{\alpha_1\alpha_2\ldots\alpha_n}(t) = (-i)^n \prod_{l=1}^{n} \left(\frac{\partial}{\partial k_{\alpha_l}} \right) \Phi(k,t) \Bigg|_{k=0} . \tag{2.19}$$

Each cumulant of order n is a combination of moments of order $l \leq n$, as can be seen by substitution of (2.17) and (2.18) into (2.19). We get for the first cumulants

$$\begin{aligned}
c^{(1)}_{\alpha} &= m^{(1)}_{\alpha} \\
c^{(2)}_{\alpha\beta} &= m^{(2)}_{\alpha\beta} - m^{(1)}_{\alpha} m^{(1)}_{\beta} \\
c^{(3)}_{\alpha\beta\gamma} &= m^{(3)}_{\alpha\beta\gamma} - m^{(2)}_{\alpha\beta} m^{(1)}_{\gamma} - m^{(2)}_{\beta\gamma} m^{(1)}_{\alpha} - m^{(2)}_{\gamma\alpha} m^{(1)}_{\beta} + 2 m^{(1)}_{\alpha} m^{(1)}_{\beta} m^{(1)}_{\gamma} \\
&\vdots
\end{aligned} \tag{2.20}$$

The first-order cumulants are the averages of the single components Y_α. The second-order cumulants define the covariance matrix $\widetilde{\sigma}$ with the elements

$$\widetilde{\sigma}_{\alpha\beta} = c^{(2)}_{\alpha\beta} = m^{(2)}_{\alpha\beta} - m^{(1)}_{\alpha} m^{(1)}_{\beta} . \tag{2.21}$$

The covariance is a generalized measure of the degree to which the values Y deviate from the central tendencies. In particular, for a single variable

Y, the second-order cumulant is equivalent to the variance σ^2. Higher-order cumulants contain information of decreasing significance. Especially if all higher-order cumulants vanish, we can easily deduce using (2.18) and (2.15) that the corresponding probability density $p(Y,t)$ is a Gaussian probability distribution

$$p(Y,t) = \frac{1}{\sqrt{\det 2\pi\widetilde{\sigma}}} \exp\left\{-\frac{1}{2}\left(Y - m^{(1)}\right)\widetilde{\sigma}^{-1}\left(Y - m^{(1)}\right)\right\}. \tag{2.22}$$

Note that the theorem of Marcienkiewicz [254] shows that either all but the first two cumulants vanish or there are an infinite number of nonvanishing cumulants. In other words, the cumulant generating function cannot be a polynomial of degree greater than 2.

Obviously, higher-order cumulants characterize the natural deviation from Gaussian behavior. In the case of a single variable Y, the normalized third-order cumulant $\lambda_3 = c^{(3)}/\sigma^3$ is called the skewness, while $\lambda_4 = c^{(4)}/\sigma^4$ is called excess kurtosis. The skewness is a measure of the asymmetry of the probability distribution function. For symmetric distributions, the excess kurtosis quantifies the first correction to the Gaussian behavior.

2.2 Generalized Rate Equations

2.2.1 The Formal Solution of the Liouville Equation

We can formally integrate the Liouville equation (1.9) to obtain the solution

$$\rho(\Gamma,t) = \exp\left\{-\hat{L}t\right\}\rho(\Gamma,0). \tag{2.23}$$

This expression considers the microscopic equations of motion due to the concrete structure of the Liouvillian $\hat{L}$. The operator $\exp\{-\hat{L}t\}$ is referred to as the time propagator associated with the dynamical variables of the system. For a better understanding of the meaning of the time propagator, let us expand the exponential function in powers of t:

$$\rho(\Gamma,t) = \left[1 - \hat{L}t + \frac{1}{2!}\left(\hat{L}t\right)^2 - \frac{1}{3!}\left(\hat{L}t\right)^3 + \ldots\right]\rho(\Gamma,0). \tag{2.24}$$

The right-hand side may be interpreted as a perturbative solution obtained from a successive integration of the Liouville equation. To demonstrate this, we write the Liouville equation as an integral equation. Then, we are able to construct the map

$$\rho^{(n+1)}(\Gamma,t) = \rho(\Gamma,0) - \int_0^t \hat{L}\rho^{(n)}(\Gamma,\tau)\,d\tau \tag{2.25}$$

with the initial function $\rho^{(0)}(\Gamma,t) = \rho(\Gamma,0)$. The series $\rho^{(0)}, \rho^{(1)}, \ldots, \rho^{(n)}, \ldots$ converges eventually against the solution $\rho(\Gamma,t)$ of the Liouville equation. In fact, we receive

$$\begin{aligned}
\rho^{(1)} &= \rho^{(0)} - \hat{L}\rho^{(0)}t \\
\rho^{(2)} &= \rho^{(0)} - t\hat{L}\rho^{(0)} + \frac{t^2}{2}\hat{L}^2\rho^{(0)} \\
\rho^{(3)} &= \rho^{(0)} - t\hat{L}\rho^{(0)} + \frac{t^2}{2}\hat{L}^2\rho^{(0)} - \frac{t^3}{6}\hat{L}^3\rho^{(0)} \\
&\vdots
\end{aligned} \tag{2.26}$$

As expected, the solutions $\rho^{(0)}, \rho^{(1)}, \rho^{(2)}, \ldots$ of the hierarchical system (2.25) are identical with the first terms of the expansion (2.24).

Unfortunately, for complex systems, the formal solution (2.23) is in general too complicated to be useful in practice. In order to describe the dynamical behavior of such systems, we must look for alternative ways.

2.2.2 The Nakajima–Zwanzig Equation

Obviously, knowledge of the relevant probability density $p(Y,t)$ is a sufficient presupposition for the study of complex systems on the level of the chosen relevant degrees of freedom. Our previous knowledge allows us to derive this function from the complete microscopic probability distribution function $\rho(\Gamma,t)$. For this purpose, we would have to solve the Liouville equation with all of the microscopic degrees of freedom at first. Then, in the subsequent step, we would be able to remove the irrelevant degrees of freedom from the microscopic distribution function by integration.

To avoid this unrealistic procedure, we want to answer the question of whether one can find an equation that describes the evolution of $p(Y,t)$ and contains exclusively relevant degrees of freedom.

Of course, we can also remove the relevant degrees of freedom from every given microscopic probability distribution so that we arrive at the distribution function of the irrelevant degree of freedom

$$\rho_{\text{irr}}(\Gamma_{\text{irr}}, t) = \int dY \rho(Y, \Gamma_{\text{irr}}, t), \tag{2.27}$$

where we have to consider the normalization condition

$$\int d\Gamma_{\text{irr}} \rho_{\text{irr}}(\Gamma_{\text{irr}}, t) = 1. \tag{2.28}$$

The product of the probability distributions (2.27) at the initial time t_0 and (2.2) at the time t is again a probability density

$$\widetilde{\rho}(Y, \Gamma_{\text{irr}}, t, t_0) = \rho_{\text{irr}}(\Gamma_{\text{irr}}, t_0)\, p(Y,t). \tag{2.29}$$

Of course, this surrogate probability distribution is no longer identical to the microscopic probability density $\rho(\Gamma,t)$. But the average values of any functions of relevant degrees of freedom calculated by an application of the density $\widetilde{\rho}$ remain unchanged in comparison with the use of ρ. Indeed, we get

$$p(Y,t) = \int d\Gamma_{\text{irr}} \rho(Y, \Gamma_{\text{irr}}, t) = \int d\Gamma_{\text{irr}} \widetilde{\rho}(Y, \Gamma_{\text{irr}}, t, t_0). \tag{2.30}$$

The generation of the surrogate probability distribution $\widetilde{\rho}$ is usually called a projection formalism . This procedure may be symbolically expressed by an application of a projection operator onto the probability distribution function

$$\widetilde{\rho}(Y, \Gamma_{\text{irr}}, t, t_0) = \hat{P}\rho(\Gamma, t), \tag{2.31}$$

where we have introduced the special projection operator

$$\hat{P} \ldots = \rho_{\text{irr}}(\Gamma_{\text{irr}}, t_0) \int d\Gamma_{\text{irr}} \ldots . \tag{2.32}$$

Apart from $\hat{P}$, we still need the complementary operator $\hat{Q} = 1 - \hat{P}$. Using (2.32), it is simple to demonstrate that these operators have the "idempotent" properties

$$\hat{P}^2 = \hat{P}, \qquad \hat{Q}^2 = \hat{Q}, \qquad \text{and} \qquad \hat{P}\hat{Q} = \hat{Q}\hat{P} = 0, \tag{2.33}$$

typically for all projection operators. The first equation is a direct consequence of (2.32), while the last two follow from

$$\hat{Q}^2 = 1 - 2\hat{P} + \hat{P}^2 = 1 - 2\hat{P} + \hat{P} = 1 - \hat{P} = \hat{Q} \tag{2.34}$$

and

$$\hat{Q}\hat{P} = \hat{P} - \hat{P}^2 = \hat{P} - \hat{P} = 0. \tag{2.35}$$

We return now to the question of how to describe the time-dependent evolution of the relevant probability density. To proceed, we need some information about the initial distribution at time t_0. Although we can provide a meaningful initial distribution for simple physical systems due to the realization of an arbitrary number of repeatable experiments, we must fall back on more or less accurate estimations depending on the respective level of experience if we want to describe phenomena of social or economic systems. The distribution of the relevant degrees of freedom can be fixed relatively simply: we assume that the values Y_0 of all relevant degrees of freedom are well-known at the initial time t_0. Therefore, we can write

$$p(Y, t_0) = p(Y, t_0 \mid Y_0, t_0) = \delta(Y - Y_0). \tag{2.36}$$

In this context, $p(Y, t \mid Y_0, t_0)$ means the probability density of the relevant degrees of freedom at time t, while the initial state was Y_0. In principle, the following procedure works also for all other thinkable initial distributions. We will see somewhat later that all of these cases can be mapped onto (2.36). On the other hand, we have no essential information about the irrelevant degrees of freedom.

However, we may assume that relevant and irrelevant degrees of freedom are uncorrelated, at least for the initial state. Here the idea of Bayesian statistics comes into its own. The statistical independence of relevant and irrelevant degrees can be neither verified nor rejected. It is an "a priori" assumption reflecting the degree of our belief. Considering these assumptions, the initial microscopic probability distribution can be written as

$$\rho\left(\Gamma,t_0\right)=\rho_{\mathrm{irr}}\left(\Gamma_{\mathrm{irr}},t_0\right)\delta\left(Y-Y_0\right)=\widetilde{\rho}\left(Y,\Gamma_{\mathrm{irr}},t_0,t_0\right) \tag{2.37}$$

with the property

$$\hat{P}\rho\left(\Gamma,t_0\right)=\rho\left(\Gamma,t_0\right). \tag{2.38}$$

Now, we apply the projection operator $\hat{P}$ to the Liouville equation (1.9) and obtain

$$\begin{aligned}\frac{\partial\hat{P}\rho\left(\Gamma,t\right)}{\partial t}&=-\hat{P}\hat{L}\left(\hat{P}+\hat{Q}\right)\rho\left(\Gamma,t\right)\\&=-\hat{P}\hat{L}\hat{P}\rho\left(\Gamma,t\right)-\hat{P}\hat{L}\hat{Q}\rho\left(\Gamma,t\right).\end{aligned} \tag{2.39}$$

We replace $\rho\left(\Gamma,t\right)$ in the second term of the right-hand side by the formal solution (2.23) of the Liouville equation where the initial time t_0 is taken into account. Then, we arrive at

$$\hat{P}\hat{L}\hat{Q}\rho\left(\Gamma,t\right)=\hat{P}\hat{L}\hat{Q}\mathrm{e}^{-\hat{L}(t-t_0)}\rho\left(\Gamma,t_0\right). \tag{2.40}$$

For the further treatment of this expression, we need the identity

$$\mathrm{e}^{-\hat{L}(t-t_0)}=\mathrm{e}^{-\hat{L}_1(t-t_0)}-\int_{t_0}^{t}dt'\mathrm{e}^{-\hat{L}_1(t-t')}\hat{L}_2\mathrm{e}^{-\hat{L}(t'-t_0)}, \tag{2.41}$$

where we have split the Liouvillian into two arbitrary parts, $\hat{L}_1$ and $\hat{L}_2$, via $\hat{L}=\hat{L}_1+\hat{L}_2$. This identity may be checked by the derivative with respect to the time

$$\begin{aligned}-\hat{L}\mathrm{e}^{-\hat{L}(t-t_0)}&=-\hat{L}_1\mathrm{e}^{-\hat{L}_1(t-t_0)}-\hat{L}_2\mathrm{e}^{-\hat{L}(t-t_0)}\\&\quad+\hat{L}_1\int_{t_0}^{t}dt'\mathrm{e}^{-\hat{L}_1(t-t')}\hat{L}_2\mathrm{e}^{-\hat{L}(t'-t_0)}.\end{aligned} \tag{2.42}$$

Then, substituting the integral kernel using (2.41), we obtain

$$\begin{aligned}-\hat{L}\mathrm{e}^{-\hat{L}\Delta t}&=-\hat{L}_1\mathrm{e}^{-\hat{L}_1\Delta t}-\hat{L}_2\mathrm{e}^{-\hat{L}\Delta t}+\hat{L}_1\left[\mathrm{e}^{-\hat{L}_1\Delta t}-\mathrm{e}^{-\hat{L}\Delta t}\right]\\&=-\hat{L}_2\mathrm{e}^{-\hat{L}\Delta t}-\hat{L}_1\mathrm{e}^{-\hat{L}\Delta t}\\&=-\hat{L}\mathrm{e}^{-\hat{L}\Delta t}\end{aligned} \tag{2.43}$$

with $\Delta t=t-t_0$. Thus, the identity (2.41) is proven. In particular, if we replace $\hat{L}_1$ by $\hat{L}\hat{Q}$ and $\hat{L}_2$ by $\hat{L}\hat{P}$, we get

$$\mathrm{e}^{-\hat{L}(t-t_0)}=\mathrm{e}^{-\hat{L}\hat{Q}(t-t_0)}-\int_{t_0}^{t}dt'\mathrm{e}^{-\hat{L}\hat{Q}(t-t')}\hat{L}\hat{P}\mathrm{e}^{-\hat{L}(t'-t_0)}. \tag{2.44}$$

We substitute (2.44) into (2.40) so that we obtain

$$\hat{P}\hat{L}\hat{Q}\rho(\Gamma,t) = \hat{P}\hat{L}\hat{Q}\mathrm{e}^{-\hat{L}\hat{Q}(t-t_0)}\rho(\Gamma,t_0) - \int_{t_0}^{t} dt'\hat{P}\hat{L}\hat{Q}\mathrm{e}^{-\hat{L}\hat{Q}(t-t')}\hat{L}\hat{P}\mathrm{e}^{-\hat{L}(t'-t_0)}\rho(\Gamma,t_0). \quad (2.45)$$

The first addend on the right-hand side disappears. This property follows from a Taylor expansion of the exponential function. The expansion is apparently an infinite series, but by (2.38) we know that all coefficients must vanish identically as a result of (2.33). To go further, we write the integral kernel in a more symmetric form. Considering $\hat{Q} = \hat{Q}^2$, we conclude that

$$\begin{aligned}\hat{Q}\mathrm{e}^{-\hat{L}\hat{Q}\tau} &= \hat{Q}\left[1 - \tau\hat{L}\hat{Q} + \frac{\tau^2}{2}\hat{L}\hat{Q}\hat{L}\hat{Q} + ...\right] \\ &= \hat{Q}\left[1 - \tau\hat{Q}\hat{L}\hat{Q} + \frac{\tau^2}{2}\hat{Q}\hat{L}\hat{Q}^2\hat{L}\hat{Q} + ...\right]\hat{Q} = \hat{Q}\mathrm{e}^{-\hat{Q}\hat{L}\hat{Q}\tau}\hat{Q}. \quad (2.46)\end{aligned}$$

From (2.23), we see that

$$\hat{P}\hat{L}\hat{Q}\rho(\Gamma,t) = -\int_{t_0}^{t} dt'\hat{P}\hat{L}\hat{Q}\mathrm{e}^{-\hat{Q}\hat{L}\hat{Q}(t-t')}\hat{Q}\hat{L}\hat{P}\rho(\Gamma,t'), \quad (2.47)$$

and coming back to (2.39), we obtain

$$\frac{\partial\hat{P}\rho(\Gamma,t)}{\partial t} = -\hat{P}\hat{L}\hat{P}\rho(\Gamma,t) + \int_{t_0}^{t} dt'\hat{P}\hat{L}\hat{Q}\mathrm{e}^{-\hat{Q}\hat{L}\hat{Q}(t-t')}\hat{Q}\hat{L}\hat{P}\rho(\Gamma,t'). \quad (2.48)$$

When this relationship is integrated over all irrelevant degrees of freedom, we obtain a closed linear integrodifferential equation for the probability distribution function of the relevant degrees of freedom. Considering (2.32), we get

$$\begin{aligned}\frac{\partial p(Y,t)}{\partial t} &= -\int d\Gamma_{\mathrm{irr}}\left[\hat{L}\rho_{\mathrm{irr}}(\Gamma_{\mathrm{irr}},t_0)\right]p(Y,t) \\ &+ \int_{t_0}^{t} dt' \int d\Gamma_{\mathrm{irr}} \\ &\times\left[\hat{L}\hat{Q}\mathrm{e}^{-\hat{Q}\hat{L}\hat{Q}(t-t')}\hat{Q}\hat{L}\rho_{\mathrm{irr}}(\Gamma_{\mathrm{irr}},t_0)\right]p(Y,t') \quad (2.49)\end{aligned}$$

or, more precisely,

$$\begin{aligned}\frac{\partial p(Y,t\mid Y_0,t_0)}{\partial t} &= -\hat{M}(t_0)\,p(Y,t\mid Y_0,t_0) \\ &+ \int_{t_0}^{t} dt'\hat{K}(t_0,t-t')\,p(Y,t'\mid Y_0,t_0), \quad (2.50)\end{aligned}$$

where we have introduced the frequency operator

$$\hat{M}(t_0) = \int d\Gamma_{\mathrm{irr}} \left[\hat{L}\rho_{\mathrm{irr}}(\Gamma_{\mathrm{irr}}, t_0) \right] \tag{2.51}$$

and the memory operator

$$\hat{K}(t_0, t - t') = \int d\Gamma_{\mathrm{irr}} \left[\hat{L}\hat{Q} \exp\left\{ -\hat{Q}\hat{L}\hat{Q}(t - t') \right\} \hat{Q}\hat{L}\rho_{\mathrm{irr}}(\Gamma_{\mathrm{irr}}, t_0) \right]. \tag{2.52}$$

This equation is called the Nakajima–Zwanzig equation or the generalized rate equation. The Nakajima–Zwanzig equation is still a proper relation, although it apparently describes only the evolution of the relevant probability distribution function. However, the complete dynamics of the irrelevant degrees of freedom, including their interaction with the relevant degrees of freedom in particular, is hidden in the memory operator.

The dependency of the operators $\hat{M}$ and $\hat{K}$ on the initial time t_0 is a remarkable property and reflects the fact that a complex system does not necessarily have to be in a stationary state. Therefore, completely different developments of the probability density $p(Y, t \mid Y_0, t_0)$ may be observed for the same system and for the same initial conditions but for different initial times.

The Nakajima–Zwanzig equation allows the prediction of the further evolution of the relevant probability distribution function, presupposing that we are able to determine the exact mathematical structure of the frequency and memory operators. In principle, we are also able to derive more general evolution equations than the Nakajima–Zwanzig equation (e.g., by use of time-dependent projectors or projection operators that depend even on the relevant probability distribution function). But then the useful convolution property is lost, which characterizes the memory term in (2.50). Additionally, all evolution equations obtained by projection formalisms are physically equivalently and mathematically accurate so that also from this point of view none of the thinkable evolution equations possesses a possible preference.

The main problem is, however, the determination of the operators $\hat{M}$ and $\hat{K}$. The complete determination of these quantities equals the solution of the Liouville equation. Consequently, this way is unsuitable for systems with a sufficiently high degree of complexity. But we can try to approach these operators of the Nakajima–Zwanzig equation in a heuristic manner using empirical experiences and mathematical considerations. Physical intuition plays an important role at several stages of this approach. Furthermore, especially for economic systems, we must take into account a lot of economical, technological, and social facts.

In this way, one can combine certain model conceptions and real observations and arrive at a comparatively reasonable approximation of the accurate evolution equation. Here it also becomes important what projection formalism one used. However, for the majority of economic problems, (2.50) is quite a suitable equation that gives us the opportunity for further progress.

2.3 Combined Probabilities

2.3.1 Conditional Probability

In the future, we will usually consider the space of the relevant degrees of freedom. Therefore, we will abandon an extra designation of all quantities that are related to this space. We speak now of an N-dimensional state Y and of the corresponding state space instead of relevant degrees of freedom and their corresponding subspace of dimension N_{rel}. Only if the possibility of a mistake exists will we use the old notation.

As discussed in the previous section, $p(Y,t \mid Y_0,t_0)$ is the probability density that the system in the state Y_0 at time t_0 will be in the state Y at time $t > t_0$. Hence,

$$P(R,t \mid Y_0,t_0) = \int_R dY p(Y,t \mid Y_0,t_0) \tag{2.53}$$

is the probability that the system occupies an arbitrary state of the region R at time t if the system was in the state Y_0 at time t_0. This is a special kind of conditional probability that is directly related to the time development of a complex system.

More generally, the conditional probability may be defined in the language of set theory. Here $P(A \mid B)$ is the probability that an event contained in the set A appears under the condition that we know it was also contained in the set B.

In particular, we can interpret A as the set of all trajectories of the system that touch the region R at time t, while B is the set of trajectories that go at time t_0 through the point Y_0. In this sense, each trajectory is an event. Both A and B are subsets of the set Ω of all trajectories. Then, $P(A \mid B) = P(R,t \mid Y_0,t_0)$ may be understood as the probability that any trajectory of B belongs also to A. In particular, we therefore receive the normalization condition $P(\Omega \mid B) = 1$, which allows us to conclude that

$$\int dY p(Y,t \mid Y_0,t_0) = 1. \tag{2.54}$$

Statistical independence means $P(A \mid B) = P(A)$ (i.e., the knowledge that one event occurs in B does not change the probability that it occurs in A). If $P(A \mid B) > P(A)$, we say that A and B are positively correlated, while $P(A \mid B) < P(A)$ corresponds to a negative correlation between A and B.

2.3.2 Joint Probability

Let us now consider an event that is an element of the set A as well as of the set B. Then, the event is contained also in $A \cap B$. The probability $P(A \cap B)$ is called the joint probability that the event is contained in both classes. Conditional probabilities, the joint probabilities, and the usual probabilities or unconditional probabilities become connected very naturally as

$$P(A \cap B) = P(A \mid B)P(B) = P(B \mid A)P(A). \tag{2.55}$$

This representation allows a natural definition of statistically independent events. Obviously, statistical independence requires simply $P(A \cap B) = P(A)P(B)$ and therefore $P(A \mid B) = P(A)$ and $P(B \mid A) = P(B)$. For example, the probability that a complex system stays in the infinitesimal small volume dY at time t and was in the volume dY_0 at the initial time t_0 is a typical problem to consider. The corresponding (infinitesimal) joint probability may be written as $dP(Y, t; Y_0, t_0) = p(Y, t; Y_0, t_0)\, dY dY_0$ with

$$p(Y, t; Y_0, t_0) = p(Y, t \mid Y_0, t_0)\, p(Y_0, t_0). \tag{2.56}$$

Suppose that we know all sets B_i that could condition the appearance of an event in the set A. The B_i should be mutually exclusive, $B_i \cap B_j = \emptyset$ for all $i \neq j$, and exhaustive, $\bigcup_i B_i = \Omega$. Thus, we obtain

$$P(A) = P(A \cap \Omega) = P\left(A \cap \bigcup_i B_i\right) = P\left(\bigcup_i (A \cap B_i)\right). \tag{2.57}$$

If we take into account $(A \cap B_i) \cap (A \cap B_j) = \emptyset$, we obtain due to (1.4) and (2.55)

$$P(A) = \sum_i P(A \cap B_i) = \sum_i P(A \mid B_i)\, P(B_i). \tag{2.58}$$

This general relation specifies immediately in the case of the probability density to

$$p(Y, t) = \int p(Y, t \mid Y_0, t_0)\, p(Y_0, t_0)\, dY_0. \tag{2.59}$$

Because of the symmetry $P(A \cap B) = P(B \cap A)$, we get also

$$p(Y_0, t_0) = \int p(Y, t \mid Y_0, t_0)\, p(Y_0, t_0)\, dY. \tag{2.60}$$

Due to (2.54), the last equation is a simple identity. Equation (2.56) permits in particular the extension of the initial condition (2.36) on any probability distributions.

This constitutes a warning that it is always preferable to represent each joint probability distribution function $p(Y, t; Z, \tau)$ in the form (2.56). If we want to generally determine this joint probability for $t > \tau > t_0$, then we must calculate the integral

$$p(Y, t; Z, \tau) = \int dY_0 p(Y, t \mid Z, \tau; Y_0, t_0)\, p(Z, \tau \mid Y_0, t_0)\, p(Y_0, t_0) \tag{2.61}$$

in which the conditional probability $p(Y, t \mid Z, \tau; Y_0, t_0)$ occurs. The reason for the more complicated structure consists in the fact that the deterministic character of the microscopic dynamics is possibly partially conserved on the level of the relevant degrees of freedom. Hence, it remains a certain memory

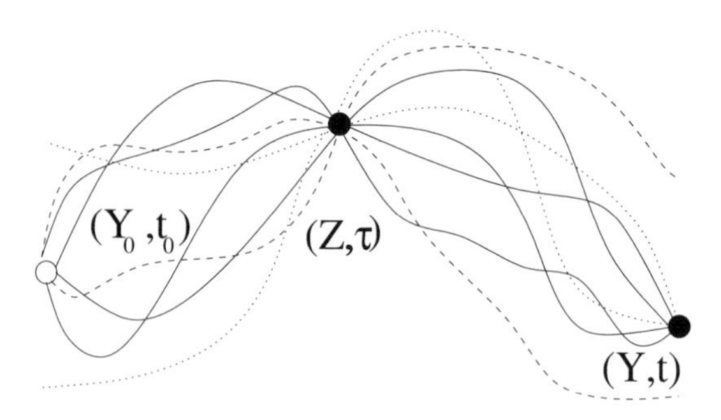

Fig. 2.1. Possible contributions to the joint probability density $p(Y, t; Z, \tau; Y_0, t_0)$. The integration over all positions Y_0 leads to $p(Y, t; Z, \tau)$. Only the full trajectories contribute to the conditional probability density $p(Z, \tau \mid Y_0, t_0)$ as well as to the conditional probability density $p(Y, t \mid Z, \tau; Y_0, t_0)$. These events form, together with the probability density $p(Y_0, t_0)$, the joint probability $p(Y, t; Z, \tau; Y_0, t_0)$. The dashed curves are also contained in $p(Z, \tau \mid Y_0, t_0)$ but not in $p(Y, t; Z, \tau)$. Roughly speaking, they are filtered out due to conditional probability $p(Y, t; \mid Z, \tau; Y_0, t_0)$ in the expression $p(Y, t; Z, \tau; Y_0, t_0) = p(Y, t; \mid Z, \tau; Y_0, t_0)p(Z, \tau \mid Y_0, t_0)p(Y_0, t_0)$. On the other hand, the product $p(Y, t; \mid Z, \tau)p(Z, \tau \mid Y_0, t_0)p(Y_0, t_0)$ considers also the dotted lines, which contribute particularly to $p(Y, t \mid Z, \tau)$. Thus, we have to expect $p(Y, t; \mid Z, \tau; Y_0, t_0) \neq p(Y, t; \mid Z, \tau)$. The equivalence between both quantities holds only if no memory effect appears.

of the initial information, which is, for instance, expressed by the appearance of the memory kernel $\hat{K}(t - t')$ in the Nakajima–Zwanzig equation. This effect indicates a possible feedback between the relevant degrees of freedom via the hidden irrelevant degrees of freedom. Only if this feedback disappears does $p(Y, t \mid Z, \tau; Y_0, t_0) = p(Y, t \mid Z, \tau)$ apply and the simpler relationship $p(Y, t; Z, \tau) = p(Y, t \mid Z, \tau)\, \rho(Z, \tau)$ become valid for arbitrary points in time (see Figure 2.1).

On the other hand, (2.56) is always valid. The correctness of this relation is justified by the fact that relevant and irrelevant degrees of freedom are assumed to be initially uncorrelated as well as by the fact that even former information is unknown. We point out again that this assumption has to be understood in the sense of the Bayesian definition of statistics.

2.4 Markov Approximation

We once again return to the problem of the selection of relevant degrees of freedom. If we define the relevant degrees of the complex system in such a way that all of these variables change relatively slowly compared with the irrelevant degrees of freedom, then the memory kernel (2.52) of the Nakajima–Zwanzig equation may approach

$$\hat{K}(t_0, t - t') = \hat{K}(t_0)\, \delta(t - t')\,. \tag{2.62}$$

This representation is called the Markov approximation. The assumption of such a separation between slow relevant timescales and fast irrelevant timescales is at least an appropriate approximation for many complex systems. But it should be remarked that there is really no such thing as a system with Markov character. If we observe the system on a very fine timescale, the immediate history will almost certainly be required to predict the probabilistic development. In other words, there is a certain characteristic time during which the previous history is important. However, systems whose memory time is so small may be, on the timescale on which we carry out observations, assumed to be Markov-like systems. We substitute (2.62) in the Nakajima–Zwanzig equation (2.50) to get

$$\frac{\partial p\left(Y,t \mid Y_0,t_0\right)}{\partial t} = -\hat{L}_{\mathrm{Markov}} p\left(Y,t \mid Y_0,t_0\right) \tag{2.63}$$

with the Markovian $\hat{L}_{\mathrm{Markov}} = \hat{M}\left(t_0\right) - \hat{K}(t_0)$. However, we know also about situations where a part of the irrelevant degrees of freedom is considerably slower than the relevant degrees of freedom and only the remaining part of the irrelevant degrees of freedom contributes to the fast dynamics. In these cases, it seems to be more favorable to derive the evolution equation for the probability density $p\left(Y,t \mid Y_0,t_0\right)$ by the use of time-dependent projectors capturing the effects of the slow irrelevant dynamics.

Such a generalization basically changes nothing in the general procedure of the separation of the timescales except for the occurrence of an explicit time dependence of the operator $\hat{L}_{\mathrm{Markov}}\left(t\right)$. Therefore, we can use the Markov approximation also for these problems. However, the concept of the separation of timescales fails or becomes uncontrolled if a suitable set of irrelevant degrees of freedom offers characteristic timescales similar to those of the relevant degrees of freedom. By assuming an infinitesimal time interval dt, we obtain from (2.63)

$$p\left(Y,t+dt \mid Y_0,t_0\right) = \left[1 - \hat{L}_{\mathrm{Markov}}\left(t\right) dt\right] p\left(Y,t \mid Y_0,t_0\right). \tag{2.64}$$

In general, we may express the operator $1 - \hat{L}_{\mathrm{Markov}}\left(t\right) dt$ by an integral representation

$$p\left(Y,t+dt \mid Y_0,t_0\right) = \int dZ U_{\mathrm{Markov}}\left(Y,t+dt \mid Z,t\right) p\left(Z,t \mid Y_0,t_0\right). \tag{2.65}$$

We multiply (2.65) with the initial distribution function $p\left(Y_0,t_0\right)$ and integrate over all configurations Y_0. Considering (2.59), we get

$$p\left(Y,t+dt\right) = \int dZ U_{\mathrm{Markov}}\left(Y,t+dt \mid Z,t\right) p\left(Z,t\right). \tag{2.66}$$

Thus, the integral kernel $U_{\mathrm{Markov}}\left(Y,t+dt \mid Z,t\right)$ can be interpreted as the conditional probability density $p\left(Y,t+dt \mid Z,t\right)$ for a transition from the state Z at time t to the state Y at time $t+dt$. This constitutes a further explanation. We remember that (2.61) requires the more general relation

$$p(Y, t + dt) = \int dY_0 \int dZ p(Y, t + dt \mid Z, t; Y_0, t_0) \times p(Z, t \mid Y_0, t_0) \, p(Y_0, t_0) \,. \tag{2.67}$$

A simple comparison between (2.66) and (2.67) leads to the necessary condition $p(Y, t + dt \mid Z, t; Y_0, t_0) = p(Y, t + dt \mid Z, t)$. This is simply another formulation of the Markov property. It is, even by itself, extremely powerful. In particular, this property means that we can define higher conditional and joint probabilities in terms of the simple conditional probability. To obtain a general relation between the conditional probabilities at different times, we shift the time $t \to t + dt$ in (2.65) and obtain

$$p(Y, t + 2dt \mid Y_0, t_0) = \int dZ p(Y, t + 2dt \mid Z, t + dt) \times p(Z, t + dt \mid Y_0, t_0) \,. \tag{2.68}$$

On the other hand, the transformation $dt \to 2dt$ leads to

$$p(Y, t + 2dt \mid Y_0, t_0) = \int dZ p(Y, t + 2dt \mid Z, t) \, p(Z, t \mid Y_0, t_0) \tag{2.69}$$

so that we obtain from (2.65), (2.68), and (2.69)

$$p(Y, t + 2dt \mid Z, t) = \int dX p(Y, t + 2dt \mid X, t + dt) \times p(X, t + dt \mid Z, t) \,. \tag{2.70}$$

When repeating this procedure infinitely many times, one obtains a relation for finite time differences

$$p(Y, t \mid Z, t'') = \int dX p(Y, t \mid X, t') \, p(X, t' \mid Z, t'') \,, \tag{2.71}$$

which is the Chapman–Kolmogorov equation. This equation is a rather complex nonlinear functional equation relating all conditional probabilities obtained from a given Markovian to each other.

This is a remarkable result: the conditional probability density obtained from an arbitrary Markovian must satisfy the Chapman-Kolmogorov equation. In addition, the Chapman–Kolmogorov equation is an important criterion for presence of the Markov property. Whenever empirically determined conditional probabilities satisfy (2.71), we are able to introduce the Markov property.

2.5 Generalized Fokker–Planck Equation

2.5.1 Differential Chapman–Kolmogorov Equation

The determination of the Markovian for a given process on the basis of a microscopic theory is probably excluded. Therefore, we are always dependent on empirical considerations and observations. It would be reasonable to know some rules from which the Markovian $\hat{L}_{\text{Markov}}$ could be constructed.

For all evolutionary processes with Markov properties, the parameter-free Chapman–Kolmogorov equation (2.71) is a universal relation. However, the Chapman–Kolmogorov equation has many solutions. In particular, for a given dimension N of the state Y, every solution of (2.63) must also be a solution of (2.71) independent from the special mathematical structure of the operator $\hat{L}_{\text{Markov}}$. Therefore, we could possibly use this equation to obtain information about the general mathematical structure of $\hat{L}_{\text{Markov}}$. To do so, we follow Gardiner [143] and define the subsequent quantities for all $\varepsilon > 0$:

$$A_\alpha (Z,t) = \lim_{\delta t \to 0} \frac{1}{\delta t} \int\limits_{|Y-Z|<\varepsilon} dY p(Y, t+\delta t \mid Z, t) \, \Delta Y_\alpha + o(\varepsilon) \tag{2.72}$$

and

$$B_{\alpha\beta} (Z,t) = \lim_{\delta t \to 0} \frac{1}{\delta t} \int\limits_{|Y-Z|<\varepsilon} dY p(Y, t+\delta t \mid Z, t) \, \Delta Y_\alpha \Delta Y_\beta + o(\varepsilon), \tag{2.73}$$

where we have used the notation $\Delta Y_\alpha = Y_\alpha - Z_\alpha$. Furthermore, we introduce

$$W(Y \mid Z; t) = \lim_{\delta t \to 0} \frac{1}{\delta t} p(Y, t+\delta t \mid Z, t) \tag{2.74}$$

for $|Y - Z| > \varepsilon$. We will see later that these quantities were chosen in a very natural way. They can be obtained directly from observations or defined by suitable model assumptions.

If we are able to build the Markovian, $\hat{L}_{\text{Markov}}$ by the exclusive use of these quantities, we have arrived at our goal. Note that possible higher-order coefficients must vanish for $\varepsilon \to 0$. For instance, the third-order quantity defined by

$$C_{\alpha\beta\gamma} (Z,t) = \lim_{\delta t \to 0} \frac{1}{\delta t} \int\limits_{|Y-Z|<\varepsilon} dY p(Y, t+\delta t \mid Z, t) \, \Delta Y_\alpha \Delta Y_\beta \Delta Y_\gamma \tag{2.75}$$

may be approximated by $|C_{\alpha\beta\gamma}| \simeq |B_{\alpha\beta}| \, \varepsilon = o(\varepsilon)$. Thus, if $B_{\alpha\beta}$ exists, the coefficient $C_{\alpha\beta\gamma}$ is of an order of magnitude $o(\varepsilon)$ and disappears for $\varepsilon \to 0$. To proceed, we consider the time evolution of the average of an arbitrary function f that is twice continuously differentiable,

$$\begin{aligned}
&\frac{\partial}{\partial t} \int dZ f(Z) \, p(Z, t \mid Y, t') \\
&= \lim_{\delta t \to 0} \frac{1}{\delta t} \int dZ f(Z) \left[p(Z, t+\delta t \mid Y, t') - p(Z, t \mid Y, t') \right] dZ \\
&= \lim_{\delta t \to 0} \frac{1}{\delta t} \iint dZ dX f(Z) \, p(Z, t+\delta t \mid X, t) \, P(X, t \mid Y, t') \\
&\quad - \lim_{\delta t \to 0} \frac{1}{\delta t} \iint dZ dX f(X) \, p(Z, t+\delta t \mid X, t) \, P(X, t \mid Y, t'),
\end{aligned} \tag{2.76}$$

where we have used the Chapman–Kolmogorov equation (2.71) in the first term and the normalization condition (2.54) to produce the corresponding terms in (2.76). Since $f(Z)$ is twice continuously differentiable, we may write

$$\begin{aligned} f(Z) = f(X) &+ \sum_{\alpha} \frac{\partial f(X)}{\partial X_{\alpha}} [Z_{\alpha} - X_{\alpha}] \\ &+ \frac{1}{2} \sum_{\alpha\beta} \frac{\partial^2 f(X)}{\partial X_{\alpha} \partial X_{\beta}} [Z_{\alpha} - X_{\alpha}] [Z_{\beta} - X_{\beta}] + R(Z, X), \end{aligned} \tag{2.77}$$

where the reminder function $R(Z, X)$ vanishes for $|X - Z| = \varepsilon \to 0$ as $o(\varepsilon^2)$. We now divide the integrals in (2.76) into two regions, $|X - Z| \leq \varepsilon$ and $|X - Z| > \varepsilon$, and substitute (2.77) into (2.76):

$$\begin{aligned} &\frac{\partial}{\partial t} \int dZ f(Z) p(Z, t \mid Y, t') \\ &= \lim_{\delta t \to 0} \frac{1}{\delta t} \iint_{|Z-X|<\varepsilon} dZ dX p(Z, t + \delta t \mid X, t) p(X, t \mid Y, t') \\ &\quad \times \left\{ \sum_{\alpha} \frac{\partial f(X)}{\partial X_{\alpha}} [Z_{\alpha} - X_{\alpha}] + \frac{1}{2} \sum_{\alpha\beta} \frac{\partial^2 f(X)}{\partial X_{\alpha} \partial X_{\beta}} [Z_{\alpha} - X_{\alpha}] [Z_{\beta} - X_{\beta}] \right\} \\ &\quad + \lim_{\delta t \to 0} \frac{1}{\delta t} \iint_{|Z-X|<\varepsilon} dZ dX R(Z, X) p(Z, t + \delta t \mid X, t) p(X, t \mid Y, t') \\ &\quad + \lim_{\delta t \to 0} \frac{1}{\delta t} \iint_{|Z-X|<\varepsilon} dZ dX f(X) p(Z, t + \delta t \mid X, t) p(X, t \mid Y, t') \\ &\quad + \lim_{\delta t \to 0} \frac{1}{\delta t} \iint_{|Z-X|\geq\varepsilon} dZ dX f(Z) p(Z, t + \delta t \mid X, t) p(X, t \mid Y, t') \\ &\quad - \lim_{\delta t \to 0} \frac{1}{\delta t} \iint dZ dX f(X) p(Z, t + \delta t \mid X, t) p(X, t \mid Y, t') . \end{aligned} \tag{2.78}$$

Let us compute the limit $\varepsilon \to 0$ line-by-line. The first term of this expression can be transformed in the following way. We take the limit $\delta t \to 0$ inside the integral to obtain, with the help of (2.72) and (2.73),

$$\begin{aligned} (1) = &\int dX p(X, t \mid Y, t') \\ &\times \left\{ \sum_{\alpha} A_{\alpha}(X, t) \frac{\partial f(X)}{\partial X_{\alpha}} + \frac{1}{2} \sum_{\alpha\beta} B_{\alpha\beta}(X, t) \frac{\partial^2 f(X)}{\partial X_{\alpha} \partial X_{\beta}} \right\} . \end{aligned} \tag{2.79}$$

The second term of (2.78) disappears for $\varepsilon \to 0$ due to $R(Z, X) \sim o(|Z - X|^2)$. The third term and the fifth term can be collected in one expression. Thus, we get

$$(3)+(5) = -\lim_{\delta t\to 0}\frac{1}{\delta t}\iint_{|Y-Z|\geq\varepsilon} dZdXf(X)\,p(Z,t+\delta t\mid X,t)$$
$$\times p(X,t\mid Y,t')\,, \tag{2.80}$$

or with the use of (2.74) and considering $\varepsilon\to 0$,

$$(3)+(5) = -\mathcal{H}\iint dZdXf(X)\,W(Z\mid X;t)\,p(X,t\mid Y,t')\,. \tag{2.81}$$

Notice that we use the symbol $\mathcal{H}$ to indicate the principal value integral. We finally get for the fourth term

$$(4) = \mathcal{H}\iint dXdZf(Z)\,W(Z\mid X;t)\,p(X,t\mid Y,t')\,. \tag{2.82}$$

We put these results together to obtain

$$\int dX\frac{\partial}{\partial t}p(X,t\mid Y,t')\,f(X) = \int dXp(X,t\mid Y,t')$$
$$\times\left\{\sum_\alpha A_\alpha(X,t)\frac{\partial f(X)}{\partial X_\alpha}+\frac{1}{2}\sum_{\alpha\beta}B_{\alpha\beta}(X,t)\frac{\partial^2 f(X)}{\partial X_\alpha\partial X_\beta}\right\}$$
$$+\mathcal{H}\iint dZdX\,[f(X)-f(Z)]\,W(X\mid Z;t)\,p(Z,t\mid Y,t')\,, \tag{2.83}$$

and after integrating by parts, we get

$$\frac{\partial}{\partial t}\int dXf(X)\,p(X,t\mid Y,t') = \int dXf(X)$$
$$\times\left\{-\sum_\alpha\frac{\partial}{\partial X_\alpha}A_\alpha(X,t)+\frac{1}{2}\sum_{\alpha\beta}\frac{\partial^2}{\partial X_\alpha\partial X_\beta}B_{\alpha\beta}(X,t)\right\}p(X,t\mid Y,t')$$
$$+\mathcal{H}\iint dZdX\,[f(X)-f(Z)]\,W(X\mid Z;t)\,p(Z,t\mid Y,t')\,. \tag{2.84}$$

Finally, we consider that we have chosen the function f to be arbitrary. We can then deduce that the conditional probability fulfills the relation

$$\frac{\partial}{\partial t}p(X,t\mid Y,t') = -\sum_\alpha\frac{\partial}{\partial X_\alpha}A_\alpha(X,t)p(X,t\mid Y,t')$$
$$+\frac{1}{2}\sum_{\alpha\beta}\frac{\partial^2}{\partial X_\alpha\partial X_\beta}B_{\alpha\beta}(X,t)p(X,t\mid Y,t') \tag{2.85}$$
$$+\mathcal{H}\int dZ\,[W(X\mid Z;t)\,p(Z,t\mid Y,t')-W(Z\mid X;t)\,p(X,t\mid Y,t')]\,.$$

This equation is the differential form of the Chapman–Kolmogorov equation, which is denoted in the literature as the forward differential Chapman–Kolmogorov equation. The right-hand side of this equation defines the general structure of the Markovian $\hat{L}_{\mathrm{Markov}}$.

If we want to specify the differential Chapman–Kolmogorov equation, we must consider that by definition the components $B_{\alpha\beta}(X,t)$ must form a positive-definite matrix and that $W(X \mid Z;t)$ must be a nonnegative function. Then, it can be shown under certain conditions that a nonnegative solution to the differential Chapman–Kolmogorov equation exists and that this solution also satisfies the Chapman–Kolmogorov equation. The conditions to be satisfied are the initial condition

$$p(X,t \mid Y,t) = \delta(X-Y), \tag{2.86}$$

which follows directly from (2.59), and any appropriate boundary conditions.

We may also derive the backward differential Chapman–Kolmogorov equations, which give the time evolution with respect to the initial variables of $p(X,t \mid Y,t')$. To do this, we consider

$$\begin{aligned}
\frac{\partial}{\partial t'} p(X,t \mid Y,t') &= \lim_{\delta t \to 0} \frac{1}{\delta t'} \left[p(X,t \mid Y,t') - p(X,t \mid Y,t'-\delta t') \right] \\
&= \lim_{\delta t \to 0} \frac{1}{\delta t'} \int dZ p(Z,t' \mid Y,t'-\delta t') p(X,t \mid Y,t') \\
&\quad - \lim_{\delta t \to 0} \frac{1}{\delta t'} \int dZ p(X,t \mid Z,t') p(Z,t' \mid Y,t'-\delta t')
\end{aligned} \tag{2.87}$$

by use of the normalization condition (2.54) in the first term and the Chapman–Kolmogorov equation (2.71) in the second term. It is easy to show that we can carry out the infinitesimal shift $p(Z,t' \mid Y,t'-\delta t') \to p(Z,t'+\delta t' \mid Y,t')$ without a noticeable change of (2.87). Hence, we get

$$\begin{aligned}
\frac{\partial}{\partial t'} p(X,t \mid Y,t') = \lim_{\delta t \to 0} \int dZ \, & \frac{p(Z,t'+\delta t' \mid Y,t')}{\delta t'} \\
& \times \left[p(X,t \mid Y,t') - p(X,t \mid Z,t') \right]
\end{aligned} \tag{2.88}$$

and therefore, using techniques similar to those used for the derivation of (2.85),

$$\begin{aligned}
\frac{\partial}{\partial t'} p(X,t \mid Y,t') &= -\sum_{\alpha} A_{\alpha}(Y,t') \frac{\partial}{\partial Y_{\alpha}} p(X,t \mid Y,t') \\
-\frac{1}{2} \sum_{\alpha\beta} & B_{\alpha\beta}(Y,t') \frac{\partial^2}{\partial Y_{\alpha} \partial Y_{\beta}} p(X,t \mid Y,t') \\
+ \mathcal{H} \int & dZ W(Z \mid Y;t') \left[p(X,t \mid Y,t') - p(X,t \mid Z,t') \right].
\end{aligned} \tag{2.89}$$

This equation is called the backward differential Chapman–Kolmogorov equation. We remark that the forward and backward equations are equivalent to each other. The main difference is which set of variables is held fixed. For the forward equation, solutions exist for $t \geq t'$ and (2.86) is the initial condition with respect to the free variables (X,t). In the case of the backward equation, we hold (X,t) fixed so that since the backward equation expresses development in $t' \leq t$, (2.86) is the final condition of (2.89).

2.5.2 Deterministic Processes

There are also in complex systems phenomena that may be described by a completely deterministic motion. If the corresponding processes possess also a Markov character, then they can be described by a differential Chapman–Kolmogorov equation with $B_{\alpha\beta} = 0$ and $W(Y \mid Z; t) = 0$. It remains the equation

$$\frac{\partial}{\partial t} p(X, t \mid Y, t') = -\sum_{\alpha} \frac{\partial}{\partial X_\alpha} A_\alpha(X, t) p(X, t \mid Y, t') . \tag{2.90}$$

The solution to this equation with the initial condition (2.86) is

$$p(X, t \mid Z, t') = \delta\left(X - \widetilde{X}(t)\right) . \tag{2.91}$$

This means that the system moves along the trajectory $\widetilde{X}(t) = \{\widetilde{x}_1(t), \widetilde{x}_2(t), ...\}$ obtained by solving the ordinary differential equations

$$\frac{d\widetilde{x}_\alpha(t)}{dt} = A_\alpha(\widetilde{X}(t), t) = A_\alpha(\{\widetilde{x}_1(t), \widetilde{x}_2(t), ...\}, t) \tag{2.92}$$

with the initial conditions

$$\widetilde{x}(t') = \{\widetilde{x}_1(t'), \widetilde{x}_2(t'), ...\} = Y. \tag{2.93}$$

The equations (2.92) are also called kinetic equations. These differential equations, however, are generally not equations of motion in the sense of classical mechanics. In particular, most kinetic equations are irreversible (i.e., they are not invariant under reversal of the time direction).

If a system is far from possible stationary states, then such equations describe a complex system mostly sufficiently. The structure of the solution (2.91) indicates the deterministic character of (2.90). To demonstrate the validity of (2.91), we point out first that for $t = t'$ the initial conditions (2.93) lead to $P(X, t \mid Z, t') = \delta(X - Z)$. The proof for all other times is best obtained by direct substitution. We see that

$$\begin{aligned}
\frac{dp(X, t \mid Y, t')}{dt} &= -\sum_{\alpha} \left[\frac{\partial}{\partial X_\alpha} \delta\left(X - \widetilde{X}(t)\right)\right] \frac{d\widetilde{x}_\alpha(t)}{dt} \\
&= \sum_{\alpha} \frac{\partial}{\partial X_\alpha} \left[\delta\left(X - \widetilde{X}(t)\right) A_\alpha(\widetilde{X}(t), t)\right] \\
&= \sum_{\alpha} \frac{\partial}{\partial X_\alpha} \left[A_\alpha(X, t) P(X, t \mid Y, t')\right]
\end{aligned} \tag{2.94}$$

leads to the expected identity. It should be remarked that the methods of characteristics can be used to obtain (2.92) from (2.90) in a direct way.

2.5.3 Fokker–Planck Equation

If we assume the quantities $W(X \mid Z; t)$ to be zero, the differential Chapman–Kolmogorov equation reduces to the generalized Fokker–Planck equation

$$\frac{\partial}{\partial t} p(X, t \mid Y, t') = -\sum_{\alpha} \frac{\partial}{\partial X_{\alpha}} A_{\alpha}(X, t) p(X, t \mid Y, t') + \sum_{\alpha\beta} \frac{1}{2} \frac{\partial^2}{\partial X_{\alpha} \partial X_{\beta}} B_{\alpha\beta}(X, t) p(X, t \mid Y, t') . \tag{2.95}$$

The functions $A_{\alpha}(X, t)$ are the components of the drift vector, and the $B_{\alpha\beta}(X, t)$ are known as components of the diffusion matrix. As mentioned above, the diffusion matrix is symmetric and positive-semidefinite as a result of the definition (2.73).

The general solution of this differential equation cannot be given explicitly. However, we can use certain similarities in the mathematical structure of the Fokker–Planck equation and the Schrödinger equation of quantum mechanics in order to transfer well-known solution methods.

But there are some important differences between both equations that have to do with the operator structure of the right-hand side. Each Schrödinger equation always requires a self-adjoint but not necessarily positive-definite Hamilton operator, while the differential operator of a Fokker–Planck equation must be positive-semidefinite but not self adjoint. In the case of constant components A_{α} and $B_{\alpha\beta}$, the Fokker–Planck equation (2.95) can be solved exactly, subject to the initial condition (2.86), and we arrive at

$$p(X, t \mid Y, t') = \frac{1}{\sqrt{\det(2\pi B \Delta t)}} \exp\left\{ -\frac{1}{2\Delta t} \sum_{\alpha\beta} \Delta X_{\alpha} B^{-1}_{\alpha\beta} \Delta X_{\beta} \right\} \tag{2.96}$$

with $\Delta X_{\alpha} = X_{\alpha} - Y_{\alpha} - A_{\alpha} \Delta t$, $\Delta t = t - t'$, and N the dimension of the state space. This is nothing other than a multivariable Gaussian probability distribution function. The initial condition appears for $\Delta t \to 0$, while the Gaussian spreads over the whole space for $\Delta t \to \infty$. The center of the Gaussian moves with the constant velocity $A = \{A_1, A_2, ...\}$.

2.5.4 The Master Equation

Finally, we consider the case $A_{\alpha} = B_{\alpha\beta} = 0$ so that we now have

$$\frac{\partial}{\partial t} p(X, t \mid Y, t') = \mathcal{H} \int dZ W(X \mid Z; t) p(Z, t \mid Y, t') - \mathcal{H} \int dZ W(Z \mid X; t) p(X, t \mid Y, t') . \tag{2.97}$$

This is a so called master (or rate) equation. The initial condition of this equation is again given by (2.86). In order to discuss the underlying processes described by master equations, we solve (2.97) approximately in first order to a small time interval δt. The short-time solution to this equation with the initial condition (2.86) is

$$p(X, t' + \delta t \mid Y, t') = \delta(X - Y)\left[1 - \mathcal{H}\int dZ W(Z \mid Y; t)\,\delta t\right] + W(X \mid Y; t)\,\delta t. \tag{2.98}$$

The first contribution corresponds to a finite probability for the system to stay at the original position Y in the state space. This probability decreases with increasing time. The probability that the system does not remain at Y is given by the second term of (2.98). Hence, a characteristic path of the system through the state space will consist of a series of discontinuous jumps whose distribution is given by $W(X \mid Y; t)$. For this reason, processes described by master equations are denoted as jump processes.

The master equation (2.97) may be specified for the case where the state space consists of discrete numbers only. Then, the master equation takes the form

$$\frac{\partial}{\partial t} p_{nn'}(t, t') = \sum_m \left[W_{nm}(t)\, p_{mn'}(t, t') - W_{mn}(t)\, p_{nn'}(t, t')\right] \tag{2.99}$$

with $W_{nm}(t) = W(n \mid m; t)$ and $p_{nm}(t, t') = p(n, t \mid m, t')$. In this representation, the concept of jump processes becomes particularly clear. But it should be noted again that pure jump processes can occur even in a continuous state space[1].

2.6 Correlation and Stationarity

2.6.1 Stationarity

The macroscopic dynamics of a complex system are stationary in a strict sense if all joint probabilities are invariant under a time shift Δt:

$$p(X_1, t_1; ..., X_n, t_n; ...) = p(X_1, t_1 + \Delta t; ..., X_n, t_n + \Delta t; ...). \tag{2.100}$$

From here, we conclude that $p(X, t) = p(X)$ so that, due to (2.55), all conditional probabilities are also invariant under a time shift.

Furthermore, the definition of stationarity implies that the operators of the Nakajima-Zwanzig equation (2.50) no longer depend on the initial time, $\hat{M}(t_0) = \hat{M}$ and $\hat{K}(t_0, t - t') = \hat{K}(t - t')$, and that the coefficients of the differential Chapman–Kolmogorov equation (2.85) are simple functions of

[1] Of course, the trajectory of a large but finite complex system has no real jumps. The appearance of jumps is the result of the Markov approximation. If we take into account the exact equation of motion, the jumps correspond to relatively fast but continuous changes of the macroscopic state during a short time period.

the states (i.e., $A_\alpha(X,t) = A_\alpha(X)$, $B_{\alpha\beta}(X,t) = B_{\alpha\beta}(X)$, and $W(X \mid Y;t) = W(X \mid Y)$). Finally, stationarity means that all moments and cumulants have constant values.

There exist several other definitions of stationary processes that are, in fact, less restrictive. For instance, an nth order stationary process arises when (2.100) holds only for joint probability distribution functions of less than $n+1$ points in time, while asymptotically stationary processes are observed only for infinitely large shifts Δt.

2.6.2 Correlation

The knowledge of moments and cumulants does not tell a great deal about the dynamics of a complex system. For instance, all moments have constant values in the special case of stationary systems. What would be of interest are measurable quantities obtained from joint probabilities (or alternatively from the conditional probabilities) that give simultaneously information about the state of the system at several points in time. The simplest quantity is the correlation function

$$\overline{f(t)g(t')} = \int dX dX' f(X) g(X') p(X,t;X',t') \tag{2.101}$$

between two arbitrary functions, f and g, of the state of the system. A special case is the autocorrelation function $\overline{f(t)f(t')}$, which considers the same quantity at different time points. Higher correlations can be constructed in a similar way:

$$\overline{f_1(t_1)...f_n(t_n)} = \int \prod_i^n \left[dX^{(i)} f_i \left(X^{(i)} \right) \right] p(X^{(1)},t_1;...;X^{(n)},t_n). \tag{2.102}$$

A very natural class of correlation functions may be obtained by identifying the functions f_i with the components X_α of the state vector. These correlation functions may be interpreted as generalized moments, $\overline{x_{\alpha_1}(t_1)...x_{\alpha_n}(t_n)}$, which approach the standard definition (2.12) for $t_1 = ... = t_n = t$. Analogously, we can design combinations of correlation functions that may be understood as a generalization of the cumulants (2.19). For instance, the generalized covariance functions are defined by

$$C_{\alpha\beta}(t,t') = \overline{x_\alpha(t)x_\beta(t')} - \overline{x_\alpha(t)}\ \overline{x_\beta(t')}, \tag{2.103}$$

which are useful to consider for processes with average values different from zero. Because of (2.100), stationarity always requires that the correlation functions be invariant under a time shift. In particular, we get for stationary processes $\overline{x_\alpha(t)x_\beta(t')} = \overline{x_\alpha(t-t')x_\beta(0)}$ and $C_{\alpha\beta}(t,t') = C_{\alpha\beta}(t-t',0)$, which is simply denoted as $C_{\alpha\beta}(t-t')$.

Finally, we introduce the so-called correlation time. For the sake of simplicity, here we concentrate on the discussion of the autocovariance function $C(t-t')$ of a stationary process in a one-dimensional state space. The symmetry of this function requires immediately

$$C(t-t') = C(t'-t) = C(|\, t-t' \,|). \tag{2.104}$$

One possibility that provides a measure for the corresponding correlation time τ_c is the integral

$$\tau_c = C^{-1}(0) \int_0^\infty C(t)dt. \tag{2.105}$$

This definition is independent of the precise functional form of the autocovariance function. The correlation time τ_c may have a finite, infinite, or indeterminate value.

The last case is more or less irrelevant for the majority of complex systems. It corresponds, for instance, to periodic autocovariance functions (e.g., $C(t) \sim \cos(\omega t)$). A divergent timescale $\tau_c \to \infty$ indicates the existence of a dominant correlation. Processes characterized by such an autocorrelation function are said to be long-range correlated. For instance, $C(t) \sim t^{-a}$ with $0 < a < 1$ is a typical autocovariance function of long-range correlated processes.

When the integral (2.105) is finite, the corresponding process shows a short-range correlated behavior. In this case, the values of the observed quantities are practically uncorrelated if sequentially realized measurements are separated by a timescale sufficiently longer than τ_c. An important example is the autocovariance of the Ornstein–Uhlenbeck process, $C \sim \exp\{-t/\tau_c\}$, which is often used to model a realistic noise signal.

2.6.3 Spectra

In order to characterize a stationary process, it is very natural to calculate the Fourier transform for the covariance functions $C_{\alpha\beta}(t)$:

$$S_{\alpha\beta}(\omega) = \int_{-\infty}^{+\infty} dt C_{\alpha\beta}(t) \exp\{i\omega t\}. \tag{2.106}$$

Due to the symmetry property $C_{\alpha\beta}(t) = C_{\beta\alpha}(-t)$, the Fourier transforms are self-adjoint; $S_{\alpha\beta}(\omega) = S^*_{\beta\alpha}(\omega)$. The $S_{\alpha\beta}(\omega)$ are called the spectral functions of the underlying processes. In addition to the classification of correlation processes introduced above, the same properties might be investigated in the frequency domain.

To this end, we consider again a stationary process in a one-dimensional state space. We obtain from (2.106) and (2.105) the important relation $\tau_c = S(0)$. Therefore, we conclude that a convergent behavior of $S(\omega)$ in the low-frequency regions indicates a short-range correlation, while any kind of divergence is related to long-range correlations.

2.7 Stochastic Equations of Motion

2.7.1 The Mori–Zwanzig Equation

At the beginning of the book, we already mentioned that the microscopic mechanical (or quantum-mechanical) equations of motion and the Liouville equation are equivalent representations of a given system. Subsequently, we demonstrated that the Liouville equation could be reduced to a Nakajima–Zwanzig equation, which contains only the relevant degrees of freedom representing a suitable description of the system on a more or less macroscopic level. It is reasonable to ask whether one can also reduce the complicated system of microscopic mechanical equations of motion to a macroscopic description. To this end, we introduce a set of linearly independent, differentiable functions $G_\alpha(t)$ ($\alpha = 1, ..., M$) that are assumed to be functions of the microscopic state $\Gamma = \{q_1, ..., p_N\}$, $G_\alpha(t) = G_\alpha(\Gamma(t))$. All of these functions are denoted as relevant variables representing macroscopically observable or measurable quantities of interest. The time evolution of each G_α is ruled by the microscopic equations of motion

$$\frac{dG_\alpha}{dt} = \sum_i^N \left[\frac{\partial G_\alpha}{\partial q_i} \dot{q}_i + \frac{\partial G_\alpha}{\partial p_i} \dot{p}_i \right] = \sum_i^N \left[\frac{\partial G_\alpha}{\partial q_i} \frac{\partial H}{\partial p_i} - \frac{\partial G_\alpha}{\partial p_i} \frac{\partial H}{\partial q_i} \right] \tag{2.107}$$

where we have used Hamilton's equations (1.1) describing the evolution of all $2N$ microscopic coordinates q_i and momenta p_i. From (2.107), we get

$$\frac{dG_\alpha}{dt} = \hat{L} G_\alpha, \tag{2.108}$$

where the Liouvillian $\hat{L}$ is defined as in (1.10). Equation (2.108) is formally integrated to yield

$$G_\alpha(t) = \exp\{\hat{L}(t - t_0)\} G_\alpha^0, \tag{2.109}$$

and the $G_\alpha^0 = G_\alpha(\Gamma_0)$ are fixed by the microscopic initial state $\Gamma_0 = \Gamma(t_0)$. In other words, the relevant quantities $G_\alpha(t)$ of a given system are unique functions of the initial state, $G_\alpha(t) = G_\alpha(t, \Gamma_0)$. To proceed, we should eliminate all irrelevant degrees of freedom contained in the Liouvillian by the use of an appropriate projection formalism.

A convenient starting point for this intention is the introduction of a scalar product (A, B). As is easily verified, the scalar product properties are satisfied by identifying, for instance, the scalar product with the average

$$(A, B) = \int d\Gamma_0 A(\Gamma_0) B(\Gamma_0) p(\Gamma_0, t_0) \tag{2.110}$$

considering the probability distribution function at the initial time t_0. This representation has the advantage that the scalar products

$$(G_\alpha(t), G_\beta(t')) = \int d\Gamma_0 G_\alpha(t, \Gamma_0) G_\beta(t', \Gamma_0) p(\Gamma_0, t_0) \tag{2.111}$$

are identical to the correlation functions, $(G_\alpha(t), G_\beta(t')) = \overline{G_\alpha(t) G_\beta(t')}$. In order to interpret (2.111) we must consider that each microscopic initial state Γ_0 defines a unique trajectory of the system through the phase space, while the statistical weight of each trajectory is given by $p(\Gamma_0, t_0)\, d\Gamma_0$. We remark that many other possibilities defining a suitable scalar product exist. For the following derivation, the precise structure of the scalar product is of secondary interest.

The central point is the introduction of an appropriate projection operator $\hat{P}$. We use the definition

$$\hat{P} = \sum_{\alpha\beta} G_\alpha^0 H_{\alpha\beta} (G_\beta^0, ...), \tag{2.112}$$

where the $H_{\alpha\beta}$ are the components of the inverse of the $M \times M$ matrix formed by the scalar products (G_α^0, G_β^0). Obviously, we get $\hat{P} G_\alpha^0 = G_\alpha^0$ and therefore $\hat{P}^2 G_\alpha^0 = G_\alpha^0$. Thus, the projection operator and the corresponding complementary operator $\hat{Q} = 1 - \hat{P}$ fulfill the relations (2.33). Let us now consider the formal solution (2.109) of the equation of motion and insert the identity operator $\hat{P} + \hat{Q}$ after the propagator $\exp\{\hat{L}(t - t_0)\}$. We obtain

$$\begin{aligned} \frac{dG_\alpha(t)}{dt} &= \exp\{\hat{L}(t - t_0)\} \left(\hat{P} + \hat{Q} \right) \hat{L} G_\alpha^0 \\ &= \exp\{\hat{L}(t - t_0)\} \hat{P} \hat{L} G_\alpha^0 + \exp\{\hat{L}(t - t_0)\} \hat{Q} \hat{L} G_\alpha^0 . \end{aligned} \tag{2.113}$$

The first term can be written as

$$\exp\{\hat{L}(t - t_0)\} \hat{P} \hat{L} G_\alpha^0 = \sum_{\beta\gamma} \Omega_{\alpha\gamma} G_\gamma(t), \tag{2.114}$$

where we have introduced the $M \times M$ frequency matrix $\Omega_{\alpha\gamma}$,

$$\Omega_{\alpha\gamma} = \sum_\beta H_{\gamma\beta} (G_\beta^0, \hat{P} \hat{L} G_\alpha^0). \tag{2.115}$$

The second term in equation (2.113) can be rearranged by using the identity

$$\mathrm{e}^{\hat{L}(t-t_0)} = \mathrm{e}^{\hat{L}_1(t-t_0)} + \int_{t_0}^{t} dt' \mathrm{e}^{\hat{L}(t-t')} \hat{L}_2 \mathrm{e}^{\hat{L}_1(t'-t_0)} \tag{2.116}$$

with $\hat{L} = \hat{L}_1 + \hat{L}_2$. This identity may be checked in a similar way as (2.41) by a derivative with respect to the time. If we replace $\hat{L}_1$ by $\hat{Q}\hat{L}$ and $\hat{L}_2$ by $\hat{P}\hat{L}$, we arrive at

$$\begin{aligned} \mathrm{e}^{\hat{L}(t-t_0)} \hat{Q} \hat{L} G_\alpha^0 &= \mathrm{e}^{\hat{Q}\hat{L}(t-t_0)} \hat{Q} \hat{L} G_\alpha^0 \\ &\quad + \int_{t_0}^{t} dt' \mathrm{e}^{\hat{L}(t-t')} \hat{P} \hat{L} \mathrm{e}^{\hat{Q}\hat{L}(t'-t_0)} \hat{Q} \hat{L} G_\alpha^0 . \end{aligned} \tag{2.117}$$

Because of the properties (2.33), the first contribution can be transformed into

$$f_\alpha(t) = \mathrm{e}^{\hat{Q}\hat{L}(t-t_0)}\hat{Q}\hat{L}G^0_\alpha = \hat{Q}\mathrm{e}^{\hat{Q}\hat{L}(t-t_0)}\hat{Q}\hat{L}G^0_\alpha. \tag{2.118}$$

This quantity is referred to as the fluctuating force or the residual force. By construction, the time evolution of $f_\alpha(t)$ from its initial value $\hat{Q}\hat{L}G^0_\alpha$ is ruled by the anomalous propagator $\exp\{\hat{Q}\hat{L}(t-t_0)\}$ rather than by the usual one, $\exp\{\hat{L}(t-t_0)\}$. The presence of the complementary projection operator $\hat{Q}$ has the important consequence that

$$\left(G^0_\beta, f_\alpha(t)\right) = 0 \ . \tag{2.119}$$

From a geometrical point of view, the fluctuating forces $f_\alpha(t)$ are orthogonal to all initial relevant quantities G^0_β at all times. In other words, the forces evolve in a subspace intrinsically different from the one spanned by the set G^0_β. The first term on the right-hand side of (2.117) can be transformed into a more convenient form. We write

$$\begin{aligned}
&\mathrm{e}^{\hat{L}(t'-t_0)}\hat{P}\hat{L}\mathrm{e}^{\hat{Q}\hat{L}(t-t')}\hat{Q}\hat{L}G^0_\alpha \\
&= \mathrm{e}^{\hat{L}(t'-t_0)}\sum_{\gamma\beta} G^0_\gamma H_{\gamma\beta}(G^0_\beta, \hat{L}\hat{Q}\mathrm{e}^{\hat{Q}\hat{L}(t-t')}\hat{Q}\hat{L}G^0_\alpha) \\
&= \sum_{\beta\gamma} G_\gamma(t') H_{\gamma\beta}(G^0_\beta, \hat{L}\hat{Q}\mathrm{e}^{\hat{Q}\hat{L}(t-t')}\hat{Q}\hat{L}G^0_\alpha),
\end{aligned} \tag{2.120}$$

where we have used (2.109). As a result, the equation of motion (2.113) can be written as

$$\frac{dG_\alpha(t)}{dt} = \sum_\gamma \left[\Omega_{\alpha\gamma}G_\gamma(t) + \int_{t_0}^{t} dt' K_{\alpha\gamma}(t-t')G_\gamma(t')\right] + f_\alpha(t), \tag{2.121}$$

where we have introduced the quantity

$$\begin{aligned}
K_{\alpha\gamma}(t-t') &= \sum_\beta H_{\gamma\beta}(G^0_\beta, \hat{L}\hat{Q}\mathrm{e}^{\hat{Q}\hat{L}(t-t')}\hat{Q}\hat{L}G^0_\alpha) \\
&= \sum_\beta H_{\gamma\beta}(G^0_\beta, \hat{L}f_\alpha(t-t')),
\end{aligned} \tag{2.122}$$

which is referred to as the memory matrix. It should be pointed out that both the frequency matrix and the memory matrix still depend on the initial time t_0. Equation (2.121) is called the generalized Langevin equation or the Mori–Zwanzig equation. No approximation has been taken into account in the previous derivation, so (2.121) is still equivalent to the mechanical equations of motion (2.107).

A special situation occurs in the case of stationarity. Then, we obtain

$$K_{\alpha\gamma}(t-t') = -\sum_\beta H_{\gamma\beta}(f_\beta(t'), f_\alpha(t)), \tag{2.123}$$

where we have used the relation $(G^0_\beta, \hat{L} f_\alpha(t-t')) = -\ (f_\beta\,(t')\,, f_\alpha(t))$, which can be checked straightforwardly. For the sake of simplicity, we set $t_0 = 0$. Then, due to the stationarity, we obtain

$$(f_\beta(t'), f_\alpha(t)) = (f_\beta(0), f_\alpha(t-t')) = (\hat{Q}\hat{L}G^0_\beta, f_\alpha(t-t')). \tag{2.124}$$

It is easily demonstrated that

$$(\hat{Q}\hat{L}G^0_\beta, f_\alpha(t-t')) = (\hat{L}G^0_\beta, \hat{Q}f_\alpha(t-t')) = (\hat{L}G_\beta(0), f_\alpha(t-t')) \tag{2.125}$$

and therefore $(f_\beta(t'), f_\alpha(t)) = (\hat{L}G_\beta\,(t')\,, f_\alpha(t))$. Thus, we get the desired relation

$$\begin{aligned}(f_\beta(t'), f_\alpha(t)) &= \frac{d}{dt'}(G_\beta\,(t')\,, f_\alpha(t)) = \frac{d}{dt'}(G_\beta\,(0)\,, f_\alpha(t-t')) \\ &= -\frac{d}{dt}(G_\beta\,(0)\,, f_\alpha(t-t')) = -\frac{d}{dt}(G_\beta\,(t')\,, f_\alpha(t)) \\ &= -(G_\beta\,(t')\,, \hat{L}f_\alpha(t)) = -(G^0_\beta, \hat{L}f_\alpha(t-t')),\end{aligned} \tag{2.126}$$

where several times we have applied the stationarity condition and the formal equation of motion (2.108), which is valid, of course, for all dynamic quantities of the system.

From (2.121), it is straightforward to obtain the corresponding equation for the correlation functions $\overline{G_\alpha(t)G_\beta(t_0)}$. Exploiting the orthogonality of the residual forces and the relevant quantities (2.119), we obtain

$$\begin{aligned}\frac{d}{dt}\overline{G_\alpha(t)G_\beta(t_0)} &= \sum_\gamma \Omega_{\alpha\gamma}\overline{G_\gamma(t)G_\beta(t_0)} \\ &\quad + \sum_\gamma \int_{t_0}^{t} dt' K_{\alpha\gamma}(t-t')\overline{G_\gamma(t')G_\beta(t_0)}.\end{aligned} \tag{2.127}$$

This equation is still linear in the correlation functions and may be solved by standard methods. Equation (2.127) can also be derived from the Nakajima–Zwanzig equation (2.50) in similar form.

The remaining problem is the specification of the frequency and memory matrices. The definitions (2.115) and (2.122) are in general too complicated to be useful in practice. In particular, if we want to describe phenomena in financial markets or social systems with equations of type (2.121) and (2.127), we require alternative methods in order to approximate these quantities. Physical intuition and empirical economic, social, and psychological knowledge play very important roles at several stages of these approaches.

In a correct framework, the results of the formalism are particularly rewarding because of their simple mathematical form. Of course, the results obtained cannot be claimed to be the output of a real theory firmly rooted in microscopic intuition and reasoning, but the general structure of (2.121) and (2.127) is motivated by universal principles of theoretical physics.

2.7.2 Separation of Timescales

Suppose that we have included in the set $\{G_\alpha\}$ all of the dynamical variables with a time dependence much slower than any microscopic timescale predictable from the Liouvillian. These relevant quantities determine substantially the macroscopic behavior of the system. Since the projection formalism gives no particular hint of a preference of the set of these slow variables, we have to deal with problems that occurred also with the introduction of the Markov approximation of the Nakajima–Zwanzig equation. Especially, the choice of which variables are actually slow is largely guided by the problem in mind.

After we determine the slow quantities as relevant variables, we can assume that the projection formalism collects the fast dynamics more or less in the residual forces due to the very complicated time dependence ruled by the anomalous propagator $\exp\{\hat{Q}\hat{L}(t-t')\}$. From a macroscopic point of view, the residual forces behave apparently as random functions. As a consequence, all of the elements of the memory matrix (2.123) are likely to be characterized by decay times considerably shorter than those associated with the elements $\overline{G_\alpha(t)G_\beta(t')}$ of the correlation matrix.

Thus, we may assume that over the characteristic timescales of $\overline{G_\alpha(t)G_\beta(t')}$, the decay time of the memory matrix is so short that $K_{\alpha\gamma}(t-t')$ may be approximately written as

$$K_{\alpha\gamma}(t-t') = K^0_{\alpha\gamma}\delta(t-t'). \tag{2.128}$$

This estimation is again called the Markov approximation or the separation of timescales. The representation (2.128) requires that the residual force correlations be of a δ-type, $\overline{f_\alpha(t)f_\beta(t')} \sim \delta(t-t')$, and furthermore that the condition $\bar{f}_\alpha(t) = 0$ holds, which can always be satisfied after realizing the shifts $f_\alpha \to f_\alpha - \bar{f}_\alpha$ and the corresponding changes of G_α. As a consequence of (2.128), the Mori–Zwanzig equations (2.121) now read

$$\frac{dG_\alpha(t)}{dt} = \sum_\gamma \tilde{\Omega}_{\alpha\gamma}G_\gamma(t) + f_\alpha(t) \tag{2.129}$$

with $\tilde{\Omega}_{\alpha\gamma} = \Omega_{\alpha\gamma} + K^0_{\alpha\gamma}$. It is seen that the separation of timescales yields a complete loss of memory effects in the Mori–Zwanzig equations. The system of ordinary linear differential equations (2.129) is a linearized version of a set of so-called Langevin equations. In particular, the Markov approximation reduces (2.127) to

$$\frac{d}{dt}\overline{G_\alpha(t)G_\beta(t_0)} = \sum_\gamma \tilde{\Omega}_{\alpha\gamma}\overline{G_\gamma(t)G_\beta(t_0)}, \tag{2.130}$$

representing a simple homogeneous system of linear differential equations with constant coefficients.

2.7.3 Wiener Process

Let us now study the properties of the trajectories of the so-called normalized Wiener process $W(t)$. This process satisfies a Fokker–Planck equation (2.95) in which there is only one variable W, the drift coefficient is zero, and the diffusion coefficient is 1:

$$\frac{\partial}{\partial t} p\left(W, t \mid W^{\prime}, t^{\prime}\right)=\frac{1}{2} \frac{\partial^{2}}{\partial W^{2}} p\left(W, t \mid W^{\prime}, t^{\prime}\right) . \tag{2.131}$$

All trajectories connecting the state W' at time t' with the state W at time t contribute to the conditional probability density $p(W, t \mid W, t')$. In order to be able to discuss the properties of these trajectories, we have to solve (2.131) under the initial condition $p(W, t' \mid W', t') = \delta(W - W')$. This is a standard procedure leading to the well-known Gaussian

$$P\left(W, t \mid W^{\prime}, t^{\prime}\right)=\frac{1}{\sqrt{2 \pi\left(t-t^{\prime}\right)}} \exp \left\{-\frac{\left(W-W^{\prime}\right)^{2}}{2\left(t-t^{\prime}\right)}\right\} . \tag{2.132}$$

Thus, if the process has arrived at the state W' at time t', the averaged state at time $t > t'$ is given by

$$\bar{w}=\int W p\left(W, t \mid W^{\prime}, t^{\prime}\right) d W=W^{\prime}, \tag{2.133}$$

while the variance (2.11) becomes

$$\sigma^{2}=\int_{-\infty}^{\infty} d W\,[W-\bar{w}]^{2}\, p\left(W, t \mid W^{\prime}, t^{\prime}\right) d W=t-t^{\prime} . \tag{2.134}$$

It is easy to see that, for $\delta t = t - t' \to 0$, equation (2.134) yields $|W - W'| \sim \sqrt{\delta t} \to 0$ and $|dW(t)/dt| \sim |W - W'|/\delta t \to \infty$. Therefore, each trajectory of a Wiener process is a continuous path but not a differentiable path.

Let us now determine the autocorrelation function of the Wiener process on the condition that the initial value of the process is $W_0 = W(t_0)$. The corresponding joint probability density is given by

$$p\left(W, t ; W^{\prime}, t^{\prime} | W_{0} t_{0}\right)=p\left(W, t \mid W^{\prime}, t^{\prime}\right) p\left(W^{\prime}, t^{\prime} \mid W_{0}, t_{0}\right) \tag{2.135}$$

so that

$$\begin{aligned} \overline{w(t) w\left(t^{\prime}\right)} &=\iint W W^{\prime} p\left(W, t \mid W^{\prime}, t^{\prime}\right) p\left(W^{\prime}, t^{\prime} \mid W_{0}, t_{0}\right) d W d W^{\prime} \\ &=\min \left(t-t_{0}, t^{\prime}-t_{0}\right)+W_{0}^{2} . \end{aligned} \tag{2.136}$$

We conclude that the autocovariance function of the Wiener process is given by $C(t, t') = \min(t - t_0, t' - t_0)$. As a final point, we should note that infinitesimal changes $dW(t) = W(t + dt) - W(t)$ satisfy

$$\overline{d W(t)}=0 \quad \text { and } \quad \overline{d W(t)^{2}}=d t \tag{2.137}$$

due to (2.133) and (2.134), while (2.136) leads to

$$\overline{dW(t)dW(t')} = 0 \qquad \text{for} \qquad t \neq t'. \tag{2.138}$$

Higher orders vanish as $\overline{dW(t)^f} \sim dt^{f/2} = o(dt)$ for $f > 2$. The simplest way of characterizing these results is to say that $dW(t)$ is an infinitesimal element of order 1/2 (i.e., $dW(t) \sim \sqrt{dt}$) and that in calculating differentials, infinitesimal elements of order higher than 1 are discarded so that $dW(t)^{2+n} \sim dt^{1+n/2} \to 0$ for all $n > 0$.

From here, we obtain the important result that the stochastic fluctuation of $dW(t)$ causes $\overline{dW(t)/dt} = 0$, while $\overline{|dW(t)/dt|} \sim dt^{-1/2}$ diverges.

2.7.4 Stochastic Differential Equations

The linearity of the Langevin equation (2.129) is a consequence of the projection formalism introduced in the previous sections. In many practical cases, we have to deal with nonlinear Langevin equations. These equations may be derived in a more or less intuitive manner, but they are only rarely based on a real theoretical framework. However, in the case of Markov processes, the Langevin equations can be obtained from the corresponding Fokker–Planck equations.

To proceed, we now consider a system of stochastic differential equations that generalizes the linear system (2.129):

$$\dot{Y}_\alpha(t) = a_\alpha(Y(t)) + \sum_{k=1}^{R} b_{\alpha,k}(Y(t))\eta_k(t). \tag{2.139}$$

Here, $a_\alpha(Y)$ and $b_{\alpha,k}(Y)$ are differentiable functions of the N-dimensional state vector Y, while the $\eta_k(t)$ ($k = 1, ..., R$) are linearly independent stochastic functions.

Equations of such a type are also denoted as Langevin equations. In principle, these equations can be derived formally from (2.129) in a heuristic way. To do this, we take into account a set of N relevant quantities G_α. These relevant quantities may be specified as functions of the state vector Y, $G_\alpha(t) = G_\alpha(Y(t))$. We substitute (2.139) into $\dot{G}_\alpha = \sum_\beta (\partial G_\alpha/\partial Y_\beta)\dot{Y}_\beta$ and compare the result with (2.129). This allows us to identify

$$\sum_\beta \frac{\partial G_\alpha}{\partial Y_\beta} a_\beta = \sum_\beta \tilde{\Omega}_{\alpha\beta} G_\beta \quad \text{and} \quad f_\alpha(t) = \sum_{\beta,k} \frac{\partial G_\alpha}{\partial Y_\beta} b_{\beta,k}\eta_k(t). \tag{2.140}$$

The first equation defines the functions $a_\alpha(Y)$, while the second one requires a further explanation. The fluctuation forces $f_\alpha(t)$ are assumed in the context of the Markov approximation to be fast-varying quantities with more or less stochastic character, but they can be structured nevertheless from the relatively slow relevant quantities and the fast irrelevant variables. Even the macroscopically uncontrollable dynamics of the irrelevant degrees of freedom are the reason for the apparently stochastic behavior of the fluctuation forces $f_\alpha(t)$. Therefore, the separation

$$f_\alpha(t) = \sum_{k=1}^{R} \tilde{B}_{\alpha,k}(Y)\eta_k(t) \tag{2.141}$$

of the fluctuation forces into bilinear combinations of independent stochastic functions $\eta_k(t)$ that are assumed to be exclusively controlled by the dynamics of the irrelevant degrees of freedom and systematic terms $\tilde{B}_{\alpha,k} = \sum_\beta (\partial G_\alpha / \partial Y_\beta) b_{\beta,k}$ controlled by the relevant quantities and the state vector, respectively, is a natural ansatz that is not in disagreement with the requirements of the projection formalism.

It should be noted that (2.141) is only an obvious assumption, which is perhaps supported by empirical experience. Equation (2.141) cannot yet be generally derived in the framework of a closed theory.

As mentioned above, the Markov approximation (2.128) requires

$$\overline{f_\alpha(t) f_\beta(t')} \sim \delta(t-t') \quad \text{and} \quad \bar{f}_\alpha(t) = 0. \tag{2.142}$$

This δ-function character is transferred also to the stochastic functions $\eta_k(t)$. In general, we are able to specify the stochastic functions to

$$\overline{\eta_k(t)\eta_{k'}(t')} = \delta_{kk'}\delta(t-t') \quad \text{and} \quad \bar{\eta}_k(t) = 0 \tag{2.143}$$

by a suitable choice of the functions $\tilde{B}_{\alpha,k}(Y)$ in (2.141).

Let us return to the discussion of the stochastic differential equation (2.139). Unfortunately, this equation as it stands has no meaning, and we do not know how to deal with it. The reason is that the δ-character of the correlation functions (2.143) causes jumps in the state vector $Y(t)$ such that the value of $Y(t)$ at time t is not well-defined. Basically, the problem comes from the fact that the stochastic functions $\eta_k(t)$ change substantially during an infinitesimally small time interval so that the equation does not specify what value of $b_{\alpha,k}(Y)$ should be used in the product $b_{\alpha,k}(Y(t))\eta_k(t)$.

For a better understanding of the problem, we divide the time axis into infinitesimally small subintervals of length dt by means of partitioning points t_i with $t_{i+1} = t_i + dt$ and define intermediate points τ_i such that $t_i < \tau_i < t_{i+1}$. Then, we obtain from (2.139)

$$Y_\alpha(t_{i+1}) = Y_\alpha(t_i) + a_\alpha(Y(t_i))dt + \sum_k \int_{t_i}^{t_{i+1}} b_{\alpha,k}(Y(t'))\eta_k(t')dt' \tag{2.144}$$

or, due to the mean value theorem,

$$Y_\alpha(t_{i+1}) = Y_\alpha(t_i) + a_\alpha(Y(t_i))dt + \sum_k b_{\alpha,k}(Y(\tau_i)) \int_{t_i}^{t_{i+1}} \eta_k(t')dt'. \tag{2.145}$$

The main problem comes from the integral

$$dW_k(t_i) = \int_{t_i}^{t_{i+1}} \eta_k(t')dt'. \tag{2.146}$$

It is easily seen that the application of (2.143) leads to

$$\overline{dW_k(t_i)dW_{k'}(t_j)} = \delta_{kk'}\delta_{ij}dt \quad \text{and} \quad \overline{dW_k(t_i)} = 0. \tag{2.147}$$

This is nothing other than the mean and the correlation of infinitesimal changes of independent Wiener processes $W_k(t)$ ($k = 1, ..., R$). Now, we can rewrite (2.145) to obtain

$$Y_\alpha(t_{i+1}) = Y_\alpha(t_i) + a_\alpha(Y(t_i))dt + \sum_k b_{\alpha,k}(Y(\tau_i))dW_k(t_i). \tag{2.148}$$

However, we can evaluate $b_{\alpha,k}(Y(\tau_i))$ at an arbitrary intermediate time τ_i. It is clear that the choice of this intermediate time has important consequences for numerical simulations of stochastic processes. Two general concepts were established.

In the Ito interpretation, the value of $Y(\tau_i)$ is taken before the jump. This means explicitly

$$Y_\alpha(t_{i+1}) = Y_\alpha(t_i) + a_\alpha(Y(t_i))dt + \sum_k b_{\alpha,k}(Y(t_i))dW_k(t_i). \tag{2.149}$$

On the other hand, in the Stratonovich interpretation, we take the mean of $Y(t)$ before and after the jump so that $Y(\tau_i) = (Y(t_{i+1}) + Y(t_i))/2$; namely

$$\begin{aligned} Y_\alpha(t_{i+1}) = {} & Y_\alpha(t_i) + a_\alpha(Y(t_i))dt \\ & + \sum_k b_{\alpha,k}\left(\frac{Y(t_i) + Y(t_{i+1})}{2}\right) dW_k(t_i). \end{aligned} \tag{2.150}$$

It is known that the same stochastic process occurs for Ito stochastic differential equations and Stratonovich stochastic differential equations if the conditions

$$a_\alpha^{\text{Stratonovich}} = a_\alpha^{\text{Ito}} - \frac{1}{2}\sum_{\beta,k} b_{\beta,k}^{\text{Ito}} \frac{\partial b_{\alpha,k}^{\text{Ito}}}{\partial Y_\beta} \tag{2.151}$$

and

$$b_{\alpha,k}^{\text{Stratonovich}} = b_{\alpha,k}^{\text{Ito}} \tag{2.152}$$

are satisfied. Finally, we remark that, independently of the interpretation of the stochastic differential equation (2.139), the coefficients $a_\alpha(Y)$ and $b_{\alpha,k}(Y)$ can be extended to explicitly time-dependent functions $a_\alpha(Y,t)$ and $b_{\alpha,k}(Y,t)$. Such an extension is motivated above all by the fact that possibly a part of the irrelevant variables possesses relatively slow timescales on the order of magnitude of the characteristic time of the relevant quantities.

2.7.5 Ito's Formula and the Fokker–Planck Equation

Let us consider an arbitrary differentiable function $f(Y)$, where $Y = Y(t)$ is the solution of the Ito stochastic differential equation (2.149). For the sake of simplicity, we consider only one relevant degree of freedom and only one Wiener process. Then, the Ito differential equation reads

$$dY = a(Y,t)dt + b(Y,t)dW(t). \tag{2.153}$$

The differential df is defined as

$$df = f(Y(t+dt)) - f(Y(t)) = f(Y+dY) - f(Y). \tag{2.154}$$

We expand (2.154) to second order in dY,

$$df = \frac{\partial f(Y)}{\partial Y} dY + \frac{1}{2}\frac{\partial^2 f(Y)}{\partial Y^2} dY^2 + o\left(dY^2\right), \tag{2.155}$$

and replace dY by (2.153). Considering $dW \sim dt^{1/2}$, we expand (2.155) up to first order in dt,

$$df = \frac{\partial f(Y)}{\partial Y}\left[a(Y,t)dt + b(Y,t)dW\right] + \frac{1}{2}\frac{\partial^2 f(Y)}{\partial Y^2}\left[b(Y,t)dW\right]^2, \tag{2.156}$$

where all other terms have been discarded since they are of higher order. This relation is known as Ito's formula.

Now, we perform the average with respect to all realizations of the Wiener process $W(t)$ and obtain

$$\frac{\overline{df}}{dt} = \overline{\frac{\partial f(Y)}{\partial Y} a(Y,t)} + \frac{1}{2}\overline{\frac{\partial^2 f(Y)}{\partial Y^2} b^2(Y,t)}. \tag{2.157}$$

For the determination of the averages, we used the Ito calculus, especially (2.147) and the property that the value of $Y(t)$ is taken before the jump $dW(t)$. The last remark means that the actual value of $Y(t)$ and the value of the subsequent jump $dW(t)$ are statistically independent.

On the other hand, $Y(t)$ has the conditional probability density $p(Y,t \mid Y_0,t_0)$. If the evolution starts from the initial state $Y(t_0) = Y_0$, then the averages are given by

$$\frac{\overline{df}}{dt} = \int \frac{\partial}{\partial t} p(Y,t \mid Y_0,t_0) f(Y) dY \tag{2.158}$$

and

$$\begin{aligned} \overline{\frac{\partial f(Y)}{\partial Y} a(Y,t)} &= \int \frac{\partial f(Y)}{\partial Y} a(Y,t) p(Y,t \mid Y_0,t_0) dY \\ &= -\int f(Y) \frac{\partial}{\partial Y}\left[a(Y,t)p(Y,t \mid Y_0,t_0)\right] dY \end{aligned} \tag{2.159}$$

and

$$\begin{aligned} \overline{\frac{\partial^2 f(Y)}{\partial Y^2} b^2(Y,t)} &= \int \frac{\partial^2 f(Y)}{\partial Y^2} b^2(Y,t) P(Y,t \mid Y_0,t_0) dY \\ &= \int f(Y) \frac{\partial^2}{\partial Y^2}\left[b^2(Y,t)P(Y,t \mid Y_0,t_0)\right] dY. \end{aligned} \tag{2.160}$$

Putting all of these results together and integrating by parts, we arrive at

$$\begin{aligned}
&\int dY\, f(Y) \frac{\partial}{\partial t} p\,(Y, t \mid Y_0, t_0) \\
&= \int dY\, f(Y) \frac{1}{2} \frac{\partial^2}{\partial Y^2} \left[b^2(Y, t) p\,(Y, t \mid Y_0, t_0) \right] \\
&\quad - \int dY\, f(Y) \frac{\partial}{\partial Y} \left[a(Y, t) p\,(Y, t \mid Y_0, t_0) \right].
\end{aligned} \tag{2.161}$$

Now, we consider that we have chosen the function $f(Y)$ to be arbitrary. Hence, we conclude that the conditional probability satisfies the equation

$$\begin{aligned}
\frac{\partial}{\partial t} p\,(x, t \mid y, t') &= \frac{1}{2} \frac{\partial^2}{\partial x^2} \left[b^2(x, t) p\,(x, t \mid y, t') \right] \\
&\quad - \frac{\partial}{\partial x} \left[a(x, t) p\,(x, t \mid y, t') \right].
\end{aligned} \tag{2.162}$$

Obviously, we get a complete equivalence between the stochastic differential equation (2.153) and the Fokker–Planck equation (2.162). This result can be generalized for the case of a system of N stochastic differential equations with R Wiener processes. The set of differential equations may be given by (2.139). The corresponding Fokker–Planck equation then reads

$$\frac{\partial}{\partial t} p\,(Y, t \mid Y_0, t_0) = \frac{1}{2} \sum_{\alpha, \beta} \frac{\partial^2}{\partial Y_\alpha \partial Y_\beta} \left[B_{\alpha\beta}(Y, t) p\,(Y, t \mid Y_0, t_0) \right] \tag{2.163}$$

$$- \sum_{\alpha} \frac{\partial}{\partial Y_\alpha} \left[A_\alpha(Y, t) p\,(Y, t \mid Y_0, t_0) \right] \tag{2.164}$$

with

$$B_{\alpha\beta}(Y, t) = \sum_{k=1}^{\mu} b_{\alpha,k}(Y, t) b_{\beta,k}(Y, t) \quad \text{and} \quad A_\alpha(Y, t) = a_\alpha(Y, t). \tag{2.165}$$

A similar connection between the Stratonovich stochastic differential equations and a corresponding Fokker–Planck equation can be obtained by application of the converting rules (2.151) and (2.152). Thus, the system of stochastic differential equations (2.139) in the Stratonovich interpretation is related to the Fokker–Planck equation

$$\begin{aligned}
&\frac{\partial}{\partial t} p\,(Y, t \mid Y_0, t_0) \\
&= \frac{1}{2} \sum_{\alpha, \beta, k} \frac{\partial}{\partial Y_\alpha} \left[b_{\alpha,k}(Y, t) \frac{\partial}{\partial Y_\beta} \left(b_{\beta,k}(Y, t) p\,(Y, t \mid Y_0, t_0) \right) \right] \\
&\quad - \sum_{\alpha} \frac{\partial}{\partial Y_\alpha} \left[a_\alpha(Y, t) p\,(Y, t \mid Y_0, t_0) \right].
\end{aligned} \tag{2.166}$$

Finally, it should be noted that the connection between the stochastic differential equations and Fokker–Planck equations allows us to create representative trajectories for a given Fokker–Planck equation by numerical simulations.

3. Financial Markets

3.1 Introduction

3.1.1 Finance and Financial Mathematics Versus Econophysics

When economists, finance mathematicians, and physicists are dealing with a financial problem or wish to understand an economic issue related to business investment, operations, or financing, a wide variety of ideas and techniques are available to generate quantitative answers.

The mathematical way starts from well-defined hypotheses that consider more or less idealized economic rules and specific initial and boundary conditions. This input may be obtained from an empirical analysis of financial data or from other quantitative economic investigations. Actually, this part is an intermediate field between financial mathematics and financial management because it requires economic experience to decide which data are important in the context of the problem and what is the order of their significance. The exact solution of a financial problem formulated hypothetically is the intrinsic power of financial mathematics, which developed into an extensive field over the last four decades. Financial mathematics is, however, restricted by the capacity of the methodology to be applied. Especially, mathematics offers only solutions that have to be understood as recommendations supporting decisions in a company's investments or in financial operations or various transactions.

Apart from providing specific numerical answers using the tools of financial mathematics, the economic approach to financial problems depend significantly on the points of view of the parties involved, on the relative importance of the issue, and on the nature and reliability of the information available. The central aim of finance is to answer concrete situations with concrete decisions.

Therefore, finance may be interpreted as the art of asking significant questions and giving meaningful answers to these significant questions. Such questions include the following: Have the problem and its relative importance in the overall business context been clearly spelled out, including the relevant alternatives to be considered? Which specific factors, relationships, and trends are likely to be helpful in analyzing the special problem? What is the order of their importance, and in what sequence should they be addressed?

How precise an answer is necessary in relation to the importance of the problem itself? Would additional refinement be worth the effort? How reliable are the available data, and how is this uncertainty likely to affect the range of results?

These questions have purely economic character. In order to be able to give a sufficient answer, the economist employs a whole line of experiences in addition to possible mathematical solutions delivered from financial mathematics. All of this allows us to say that an economist is mainly interested in finding a rational approach to a given financial problem. However, both the economist and the financial mathematician will have difficulties in combining the specific dynamics of financial data with general properties of complex systems.

The physical way offers, however, the possibility of describing financial phenomena in a universal theoretical framework. There is the hope that the physical progress in investigation of complex systems with other apparent behavior allows a deeper insight also into the dynamics of the financial markets. Results that would be obtained only from a direct observation of finance dynamics can possibly also be derived from discoveries in other fields of physics.

The financial market is a complex system from a physical point of view. In such a system, the rates of stocks and other asset prices are characterized as relevant degrees of freedom. All other degrees of freedom are irrelevant quantities. The intention to want to describe the evolution of the share prices conceals itself behind this division, whereas, for instance, the mental states of the traders are interpreted as more or less uninteresting (i.e., irrelevant) information for financial transactions.

The total number of quantitatively available financial quantities such as shares, trading volumes, and funds is approximately of an order of magnitude of 10^8–10^{10}. This number is extremely small compared to the size of the irrelevant set containing the degrees of freedom with direct or indirect contact with the underlying structures of the financial market. For instance, all atoms that are involved in the raising of awareness of traders, politicians, employers and employees, in the formation of the climate, and the structure and function of production plants, traffic systems, and communication networks contribute to the irrelevant degrees of freedom. Every individual transaction basically leads to a small change in the market value of a stock. But a gigantic number of processes is hidden behind every transaction that can be formally traced back up to the microscopic level.

We have discussed in the previous chapter that the dynamics of the irrelevant microscopic degrees of freedom may be formally eliminated by application of a suitable projection formalism. It was demonstrated that one obtains probabilistic equations describing such a system on the relevant macroscopic level.

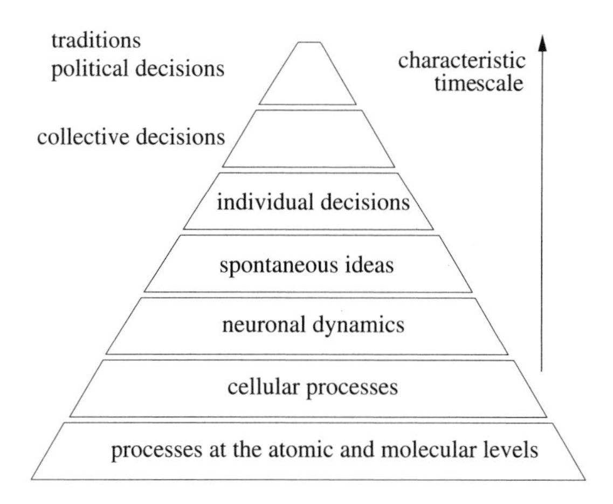

Fig. 3.1. A schematic representation of processes contributing to the complex structure of financial decisions.

At first, it seems reasonable to describe financial markets within the framework of a Markov approximation. This suggestion is supported by the fact that the characteristic timescale of financial processes is between 1 and 10^8 sec, while the effective timescale of the microscopic interactions is of an order of magnitude of 10^{-14} sec. However, this assumption is only a part of the truth. Although the irrelevant degrees of freedom are based on atomic movements, these are not exclusively chaotic but show a high collectivity and therefore a complicated hierarchy; see Figure 3.1. Especially, only the lower levels of this hierarchy are fast compared to price fluctuations of shares. Rather, the exchange rates affect the decisions of the traders, their psychological state, and their pursued trading strategy. That leads to a feedback mechanism that can modify the Markov character considerably.

3.1.2 Scales in Financial Data

In the natural sciences, especially in physics, the problem of reference units is considered basic to all experimental and theoretical work. Efforts are continually made to find the optimal reference units of a given problem and to improve the accuracy of their determination. The system of the physical units is permanently improved in order to eliminate even the smallest deviations. Unfortunately, we observe another situation in finance. The scales used here are often given in units that are themselves fluctuating in time.

The most important quantity is the price, which is indicated in units of a home currency or a key currency. However, the values of the national currencies are not constant in time. The ratios of the individual currencies among each other as well as the prices of commodities such as gold or diamonds show substantial fluctuations. The causes for these fluctuations

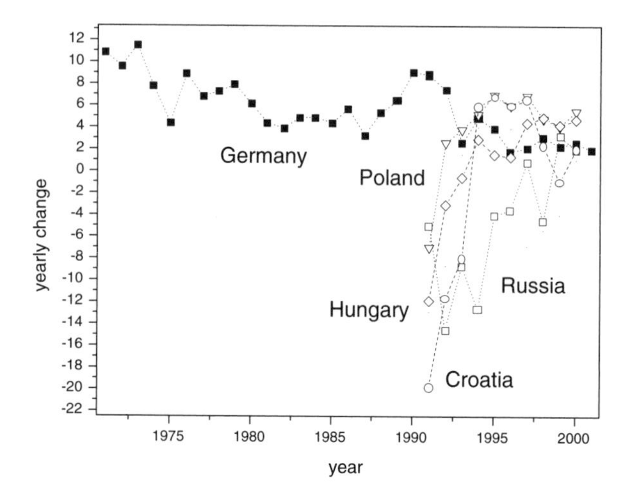

Fig. 3.2. Annual percentage change of the gross domestic product for Germany and several eastern European countries. From Statistisches Bundesamt Deutschland 2002

are miscellaneous and cannot usually be separated in detail. Possible causes are, for example, fluctuations in the global currency markets, inflation, or national and global economic growth or economic recessions. For instance, Figure 3.2 shows the annual percentage change of the German gross domestic product in comparison with those of some eastern European countries. Obviously, economic growth is neither a constant nor is it a time-dependent function that shows the same behavior for all countries.

Another problem is the choice of an appropriate timescale to use for analyzing financial data. The physical time is well-defined, but the traditional stock exchanges close aperiodically overnight and on weekends [131] and more or less randomly during holidays. The difficulty consists in the fact that we do not know how we have to handle the discontinuances and a possible arrival of information during this time. Similar problems appear in electronic stock markets. Although these markets are active 24 hours per day, the social environment and several biological cycles push the market activity to a permanent change of intensity in each financial region of the world.

Another possible timescale describing financial time series is the number of transactions. This scale eliminates the effect of the randomly distributed time intervals elapsing between transactions. Another source of randomness, the volume of transactions, still remains. It should be remarked that other definitions of timescales consider also the influence of the trading volume. For instance, the time index of the number of effective transactions occurring in the market is such a measure. This timescale is not affected by the actual trading activity but differs for different shares.

The trading time is the time that elapses during open market hours. This timescale depends on the local stock exchange. Furthermore, price changes and the release of relevant information during the night lead to jumps at the opening and therefore to a possible misfit of data. However, the trading time is the most common choice in many research studies and is also used in the context of this chapter as an underlying timescale.

It should be remarked that financial time series are discontinuous. However, we can formally extend this series to continuous functions. In practice, this is performed by the application of various interpolation techniques. Such an apparently academic continuation is a natural way in the framework of our general approach. Obviously, behind this artificial extension is concealed the fact that the microscopic processes contributing to the price formation are continuous. The problem is, that we neither know the values of the microscopic variables at a certain time nor can we calculate the actual prices from these data. From our macroscopic point of view, new information about the price is only obtainable after a transaction. This fine philosophical difference is insignificant for many practical and theoretical applications. Nevertheless, sometimes the use of the concept of continuous series proves expedient, in particular for the investigation of relatively short time periods.

3.1.3 Measurement of Price Fluctuations

Let us define $X_\alpha(t)$ as the price of a financial asset α at time t. Then, we may ask which is the appropriate variable describing the stochastic behavior of the price fluctuations. The simplest choice is the introduction of the price changes

$$\delta X_\alpha(t, \delta t) = X_\alpha(t + \delta t) - X_\alpha(t), \tag{3.1}$$

where δt is a well-defined interval –the time horizon– of the time series. The merit of this approach is that (3.1) is a simple linear relation. This means in particular that the price changes are additive:

$$\delta X_\alpha(t, \delta t_1 + \delta t_2) = \delta X_\alpha(t + \delta t_1, \delta t_2) + \delta X_\alpha(t, \delta t_1). \tag{3.2}$$

Unfortunately, the definition (3.1) is seriously affected by possible changes in money scales due to possible fluctuations in the global currency markets or inflation effects. Furthermore, the strength of the fluctuations $\delta X_\alpha(t)$ depends seriously on the order of magnitude of the actual price of the asset $X_\alpha(t)$; see Figure 3.3. Alternatively, one can analyze deflated price changes

$$\delta \widetilde{X}_\alpha(t, \delta t) = g(t + \delta t) X_\alpha(t + \delta t) - g(t) X_\alpha(t), \tag{3.3}$$

where $g(t)$ is a deflation factor that considers the effects of inflation and possible fluctuations in the growth rate of the economy. The factor $g(t)$

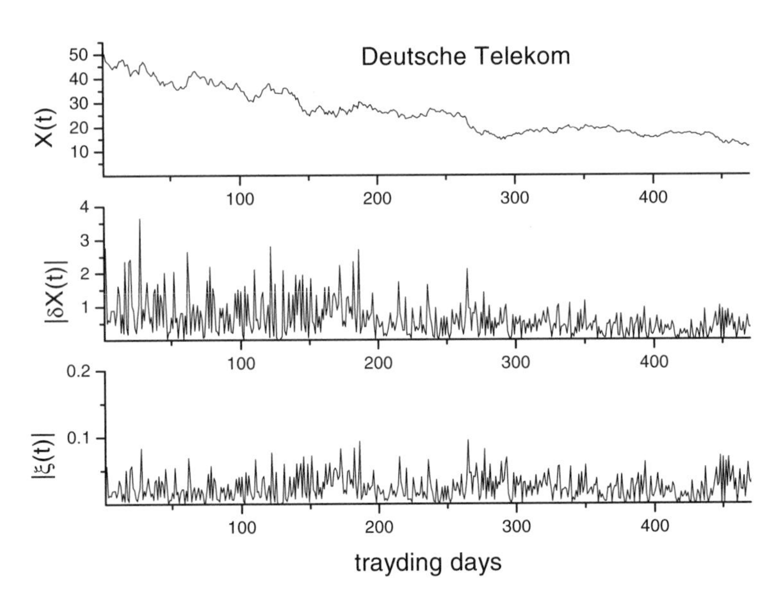

Fig. 3.3. Daily prices $X(t)$, absolute values of the price differences $|\,\delta X(t)\,|$, and logarithmic price changes $|\,\xi(t)\,|$ (Deutsche Telekom, reference time period 11/00–07/02). Intuitively, the price changes $\delta X(t)$ seem to be dependent on the daily prices $X(t)$, while the logarithmic price changes are more or less independent from $X(t)$.

rescales the price fluctuations so that $\delta \widetilde{X}_\alpha(t)$ is given in terms of a more or less constant money scale. However, the problem remains that $\delta \widetilde{X}_\alpha(t)$ scales with $X_\alpha(t)$. Additionally, the deflation factor is practically unpredictable, and there is no unique definition of $g(t)$.

A more appropriate choice is the use of the returns

$$R_\alpha(t, \delta t) = \frac{\delta X_\alpha(t)}{X_\alpha(t)} = \frac{X_\alpha(t + \delta t)}{X_\alpha(t)} - 1. \tag{3.4}$$

The merit of this approach is that returns provide a direct percentage of gain or loss in a given time period. The return itself is related to the net yield, which is defined as

$$R^\star_\alpha(t, \delta t) = \frac{\delta X_\alpha(t) + E_\alpha(t, \delta t)}{X_\alpha(t)} = R_\alpha(t, \delta t) + \frac{E_\alpha(t, \delta t)}{X_\alpha(t)}, \tag{3.5}$$

where $E_\alpha(t, \delta t)$ represents the profits (e.g., dividends or interest) in the time period $[t, t + \delta t]$. Therefore, the return is a very natural measure describing the price fluctuations of shares. The disadvantage is that the returns are nonlinearly coupled by means of multiplication:

$$R_\alpha(t, \delta t_1 + \delta t_2) = [R_\alpha(t + \delta t_1, \delta t_2) + 1]\,[R_\alpha(t, \delta t_1) + 1] - 1. \tag{3.6}$$

To overcome this problem, we introduce the difference of the natural logarithm of the price

$$\xi_\alpha(t,\delta t) = \ln X_\alpha(t+\delta t) - \ln X_\alpha(t) = \ln \frac{X_\alpha(t+\delta t)}{X_\alpha(t)}. \tag{3.7}$$

These quantities are additive,

$$\xi_\alpha(t,\delta t_1+\delta t_2) = \xi_\alpha(t+\delta t_1,\delta t_2) + \xi_\alpha(t,\delta t_1), \tag{3.8}$$

and the approach (3.7) is organized so that the correction of the scales by deflating factors leads to an additive term

$$\widetilde{\xi}_\alpha(t,\delta t) = \ln \frac{g(t+\delta t)X_\alpha(t+\delta t)}{g(t)X_\alpha(t)} = \xi_\alpha(t,\delta t) + \ln \frac{g(t+\delta t)}{g(t)}. \tag{3.9}$$

The natural logarithmic price differences and the returns are the most commonly studied quantities in financial time series. For high-frequency data, δt is small and $|\delta X_\alpha| \ll \delta X_\alpha$. Thus, we get the important approximation $\xi_\alpha(t,\delta t) \approx R_\alpha(t,\delta t)$.

3.2 Empirical Analysis

3.2.1 Probability Distributions

From a physical point of view, a financial market is a complex system whose evolution is given by the A-dimensional vector $X(t) = \{X_1(t), ..., X_A(t)\}$ of all simultaneous observed share quotations and other asset prices. Alternatively, one can also use the vector of the logarithmic price differences $\xi(t,\delta t) = \{\xi_1(t,\delta t), ..., \xi_A(t,\delta t)\}$ in place of $X(t)$. Both vectors are equivalent representations of the set of relevant degrees of freedom on the same macroscopic level of the underlying financial system. For the sake of simplicity, we mostly use the vector $\xi(t,\delta t)$, which we transform if necessary into other representations. In the context of this book, the vector function $\xi(t,\delta t)$ is also denoted as the trajectory of the financial market in the A-dimensional space of the asset prices.

We had already generally discussed that on the macroscopic level each complex system can be described by a probabilistic theory. In this sense, the probability distribution function $p_{\delta t}(\xi,t)$ is the central quantity of a physical description of financial markets. By definition, $p_{\delta t}(\xi,t)$ is the probability density for a change of the logarithmic prices by the value ξ from the beginning to the end of the time interval $[t,t+\delta t]$. The explicitly specified control parameter δt is necessary to complete the definition of the logarithmic price differences (3.7). If we want to study the properties of the probability distribution function, we must consider that in reality only one trajectory $\xi(t,\delta t)$ is at our disposal for a statistical analysis. Each repetition in the sense of a statistical experiment with the same initial conditions and the same boundary conditions is principally excluded. The quantity $p_{\delta t}(\xi, t)$ is therefore interpreted in the context of Bayesian statistics as the hypothetical probability density that a change ξ could have taken place at time t independently of the realized event. If the event is situated still in the future, $p_{\delta t}(\xi, t)$ is the probability density with which the change ξ can be expected.

Additionally, as a generalization of $p_{\delta t}(\xi, t)$, we introduce the joint probability density $p_{\delta t}(\xi^{(N)}, t_N; \xi^{(N-1)}, t_{N-1}; ...; \xi^{(0)}, t_0)$ with $N+1$ sequenced points in time. The knowledge of this function allows formally the determination of arbitrary correlations via (2.102). Normally, one chooses the parametrization of the time in such a way that the sampling times are regularly spaced, $t_{n+1} = t_n + \delta t$. In the joint probability representation, each vector function $\xi(t, \delta t)$ between the initial time t_0 and the final time t_N or, more precisely, each series $\{\xi(t_0, \delta t), ..., \xi(t_N, \delta t)\}$ is one probabilistic event.

The reduction on lower probability distribution functions is made via integration over all noninteresting variables. By the way, this procedure may be interpreted as a discrete version of a so-called path integral considering the whole set of hypothetically allowed discrete trajectories of the financial system during the time interval $[t_0, t_N]$. Using conditional probabilities, the joint probability can be written as

$$
\begin{aligned}
p_{\delta t}\left(\xi^{(N)}, t_N; ...; \xi^{(0)}, t_0\right) = {} & p_{\delta t}\left(\xi^{(N)}, t_N \mid \xi^{(N-1)}, t_{N-1}; ...; \xi^{(0)}, t_0\right) \\
& \times p_{\delta t}\left(\xi^{(N-1)}, t_{N-1} \mid \xi^{(N-2)}, t_{N-2}; ...\right) \\
& \vdots \\
& \times p_{\delta t}\left(\xi^{(1)}, t_1 \mid \xi^{(0)}, t_0\right) p_{\delta t}\left(\xi^{(0)}, t_0\right) .
\end{aligned} \tag{3.10}
$$

The main problem is now that one must obtain all information from the only trajectory observed in reality. To this end, we need some assumptions, which will be discussed in the following.

3.2.2 Ergodicity in Financial Data

The first important prerequisite is the assumption of the validity of the ergodic hypothesis. There are different formulations of this hypothesis; for further information, see the specialized literature [16, 57, 336, 378]. Roughly speaking, the ergodic hypothesis generally requires that a system starting from an arbitrary initial state can always arrive at each final state after a sufficiently long time. In other words, the total set of all possible trajectories is topologically not separable (see Figure 3.4). Since we know from our observations, however, only one function $\xi(t, \delta t)$ for a given financial market, the assumption of ergodicity is an a priori hypothesis. This theorem can neither be proven nor disproven for such systems. But even if a repetition of the development under the same macroscopic conditions were possible and the system would arrive at areas in the state space that it would

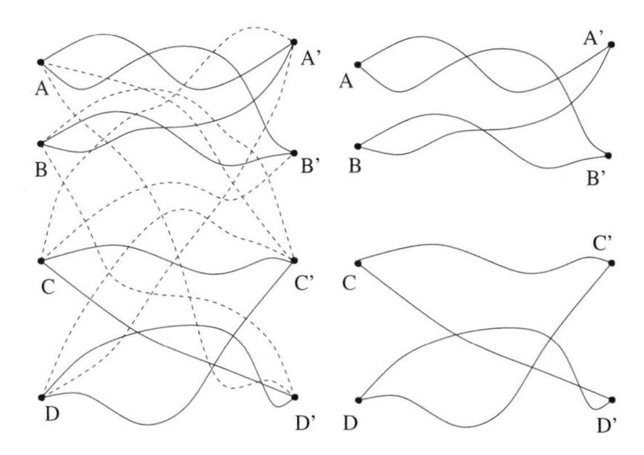

Fig. 3.4. Possible trajectories in an ergodic (left figure) and a nonergodic (right figure) system of four initial and four final states. If we possess only the empirical information about the existence of trajectories between (A, B) and (A', B'), we cannot decide from an empirical view whether the total system is ergodic or nonergodic.

never achieve during the original evolution, this new information would be irrelevant. From this point of view, the ergodicity hypothesis is an application of the principle of Occam's razor [399, 400]. This idea is attributed to the 14th century Franciscan monk William of Occam and states that entities should not be multiplied unnecessarily. The most useful statement of this principle is that the better theory of two competing theories that make exactly the same predictions is the simpler one. Occam's razor is used to cut away unprovable concepts.

3.2.3 Stationarity of Financial Markets

As a second important prerequisite for the description of financial markets, one assumes stationarity. We had already pointed out in the previous chapter that stationarity means that all probability distribution functions and all correlation functions are invariant under an arbitrary shift in time. This condition as well as the ergodic hypothesis allows us to replace the ensemble averages (2.5) and (2.102) by appropriate time averages. In particular, we expect for the correlation functions of order K

$$\overline{\prod_{k=1}^{K} \xi_{\alpha_k}(t_{n_k}, \delta t)} = \left\langle \prod_{k=1}^{K} \xi_{\alpha_k}(t_{n_k}, \delta t) \right\rangle \tag{3.11}$$

with the time-averaged correlation function

$$\left\langle \prod_{k=1}^{K} \xi_{\alpha_k}(t_{n_k}, \delta t) \right\rangle = \lim_{S \to \infty} \frac{1}{S} \sum_{m=0}^{S-1} \prod_{k=1}^{K} \xi_{\alpha_k}(t_{n_k} - m\delta t, \delta t). \tag{3.12}$$

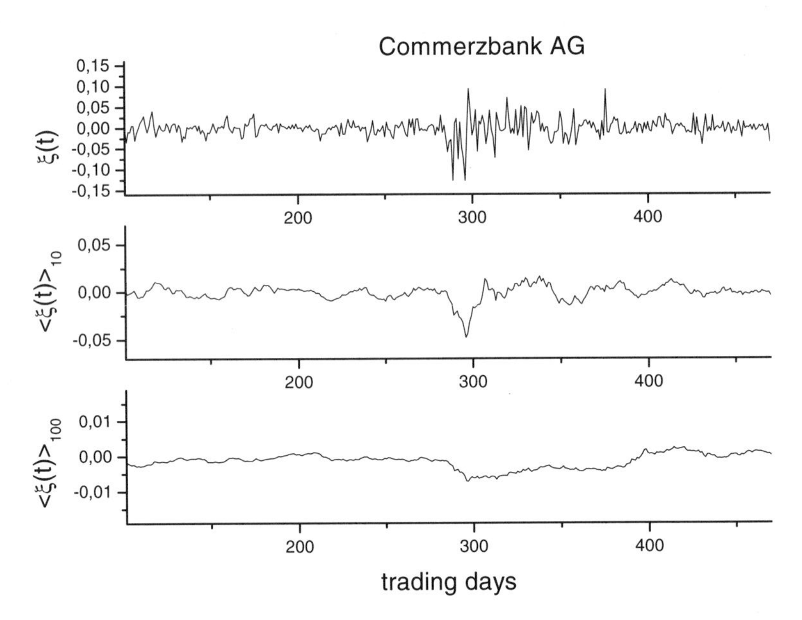

Fig. 3.5. Daily logarithmic price changes $|\xi(t)|$ and corresponding moving averages for $T = 10$ and $T = 100$ trading days. The time horizon is $\delta t = 1$ trading day (Commerzbank AG stock, reference time period 11/00–07/02).

In practice, we will attempt to determine the time-averaged values from a finite number of observations. To this end, we consider S equidistant records present in the time interval $[t - T, t]$ with $T = (S - 1)\delta t$. The heuristically constructed averages

$$\left\langle \prod_{k=1}^{K} \xi_{\alpha_k}(t_{n_k}, \delta t) \right\rangle_T = \frac{\delta t}{T + \delta t} \sum_{m=0}^{T/\delta t} \prod_{k=1}^{K} \xi_{\alpha_k}(t_{n_k} - m\delta t, \delta t) \tag{3.13}$$

are denoted also as moving averages. They are dependent on the time t, the width of the time window T, and the frequency δt^{-1} of financial observations. If we assume stationarity, then we expect that these time averages should converge to constant values for a sufficiently large number of observations. However, we must be very careful with the stationarity assumption. A typical example is the moving mean value of logarithmic price fluctuations,

$$\langle \xi_\alpha(t, \delta t) \rangle_T = \frac{\delta t}{T + \delta t} \sum_{m=0}^{T/\delta t} \xi_\alpha(t - m\delta t, \delta t). \tag{3.14}$$

If we analyze this average obtained from the data of an individual share, we find considerable fluctuations even for wide time windows and a large number of observations (see Figure 3.5). Several trends may be hidden behind these relatively slow fluctuations, which are basically influenced by macroeconomic and political phenomena rather than by the intrinsic dynamics of the financial market.

Unfortunately, it is very difficult to extract the current trend from the actual moving averages since the mean values represent only a time-delayed trend because of the asymmetric location of the time arguments with respect to the current time t. We may at least partially remove these trends by using relative fluctuations, $\widetilde{\xi}_\alpha(t,\delta t) = \xi_\alpha(t,\delta t) - \langle \xi_\alpha(t,\delta t)\rangle_T$. A typical example is the empirical second moment of the trend-corrected logarithmic price fluctuations

$$\sigma_\alpha^2(t,T,\delta t) = \left\langle \widetilde{\xi}_\alpha^2(t,\delta t) \right\rangle_T = \langle \xi_\alpha^2(t,\delta t)\rangle_T - \langle \xi_\alpha(t,\delta t)\rangle_T^2 . \tag{3.15}$$

This moment is, of course, the mean square standard deviation or the variance of the natural logarithm of the prices of the corresponding asset. The quantity is related to the so-called volatility, which is formally defined on the basis of the returns (3.4)

$$\sigma_\alpha^{\text{vola}}(t,T,\delta t) = \sqrt{\langle R_\alpha^2(t,\delta t)\rangle_T - \langle R_\alpha(t,\delta t)\rangle_T^2}. \tag{3.16}$$

The volatility is a technical term in finance that in particular evaluates the risk of a stock. For sufficiently small price fluctuations, both definitions (3.15) and (3.16) are practically equivalent. The heuristic analysis of the volatility of various asset prices shows that the thesis of stationarity is approximately valid for long elementary times δt and large time windows T (see Figure 3.6). Nevertheless, also in the case of an apparent stationarity must we be very careful in generalizing such empirical observations. If we employ small time windows but high-frequency observations, we find considerable volatility fluctuations, which are partially affected by daily cycles. Therefore, we will always speak of stationarity only within the framework of an ideal financial market.

It should be noted that the computation of moving averages is numerically executed in more extensive ways. Very popular are causal linear convolution operations [446, 447]. Such techniques can also be applied on irregularly spaced time series. As explained above, inhomogeneous time series arise from tick-by-tick data, including every quote or transaction price of the financial market. The basic idea is a very natural generalization of (3.13). We define a suitable causal kernel $\omega(t)$ and determine the moving averages via a discrete convolution

$$\left\langle \prod_{k=1}^{K} \xi_{\alpha_k}(t,\delta t) \right\rangle_\omega = \frac{\sum\limits_m \omega(t-t_m) \prod\limits_{k=1}^{K} \xi_{\alpha_k}(t_m,\delta t)}{\sum\limits_m \omega(t-t_m)}, \tag{3.17}$$

where the causality requires $\omega(t) = 0$ for $t < 0$ so that no information from the future is used. A frequently used function is the exponentially decaying kernel $\omega(t) = \exp\{-t/T\}$. The repeated application of the convolution

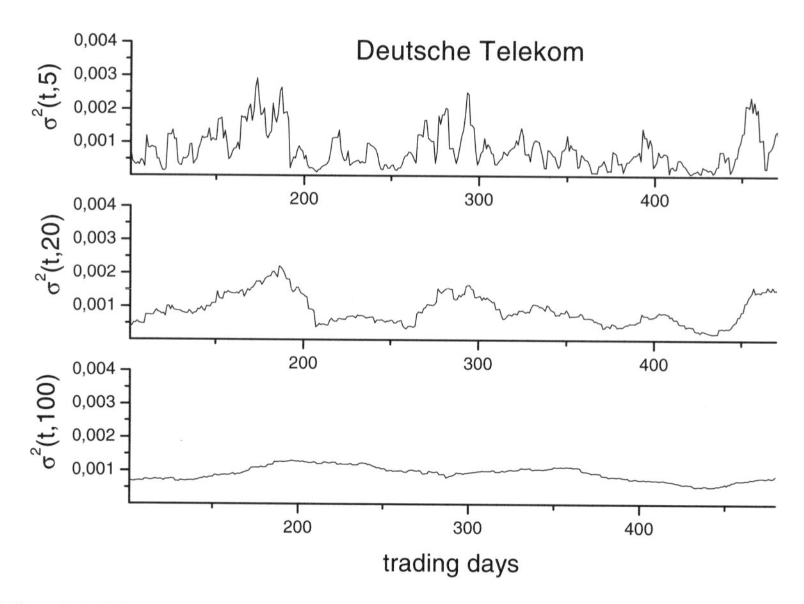

Fig. 3.6. Moving mean square standard deviation of the logarithmic price changes $\sigma^2(t, T, \delta t)$ for different time windows $T = 5$, $T = 20$, and $T = 100$ trading days. The time horizon is $\delta t = 1$ trading day (Deutsche Telekom stock, reference time period 11/00–07/02).

procedure is a method to generate new operators that may be linearly combined to make more complicated kernels. For instance, the exponential kernel allows the construction of [446]

$$\omega_n(t) = \exp\left\{-\frac{t}{T}\right\} \sum_{k=0}^{n} \frac{1}{k!} \left(\frac{t}{T}\right)^k . \tag{3.18}$$

Obviously, we find $\omega_n(t) \approx 1$ for $t \ll T$, whereas the exponential decay $\omega_n(t) \sim \exp\{-t/T\}$ dominates for $t \gg T$.

3.2.4 Markov Approximation

In principle, the joint probability (3.10) contains all necessary information for the description of a financial market on the macroscopic level using price changes at relevant quantities. Unfortunately, this function is too complicated for practical application.

However, if the logarithmic differences ξ_α of the asset prices correspond to a sufficiently long time horizon δt, then we can assume that the price fluctuations take place statistically independently. In other words, if $\delta t > \delta t_{\mathrm{Markov}}$, we may use the Markov approximation

$$p_{\delta t}\left(\xi^{(n)}, t_n \mid \xi^{(n-1)}, t_{n-1}; ...; \xi^{(0)}, t_0\right) = p_{\delta t}\left(\xi^{(n)}, t_n\right) \tag{3.19}$$

and consequently

$$p_{\delta t}\left(\xi^{(N)}, t_N; \xi^{(N-1)}, t_{N-1}; ...; \xi^{(0)}, t_0\right) = \prod_{j=0}^{N} p_{\delta t}\left(\xi^{(j)}, t_j\right). \tag{3.20}$$

In the first part of this chapter, we will regard sufficiently large time differences δt well above the Markov horizon δt_{Markov}. In this case, one can describe the development of the data of a certain financial market in the context of the theory of Markov processes.

In the opposite case (i.e., below the Markovhorizon), memory effects play an important role that cannot be neglected. We will analyze this problem in the last part of this chapter.

3.2.5 Taxonomy of Stocks

Correlation and Anticorrelation. In financial markets, many stocks are traded simultaneously. A reasonable way to detect similarities in the time evolution starts from the correlation coefficients

$$\vartheta_{\alpha\beta} = \frac{\langle \xi_\alpha(t,\delta t)\xi_\beta(t,\delta t)\rangle_T - \langle \xi_\alpha(t,\delta t)\rangle_T \langle \xi_\beta(t,\delta t)\rangle_T}{\sqrt{\langle \xi_\alpha^2(t,\delta t)\rangle_T - \langle \xi_\alpha(t,\delta t)\rangle_T^2}\sqrt{\langle \xi_\beta^2(t,\delta t)\rangle_T - \langle \xi_\beta(t,\delta t)\rangle_T^2}}. \tag{3.21}$$

These coefficients are simply the normalized components of the heuristically determined covariance matrix. With this definition, the correlation coefficients can assume values between -1 and $+1$. In the case of $\vartheta_{\alpha\beta} = 1$, the stocks α and β are completely correlated so that we obtain $\xi_\alpha(t,\delta t) = A\xi_\beta(t,\delta t) + B$ with $A > 0$ and arbitrary B. The opposite situation of a complete anticorrelation occurs for $\vartheta_{\alpha\beta} = -1$. Finally, $\vartheta_{\alpha\beta} = 0$ indicates uncorrelated changes in the time evolution of stock prices.

We may discuss the correlation coefficients in terms of a geometric representation. To this end, we introduce the normalized price fluctuations

$$\hat{\xi}_\alpha(t,\delta t) = \frac{\xi_\alpha(t,\delta t) - \langle \xi_\alpha(t,\delta t)\rangle_T}{\sqrt{\langle \xi_\alpha^2(t,\delta t)\rangle_T - \langle \xi_\alpha(t,\delta t)\rangle_T^2}} \tag{3.22}$$

so that the correlation coefficients can be written as second moments of these reduced quantities

$$\vartheta_{\alpha\beta} = \langle \hat{\xi}_\alpha(t,\delta t)\hat{\xi}_\beta(t,\delta t)\rangle_T. \tag{3.23}$$

Let us now combine the S records of the normalized price fluctuations $\hat{\xi}_\alpha(t,\delta t)$ into the S-dimensional vector $\vec{\xi}_\alpha = S^{-1/2}\left[\hat{\xi}_\alpha(t_0,\delta t), \hat{\xi}_\alpha(t_1,\delta t)..., \hat{\xi}_\alpha(t_S,\delta t)\right]$ with $S = T/\delta t + 1$ and $t_n = t - n\delta t$. Thus, the correlation coefficient is simply the scalar product $\vartheta_{\alpha\beta} = \vec{\xi}_\alpha \vec{\xi}_\beta$. The vector $\vec{\xi}_\alpha$ has unit length because of

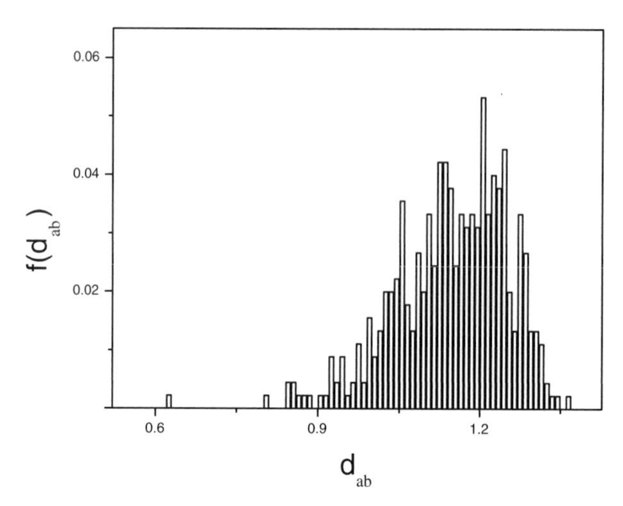

Fig. 3.7. The empirical frequency distribution function $f(d_{\alpha\beta})$ of the Euclidean distances between the stocks of the German stock index DAX (reference time interval 12/00–12/01, daily observations).

$\vec{\xi}_\alpha^{\,2} = \vartheta_{\alpha\alpha} = 1$. The Euclidean distance between the vectors $\vec{\xi}_\alpha$ and $\vec{\xi}_\beta$ is obtainable from the Pythagorean relation

$$d_{\alpha\beta}^2 = \left\| \vec{\xi}_\alpha - \vec{\xi}_\beta \right\|^2 = \vec{\xi}_\alpha^{\,2} + \vec{\xi}_\alpha^{\,2} - 2\vec{\xi}_\alpha \vec{\xi}_\beta \tag{3.24}$$

so that

$$d_{\alpha\beta} = \sqrt{2\left(1 - \vec{\xi}_\alpha \vec{\xi}_\beta\right)} = \sqrt{2\left(1 - \vartheta_{\alpha\beta}\right)}. \tag{3.25}$$

This distance is an appropriate measure to characterize the temporal synchronization of shares (see Figure 3.7). Obviously, the distance between completely correlated assets vanishes, while the distance gets its maximum value $d_{\alpha\beta} = 2$ for completely anticorrelated stocks. Due to the Euclidean character, the distances between stocks are symmetric, $d_{\alpha\beta} = d_{\beta\alpha}$, positive-semidefinite, $d_{\alpha\beta} \geq 0$ with $d_{\alpha\alpha} = 0$, and satisfy the triangular inequality, $d_{\alpha\beta} \leq d_{\alpha\gamma} + d_{\beta\gamma}$. The knowledge of the distance matrix between stocks is customarily used to decompose the set of shares into suitable subsets of closely related stocks.

Ultrametricity. The main problem with a possible decomposition procedure is the definition of a space in which the stocks are embedded. Unfortunately, the dimension of the space spanned by the vectors $\vec{\xi}_\beta$ is too large for a reasonable analysis. It is simple to check that such a space has at least the dimension $\min(A-1, S)$, where A is the total number of stocks that were taken into account. One way to obtain a simple taxonomy of the correlations between stocks, however, is the introduction of an additional hypothesis about the spatial topology.

Let us make the working hypothesis that a useful space for linking A stocks is an ultrametric space. This a posteriori hypothesis is motivated by the fact that the associated taxonomy is meaningful from an economic point of view. A good introduction to ultrametricity for physicists is given elsewhere [324]. Here, we just remind the reader that the usual triangular equation $d_{\alpha\beta} \leq d_{\alpha\gamma} + d_{\beta\gamma}$ is substitute in spaces endowed with an ultrametric distance $\hat{d}_{\alpha\beta}$ by the stronger inequality

$$\hat{d}_{\alpha\beta} \leq \max\left(\hat{d}_{\alpha\gamma}, \hat{d}_{\beta\gamma}\right) . \tag{3.26}$$

In an ultrametric space, all triangles have at least two equal sides that are larger than or equal to the third side. Ultrametric spaces provide a natural way to describe hierarchically structures of complex systems since the concept of ultrametricity is directly connected to the concept of hierarchy. A hierarchical tree is a very good way of representing an ultrametric set of objects. We remark that in the solution of the mean-field spin-glass theory, one finds an exact ultrametric structure: the equilibrium states are organized on a hierarchical tree, and if we pick up three configurations of the system and compute their distances, we find an ultrametric triangle [128, 129, 301].

The point is now to find a formalism in order to transform the metric distances into ultrametric distances. Assuming that the metric distance matrix with the components $d_{\alpha\beta}$ exists, several ultrametric spaces can be obtained by performing the given partition of the set of A shares. Among all of the possible ultrametric topologies associated with the metric distances, the subdominant ultrametric structure convinces because of its simplicity. In the presence of a metric space defined by the whole set of distances $d_{\alpha\beta}$, the subdominant ultrametric topology can be obtained by determining the minimal spanning tree.

Formally, the minimal spanning tree is a graph of $A-1$ edges connecting the A shares in such a way that the graph structure minimizes the sum over the metric edge distances. The investigation of the subdominant ultrametric topology allows us to determine in a unique way an indexed hierarchy of the A stocks. The technique of constructing a minimal spanning tree of A shares, known as Kruskal's algorithm [298, 422], is simple and direct.

In the first step, we have to find the pair of stocks separated by the smallest distance $d_{\alpha\beta}$. This pair forms a cluster of size 2, and the ultrametric distance is equivalent to the metric distance $\hat{d}_{\alpha\beta} = d_{\alpha\beta}$. We then determine the minimum of the remaining metric distances. If this distance combines another pair, we get a further cluster of size 2. But if a connection to the already existing cluster appears, the cluster size increases by 1. Furthermore, we identify all ultrametric distances between the new element and the old elements of the cluster with the smallest metric distance between the new element and the old elements. If we continue this procedure, the size of the clusters increases monotonously. Finally, we find stock pairs that combine two clusters. The ultrametric distance between all elements of the one cluster and all elements of the other cluster is then precisely the metric distance of the connecting pair.

After we have finished this procedure, we arrive at a complete matrix of ultrametric distances $\hat{d}_{\alpha\beta}$. In contrast to the matrix of metric distances, the number of different elements in the ultrametric distance matrix cannot exceed $A - 1$.

The graphic representation of the minimal spanning tree can be obtained straightforwardly. To do this, we represent each stock by a site and connect all pairs (α, β) of sites if the condition $\hat{d}_{\alpha\beta} = d_{\alpha\beta}$ is satisfied. Figure 3.8 shows the minimal spanning tree for the German stock index (DAX).

Another representation is the so-called indexed hierarchical tree, which considers the subdominant ultrametric structure more explicitly. In Figure 3.9, we show this tree associated with the minimal spanning tree of the DAX.

Random Potts Models. Another possibility for the arrangement of shares is the construction of a fixed number of groups with minimum metric distances. To do this, we assume that the number of groups is defined by $G < A$. A stock α is characterized by a group number $s_\alpha \in [1, ..., G]$ that defines the affiliation to a group. Further, let us define a functional that exclusively considers all metric distances inside the groups,

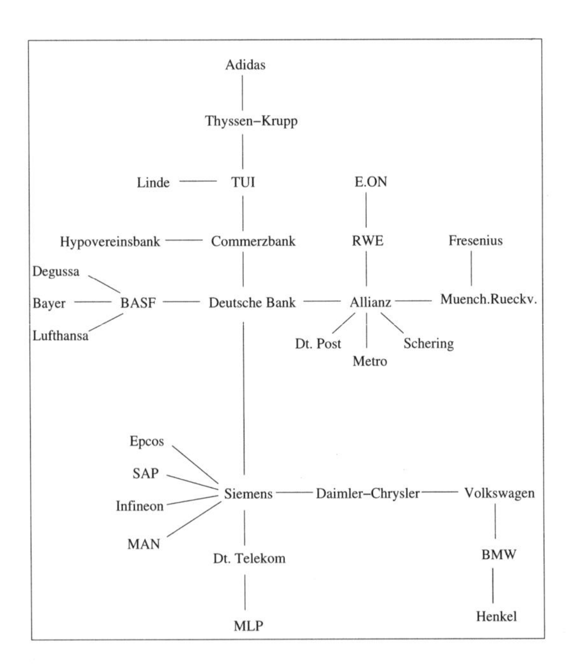

Fig. 3.8. Minimal spanning tree of the German stock index (DAX) (reference time interval 12/00–12/01, daily observations).

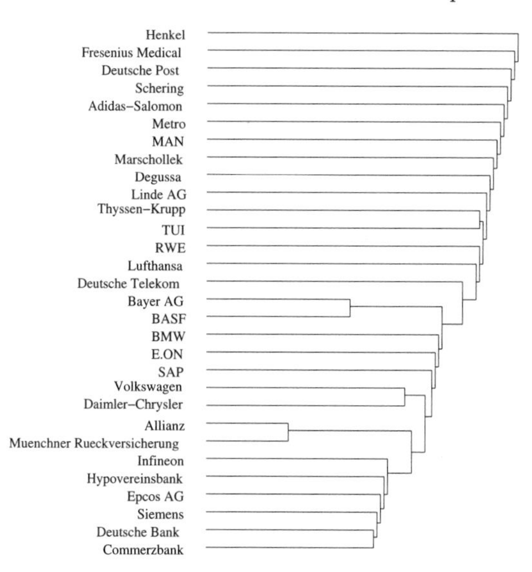

Fig. 3.9. Hierarchical tree of the German stock index (DAX) (reference time interval 12/00–12/01, daily observations). The lengths of the branches correspond to the metric distance to the nearest stock.

$$H = \sum_{\alpha \neq \beta} J_{\alpha\beta} \delta_{s_\alpha, s_\beta}, \tag{3.27}$$

with $\delta_{ss'} = 1$ for $s = s'$ and $\delta_{ss'} = 0$ for $s \neq s'$. The coupling parameters $J_{\alpha\beta} = J(d_{\alpha\beta})$ are chosen as they increase monotonously with the argument $d_{\alpha\beta}$. Supposing that the coupling parameter between the share pair (α, β) is $J_{\alpha\beta}$, the contribution to H is equal to $J_{\alpha\beta}$ if the two shares are in the same group and equal to 0 if they are in different groups. The minimization of H by rearranging the occupations should lead the groups to contain relatively synchronous stocks.

A functional of type (3.27) is usually denoted as the Hamiltonian of a Potts model [183, 314, 435] in the framework of statistical physics. Such a model is a generalization of the well-known two-state Ising model to an arbitrary number G of discrete states: $s_\alpha = 1, ..., G$ instead of $s_\alpha = \pm 1$. In the language of our problem, each Potts state corresponds to a certain stock group. The minimization of H corresponds to the determination of the ground states of the Hamiltonian of the Potts model. Because of the more or less stochastic structure of the distance matrix, there are many local minimum configurations $\left\{\widehat{s}_1^{(i)}, \widehat{s}_2^{(i)}, ..., \widehat{s}_A^{(i)}\right\}$, with $i = 1, 2, ...$, which complicate the determination of a ground state.

The Potts model is relatively sensitive to the choice of the function $J(d_{\alpha\beta})$. As an example, we study the coupling parameter $J_{\alpha\beta} = d_{\alpha\beta}^2 - c$, where c is an arbitrary number. For $c > 4$, we obtain $J_{\alpha\beta} < 0$, and the minimum

	DAX	
Deutsche Post Fresenius Medical Henkel Schering Metro	Adidas–Salomon BASF Linde Bayer Degussa	Allianz Muench.Rueckv. MLP E.ON RWE
Deutsche Telekom Epcos SAP Infineon Siemens	Lufthansa BMW MAN TUI Thyssen–Krupp	Hypovereinsbank Commerzbank Deutsche Bank Daimler–Chrysler Volkswagen

Fig. 3.10. Group structure of the DAX (reference time interval 12/00–12/01, daily observations) obtained for $c = 4$ (antiferromagnetic regime). An equipartition of the occupation per group is found.

configurations of (3.27) show a ferromagnetic behavior. This means that the global minimum [1] is one in which a particular one of the G different states s_α is more probable than the others for all shares α. In other words, for sufficiently large c, all shares are collected in one group, so that this regime is not very important for the arrangement of stocks.

	DAX	
Henkel Fresenius Medical E.ON Schering RWE MLP Deutsche Post Metro Degussa Adidas–Salomon		Allianz BASF Bayer Hypovereinsbank BMW Commerzbank Daimler–Chrysler Deutsche Bank Lufthansa Deutsche Telekom Epcos Infineon Linde MAN Muench. Rueckv. TUI SAP Siemens Thyssen–Krupp Volkswagen

Fig. 3.11. Group structure of the DAX (reference time interval 12/00–12/01, daily observations) obtained for $c = 2$ (Potts glass regime). A large main group dominates the group structure.

[1] Note that the ferromagnetic ground state is G-fold, degenerated while the ground state of the Potts glass and the antiferromagnetic regime are $G!$-fold degenerated.

For $c < 0$, there occurs an antiferromagnetic behavior. Here, the global minimum configuration is characterized by an equipartition of the G different states. Thus, the average occupation of a group is given by A/G.

The case $c \approx 2$ is related to a so-called Potts-glass. This regime offers a very complicated configuration landscape. In particular, we expect a very heterogeneous occupation of the groups. The stocks of a group are mainly correlated, whereas for stocks of different groups, anticorrelation predominates.

The analysis of the various stock markets shows that several groups are obtainable using the antiferromagnetic (Figure 3.10) or the glassy regime (Figure 3.11) of the Potts model and that these groups are homogeneous with respect to their industry sector and often also with respect to their industry subsector.

3.3 Long-Time Regime

3.3.1 The Central Limit Theorem

We want to attempt to determine the possible probability distribution function $p_{\delta t}(\xi, t)$ of the price fluctuations more closely. Up to now, we can only say that such a function must exist. A certain estimate of the probability density is possible by an application of empirical methods to financial data, but the preserved information may contain considerable errors. Therefore, we want to choose another way. To this end, we assume that we know the probability distribution function $p_{\delta t}(\xi, t)$ for price changes over a very short time horizon δt. In the following context, we designate $p_{\delta t}(\xi, t)$ also as an elementary probability distribution function. Indeed, the time horizon should be above the Markov horizon, $\delta t > \delta t_{\text{Markov}}$. Then, the joint probability for the time series $\{\xi(t_1, \delta t), ..., \xi(t_N, \delta t)\}$ is given by (3.20):

$$p_{\delta t}\left(\xi^{(N)}, t_N; \xi^{(N-1)}, t_{N-1}; ...; \xi^{(1)}, t_1\right) = \prod_{j=1}^{N} p_{\delta t}\left(\xi^{(j)}, t_j\right). \tag{3.28}$$

Let us now determine the function $p_{N\delta t}(\xi, t)$ for a price change ξ during the time interval $[t, t + N\delta t]$. Because the logarithmic price changes are additive, we get

$$p_{N\delta t}(\xi, t) = \int \prod_{j=1}^{N} d\xi^{(j)} \delta\left(\xi - \sum_{j=1}^{N} \xi^{(j)}\right) \prod_{j=1}^{N} p_{\delta t}\left(\xi^{(j)}, t_j\right) \tag{3.29}$$

with the condition $t_1 = t$. The Markov property allows us to derive the complete functional structure of the probability distribution function $p_{N\delta t}(\xi, t)$ from the sole knowledge of the elementary probability densities $p_{\delta t}\left(\xi^{(j)}, t_j\right)$. It is convenient to use the characteristic function (2.14), which is defined as the Fourier transform of the probability density. Hence, we obtain

$$\begin{aligned}\hat{p}_{N\delta t}(k,t) &= \int d\xi \exp\{ik\xi\} p_{N\delta t}(\xi,t) \\ &= \int \prod_{j=1}^{N} d\xi^{(j)} \exp\left\{ik\sum_{j=1}^{N}\xi^{(j)}\right\} \prod_{j=1}^{N} p_{\delta t}\left(\xi^{(j)},t_j\right) \\ &= \prod_{j=0}^{N} \hat{p}_{\delta t}(k,t_j). \end{aligned} \tag{3.30}$$

Let us assume for the moment that the financial market is in a steady state (i.e., the characteristic function is a time-independent quantity). Thus, (3.30) can be written as

$$\hat{p}_{N\delta t}(k) = \left[\hat{p}_{\delta t}(k)\right]^N. \tag{3.31}$$

What can we learn from this approach? To this end, we provide a naive scaling procedure to (3.31). We start from the expansion of the characteristic function in terms of cumulants. If we focus on distribution functions of single prices, we obtain

$$\hat{p}_{\delta t}(k) = \exp\left\{\sum_{n=1}^{\infty} \frac{c^{(n)}}{n!}(ik)^n\right\} \tag{3.32}$$

and, because of (3.31),

$$\hat{p}_{N\delta t}(k) = \exp\left\{\sum_{n=1}^{\infty} \frac{Nc^{(n)}}{n!}(ik)^n\right\}, \tag{3.33}$$

where k is now a simple scalar quantity instead of a vector of dimension A. Obviously, when $N \to \infty$, the quantity ξ goes to infinity with the central tendency $\bar{\xi} = Nc^{(1)}$ and the standard deviation $\sigma = \left(Nc^{(2)}\right)^{1/2}$. Since the drift can be zero or can be put to zero by a suitable shift $\xi \to \xi - \bar{\xi}$, we conclude that the relevant scale is that of the fluctuations, namely the variance σ. The corresponding range of k is simply its inverse since ξ and k are conjugate in the Fourier transform. Thus, after rescaling $k \to \hat{k}N^{-1/2}$, the cumulant expansion reads

$$\hat{p}_{N\delta t}\left(\hat{k}\right) = \exp\left\{\sum_{n=1}^{\infty} \frac{c^{(n)}N^{1-n/2}}{n!}\left(i\hat{k}\right)^n\right\}. \tag{3.34}$$

Apart from the first cumulant, we find that the second cumulant remains invariant while all higher cumulants approach zero as $N \to \infty$. Thus, only the first and second cumulants will remain for sufficiently large N, and the probability distribution function $p_{N\delta t}(\xi,t)$ approaches a Gaussian function. The result of our naive argumentation is the central limit theorem. The precise formulation of this important theorem is:

> The sum, normalized by $N^{-1/2}$ of N random independent and identically distributed states of zero mean and finite variance, is a random variable with a probability distribution function converging to the Gaussian distribution with the same variance. The convergence is to be understood in the sense of a limit in probability (i.e., the probability that the normalized sum has a value within a given interval converges to that calculated from the Gaussian distribution).

We will now give a more precise derivation of the central limit theorem. Formal proofs of the theorem may be found in probability textbooks such as Feller [101, 119, 412]. Here, we follow a more physically motivated way of Sornette [383] using the technique of the renormalization group theory.

This powerful method [444] introduced in field theory and in critical phase transitions, is a very general mathematical tool that allows one to decompose the problem of finding the collective behavior of a large number of elements on large spatial scales and for long times into a succession of simpler problems with a decreasing number of elements whose effective properties vary with the scale of observation. In the context of the central limit theorem for finance, these elements refer to the elementary price changes $\xi(t, \delta t)$.

The renormalization group theory works best when the problem is dominated by one characteristic scale that diverges at the so-called critical point. The distance to this criticality is usually determined by a control parameter, which may be identified in our special case as N^{-1}. Close to the critical point, a universal behavior becomes observable that is related to typical phenomena such as scale invariance or self-similarity. As we will see below, the form stability of the Gaussian probability distribution function is such a kind of self-similarity.

The renormalization consists of an iterative application of decimation and rescaling steps. The first step is to reduce the number of elements to transform the problem into a simpler one. We use the thesis that, under certain conditions, the knowledge of all of the cumulants is equivalent to the knowledge of the probability densityr, so we can write

$$p_{\delta t}(\xi) = f\left(\xi, c^{(1)}, c^{(2)}, ..., c^{(m)}, ...\right), \tag{3.35}$$

where f is a unique function of ξ and the infinite set of all cumulants $\{c^{(1)}, c^{(2)}, ...\}$. Every distribution function can be expressed by the same function in this way, however with differences in the infinite set of parameters.

The probability distribution function $p_{N\delta t}(\xi)$ may be the convolution of $N = 2^n$ identical distribution functions $p_{\delta t}(\xi)$. This specific choice of N is not a restriction since we are interested in the limit of large N, and how we reach this limit is irrelevant. We denote the result of the 2^n-fold convolution as

$$p_{N\delta t}(\xi) = f^{(n)}\left(\xi, c^{(1)}, c^{(2)}, ..., c^{(m)}, ...\right). \tag{3.36}$$

Furthermore, we obtain from (3.31) the general relation

$$p_{2\delta t}(\xi) = f\left(\xi, 2c^{(1)}, 2c^{(2)}, ..., 2c^{(m)}, ...\right). \tag{3.37}$$

With this knowledge, we are able to generate $p_{N\delta t}(\xi)$ also from $p_{2\delta t}(\xi)$ by a 2^{n-1}-fold convolution

$$p_{N\delta t}(\xi) = f^{(n-1)}\left(\xi, 2c^{(1)}, 2c^{(2)}, ..., 2c^{(m)}, ...\right). \tag{3.38}$$

Here, we see the effect of the decimation. The new convolution considers only 2^{n-1} price changes with respect to the new time difference $2\delta t$ instead of 2^n steps and a timescale δt. The decimation itself corresponds to the pairing due to the convolution (3.29) between two identical elementary probability distributions

$$p_{2\delta t}(\xi) = \int p_{\delta t}(\xi - \xi')\, p_{\delta t}(\xi')\, d\xi'. \tag{3.39}$$

The notation of the scale is inherent to the probability distribution function. The new elementary probability distribution function $p_{2\delta t}(\xi)$ obtained from (3.39) may display differences from the probability density from which we started. We compensate for this by the scale factor λ^{-1} for ξ. This leads to the rescaling step $\xi \to \lambda^{-1}\xi$ of the renormalization group, which is necessary to keep the reference scale.

With the rescaling of the components of the price vector ξ, the cumulants are also rescaled, and each cumulant of order m has to be multiplied by the factor λ^{-m}. This is a direct consequence of (2.19) because it demonstrates that the cumulants of order m have the dimension $|k|^{-m}$ and $|\xi|^m$, respectively. The conservation of the probabilities $p_{\delta t}(\xi)\, d\xi = p_{\delta t}(\xi')\, d\xi'$ introduces a prefactor λ^{-A} as a consequence of the change of the A-dimensional vector $\xi \to \xi'$. We thus obtain from (3.38)

$$p_{N\delta t}(\xi) = \lambda^{-A} f^{(n-1)}\left(\frac{\xi}{\lambda}, \frac{2c^{(1)}}{\lambda}, \frac{2c^{(2)}}{\lambda^2}, ..., \frac{2c^{(m)}}{\lambda^m}, ...\right). \tag{3.40}$$

The successive repetition of both decimation and rescaling leads after n steps to

$$p_{N\delta t}(\xi) = \lambda^{-nA} f^{(0)}\left(\frac{\xi}{\lambda^n}, \frac{2^n c^{(1)}}{\lambda^n}, \frac{2^n c^{(2)}}{\lambda^{2n}}, ..., \frac{2^n c^{(m)}}{\lambda^{mn}}, ...\right). \tag{3.41}$$

As mentioned above, $f^{(n)}(\xi, ...c^{(m)}, ...)$ is a function that is obtainable from a convolution of 2^n identical functions $f(\xi, ...c^{(m)}, ...)$. In this sense, we obtain the matching condition $f^{(0)} \equiv f$ so that we arrive at

$$p_{N\delta t}(\xi) = \lambda^{-nA} f\left(\frac{\xi}{\lambda^n}, \frac{2^n c^{(1)}}{\lambda^n}, \frac{2^n c^{(2)}}{\lambda^{2n}}, ..., \frac{2^n c^{(m)}}{\lambda^{mn}}, ...\right). \tag{3.42}$$

Finally, we have to fix the scale λ. We see from (3.42) that the particular choice $\lambda = 2^{1/m_0}$ makes the prefactor of the m_0th cumulant equal to 1, while all higher cumulants decrease to zero as $n = \log_2 N \to \infty$. The lower cumulants diverge with $N^{(1-m/m_0)}$, where $m < m_0$.

The only reasonable choice is $m_0 = 2$ because $\lambda = \sqrt{2}$ keeps the probability distribution function in a window with constant width. In this case, only the first cumulant may remain divergent for $N \to \infty$. As mentioned above, this effect can be eliminated by a suitable shift of ξ. Thus, we arrive at

$$\lim_{N\to\infty} p_{N\delta t}(\xi) = N^{-A/2} f\left(\frac{\xi}{\sqrt{N}}, c^{(1)}\sqrt{N}, c^{(2)}, 0, ..., 0, ...\right). \tag{3.43}$$

In particular, if we come back to the financial problem, we have thus obtained the asymptotic result that the probability distribution of price changes over large time intervals $N\delta t$ has only its two first cumulants nonzero. Hence, as discussed in subsection 2.1.5, the corresponding probability density is a Gaussian law.

If we return to the original scales, the final Gaussian probability distribution function $p_{N\delta t}(\xi)$ is characterized by the mean $\overline{\xi} = \overline{\xi}(N\delta t) = Nc^{(1)} = N\overline{\xi}(\delta t)$ and the covariance matrix $\tilde{\sigma} = \tilde{\sigma}(N\delta t) = Nc^{(2)} = N\tilde{\sigma}(\delta t)$, where $c^{(1)}$ and $c^{(2)}$ are the first two cumulants of the elementary probability density $p_{\delta t}(\xi)$. Hence, we obtain

$$\lim_{N\to\infty} p_{N\delta t}(\xi) = \frac{1}{(2\pi)^{A/2}\sqrt{\det\tilde{\sigma}}} \exp\left\{-\frac{1}{2}\left(\xi - \overline{\xi}\right)\tilde{\sigma}^{-1}\left(\xi - \overline{\xi}\right)\right\}, \tag{3.44}$$

or with the rescaled and shifted states,

$$\lim_{N\to\infty} p_{N\delta t}(\xi) = \frac{1}{(2\pi)^{A/2}\sqrt{\det c^{(2)}}} \exp\left\{-\frac{1}{2}\hat{\xi}\left[c^{(2)}\right]^{-1}\hat{\xi}\right\}. \tag{3.45}$$

The quantity $\hat{\xi}$ is simply the sum normalized by $N^{-1/2}$ of N random independent and identically distributed states $\xi(t + n\delta t, \delta t) - c^{(1)}$ of zero mean and finite variance,

$$\hat{\xi} = \frac{\xi - \overline{\xi}(N\delta t)}{\sqrt{N}} = \frac{1}{\sqrt{N}}\sum_{j=0}^{N-1}\left(\xi(t + n\delta t, \delta t) - c^{(1)}\right). \tag{3.46}$$

In other words, (3.45) is the mathematical formulation of the central limit theorem. The Gaussian distribution function itself is a fixed point of the convolution procedure in the space of functions in the sense that it is form-stable under the renormalization group approach. Notice that form stability, or alternatively self-similarity, means that the resulting Gaussian function is identical to the initial Gaussian function after an appropriate shift and a rescaling of the variables.

We remark that the convergence to a Gaussian behavior also holds if the initial variables have different probability distribution functions with finite variance of the same order of magnitude. In particular, small deviations of financial data from the stationarity condition are not so important for the long-time analysis in the frame of the central limit theorem. The generalized fixed point is now the Gaussian law (3.44) with

$$\overline{\xi} = \sum_{n=0}^{N-1} \overline{\xi}\,(t + n\delta t, \delta t) \quad \text{and} \quad \tilde{\sigma} = \sum_{n=0}^{N-1} \tilde{\sigma}\,(t + n\delta t, \delta t)\,, \tag{3.47}$$

where $\overline{\xi}\,(t + n\delta t, \delta t)$ and $\tilde{\sigma}\,(t + n\delta t, \delta t)$ are the mean trend vector and the covariance matrix, respectively, obtained from the time-dependent probability distribution function $p_{\delta t}\,(\xi, t + n\delta t)$.

Finally, it should be remarked that the two conditions of the central limit theorem may be partially relaxed. The first condition under which this theorem holds is the Markov property. This strict condition can, however, be weakened, and the central limit theorem still holds for weakly correlated variables under certain conditions. The second condition, that the variance of the variables be finite, can be somewhat relaxed to include probability functions with algebraic tails $|\xi|^{-3}$. In this case, the normalizing factor is no longer $N^{-1/2}$ but can contain logarithmic corrections.

3.3.2 Convergence Problems

As a consequence of the renormalization group analysis, the central limit theorem is applicable in a strict sense only in the limit of infinite N. But in practice, the Gaussian shape is a good approximation of the center of a probability distribution function if N is sufficiently large. It is important to realize that large deviations can occur in the tail of the probability distribution function $p_{N\delta t}\,(\xi)$, whose weight shrinks as N increases. The center is a region of width at least of the order of $\sqrt{N}$ around the average $\overline{\xi}\,(N\delta t)$.

Let us make more precise what the center of a probability distribution function means. For the sake of simplicity, we investigate states of only one component (i.e., ξ is a scalar quantity). As before, ξ is the sum of N identically distributed variables $\xi^{(j)}$ with mean $\overline{\xi}\,(\delta t) = c^{(1)}$, variance $\sigma^2\,(\delta t) = c^{(2)}$, and finite higher cumulants $c^{(m)}$. Thus, the central limit theorem reads

$$\lim_{N\to\infty} p_{N\delta t}\,(x) = \frac{1}{\sqrt{2\pi}} \exp\left\{-\frac{x^2}{2}\right\}, \tag{3.48}$$

where we have introduced the reduced variable

$$x = \sigma^{-1}\hat{\xi} = \frac{\xi - N\overline{\xi}\,(\delta t)}{\sqrt{Nc^{(2)}}}\,. \tag{3.49}$$

In order to analyze the convergence behavior for the tails [152], we start from the probability

$$P_{>}^{(N)}(z) = P^{(N)}(x > z) = \int_{z}^{\infty} p_{N\delta t}\,(x)\, dx \tag{3.50}$$

and analyze the difference $\Delta P^{(N)}(z) = P_>^{(N)}(z) - P_>^{(\infty)}(z)$, where $P_>^{(\infty)}(z)$ is simply the complementary error function due to (3.48). If all cumulants are finite, one can develop a systematic expansion in powers of $N^{-1/2}$ of the difference $\Delta P^{(N)}(z)$ [72],

$$\Delta P^{(N)}(z) = \frac{\exp\{-z^2/2\}}{\sqrt{2\pi}} \sum_{n=1}^{\infty} \left[\frac{Q_1(z)}{N^{1/2}} + \frac{Q_2(z)}{N} \ldots + \frac{Q_m(z)}{N^{m/2}} \ldots \right], \quad (3.51)$$

where the $Q_m(z)$ are polynomials in z, the coefficients of which depend on the first $m+2$ normalized cumulants of the elementary probability distribution function, $\lambda_k = c^{(k)}/\sigma^k$. The explicit form of these polynomials can be obtained from the textbook of Gnedenko and Kolmogorov [152]. The first two polynomials are

$$Q_1(z) = \frac{\lambda_3}{6}\left(1 - z^2\right) \quad (3.52)$$

and

$$Q_2(z) = \frac{\lambda_3^2}{72} z^5 + \left(\frac{\lambda_4}{24} - \frac{5\lambda_3^2}{36} \right) z^4 + \left(\frac{5\lambda_3^2}{24} - \frac{\lambda_4}{8} \right) z^3 . \quad (3.53)$$

If the elementary probability distribution function has Gaussian behavior, all of its cumulants $c^{(m)}$ of order larger than 2 vanish identically. Therefore, all $Q_m(z)$ are also zero, and the probability density $p_{N\delta t}(x)$ is Gaussian.

For an arbitrary asymmetric probability distribution function, the skewness λ_3 is nonvanishing in general and the leading correction is $Q_1(z)$. The Gaussian law is valid if the relative error $\left|\Delta P^{(N)}(z)\right| / P_>^{(\infty)}(z)$ is small compared to 1. Since the error increases with z, the Gaussian behavior first becomes observable close to the central tendency. The necessity condition $|\lambda_3| \ll N^{1/2}$ follows directly from $\left|\Delta P^{(N)}(z)\right| / P_>^{(\infty)}(z) \ll 1$ for $z \to 0$.

For large z, the approximation of $p_{N\delta t}(x)$ by a Gaussian law remains valid if the relative error remains small compared to 1. Here, we may replace the complementary error function $P_>^{(\infty)}(z)$ by its asymptotic representation $\exp\left\{-z^2/2\right\}/(\sqrt{2\pi}z)$. We thus obtain the inequality $|zQ_1(z)| \ll N^{1/2}$ leading to $\left|z^3\lambda_3\right| \ll N^{1/2}$. Because of (3.49), this relation is equivalent to the condition

$$\left|\xi - N\bar{\xi}(\delta t)\right| \ll |\lambda_3|^{-1/3} \sigma N^{2/3}. \quad (3.54)$$

This means that the Gaussian law holds in a region of order of magnitude $\left|\xi - N\bar{\xi}(\delta t)\right| \ll |\lambda_3|^{-1/3} \sigma N^{2/3}$ around the central tendency.

A symmetric probability distribution function has a vanishing skewness so that the excess kurtosis $\lambda_4 = c^{(4)}/\sigma^4$ provides the leading correction to the central limit theorem. The Gaussian law is now valid if $\lambda_4 \ll N$ and

$$\left|\xi - N\bar{\xi}(\delta t)\right| \ll |\lambda_4|^{-1/4} \sigma N^{3/4} \quad (3.55)$$

(i.e., the central region in which the Gaussian law holds is now of order of magnitude $N^{3/4}$).

Another class of inequalities describing the convergence behavior with respect to the central limit theorem was found by Berry [43] and Esséen [114]. The Berry–Esséen theorems [120] provide inequalities controlling the absolute difference $\left|\Delta P^{(N)}(z)\right|$. Suppose that the variance σ and the average

$$\eta = \int \left|\xi - \overline{\xi}(\delta t)\right|^3 p_{\delta t}(\xi)\, d\xi \tag{3.56}$$

are finite quantities. Then, the first theorem reads

$$\left|\Delta P^{(N)}(z)\right| \leq \frac{3\eta}{\sigma^3 \sqrt{N}}. \tag{3.57}$$

The second theorem is the extension to not identically distributed variables. In the language of finance, this case corresponds to a nonstationary market. Here, we have to replace the constant values of σ and η by

$$\left\langle \sigma^2(t, \delta t) \right\rangle_N = \frac{1}{N} \sum_{n=0}^{N-1} \sigma^2(t + n\delta t, \delta t) \tag{3.58}$$

and

$$\left\langle \eta(t, \delta t) \right\rangle_N = \frac{1}{N} \sum_{n=0}^{N-1} \eta(t + n\delta t, \delta t), \tag{3.59}$$

where $\sigma(t, \delta t)$ and $\eta(t, \delta t)$ are obtained from the time-dependent elementary probability distribution function $p_{\delta t}(\xi, t)$. Then, the following inequality holds:

$$\left|\Delta P^{(N)}(z)\right| \leq \frac{6 \left\langle \eta(t, \delta t) \right\rangle_N}{\left\langle \sigma^2(t, \delta t) \right\rangle_N^{3/2} \sqrt{N}}. \tag{3.60}$$

Notice that the Berry–Esséen theorems are less stringent than the results obtained from the cumulant expansion (3.51). We see that the central limit theorem gives no information about the behavior of the tails for finite N. Only the center is well-approximated by the Gaussian law. The width of the central region depends on the detailed properties of the elementary probability distribution functions.

The Gaussian probability distribution function is the fixed point, or the attractor, of a well-defined class of functions. This class is also called the basin of attraction with respect to the corresponding functional space. When N increases, the functions $p_{N\delta t}(\xi)$ become progressively closer to the Gaussian attractor. As discussed above, this process is not uniform. The convergence is faster close to the center than in the tails of the probability distribution function.

3.3.3 Fokker–Planck Equation for Financial Processes

Let us assume that the financial market is in a stationary state and we know the set of all prices at a given initial time t_0. The probability distribution function $p_{t-t_0}(\xi)$ to observe the overall logarithmic price changes ξ at $t > t_0$ with $t - t_0 \gg \delta t_{\text{Markov}}$ can be interpreted as a conditional probability

$$p(\xi, t \mid 0, t_0) = p_{t-t_0}(\xi) \,. \tag{3.61}$$

The number of elementary price changes during the time interval $\Delta t = t - t_0$ is simply given by $N = \Delta t/\delta t$. For large N and moderate price fluctuations, the probability density $p_{t-t_0}(\xi)$ is well-approximated by the Gaussian shape law (3.44)

$$p_{t-t_0}(\xi) = \frac{1}{(2\pi\Delta t)^{A/2}\sqrt{\det\left[c^{(2)}/\delta t\right]}} \times \exp\left\{-\frac{1}{2\Delta t}\left(\xi - \frac{c^{(1)}}{\delta t}\Delta t\right)\left[\frac{c^{(2)}}{\delta t}\right]^{-1}\left(\xi - \frac{c^{(1)}}{\delta t}\Delta t\right)\right\} \tag{3.62}$$

with $\Delta t = t - t_0$. Obviously, the corresponding conditional probability satisfies the initial condition $p(\xi, t_0 \mid 0, t_0) = \delta(\xi)$. The cumulants $c^{(1)}$ and $c^{(2)}$ are obtainable from the elementary probability distribution function $p_{\delta t}(\xi)$. If $\delta t \gg \delta t_{\text{Markov}}$ the function $p_{\delta t}(\xi)$ itself can be generated from probability distribution functions with significantly shorter time horizons by the convolution procedure introduced above. Therefore, we conclude that $c^{(1)} \sim \delta t$ as well as $c^{(2)} \sim \delta t$. It is reasonable to introduce the trend rate vector $\mu = c^{(1)}/\delta t$ and covariance rate matrix $\Phi = c^{(2)}/\delta t$, which are mainly independent from the timescale δt. Thus, we obtain

$$p(\xi, t \mid 0, t_0) = \frac{1}{(2\pi\Delta t)^{A/2}\sqrt{\det\Phi}} \times \exp\left\{-\frac{1}{2\Delta t}(\xi - \mu\Delta t)\,\Phi^{-1}(\xi - \mu\Delta t)\right\} . \tag{3.63}$$

This equation is equivalent to (2.96) and therefore the solution of the Fokker–Planck equation

$$\frac{\partial}{\partial t}p(\xi, t \mid 0, t_0) = \sum_{\alpha\beta}\frac{1}{2}\frac{\partial^2}{\partial\xi_\alpha\partial\xi_\beta}\Phi_{\alpha\beta}p(\xi, t \mid 0, t_0) - \sum_{\alpha}\frac{\partial}{\partial\xi_\alpha}\mu_\alpha p(\xi, t \mid 0, t_0) \tag{3.64}$$

with the initial condition $p(\xi, t_0 \mid 0, t_0) = \delta(\xi)$. In the last equation, we have explicitly emphasized the components of the A-dimensional vectors ξ and μ and the $A \times A$ matrix Φ.

The last step requires some remarks. As a consequence of the central limit theorem, the Gaussian probability distribution (3.44) is an asymptotically self-similar function under the convolution procedure discussed above. In contrast to this forward extrapolation, the construction of the Fokker–Planck equation (3.64) includes a backward extrapolation of the self-similarity of the Gaussian probability distribution function onto short times. The existence of a backward extrapolation is secured because Gaussian laws belong to the class of infinitely divisible probability distribution functions [120, 152]. But the backward procedure is not a unique formalism. Since the Gaussian probability distribution function is the limit distribution obtained from the convolution of arbitrary (usually identical) elementary probability distribution functions with finite variance, it can also be divided into several elementary probability densities in the course of the backward extrapolation.

Therefore, we require additionally the conservation of the self-similarity during the backward extrapolation to short timescales. Notice that this is only a helpful a priori model assumption, as opposed to the universal validity of the central limit theorem at long time scales.

Therefore, we cannot expect the short-time solution of the Fokker–Planck equation to conform with the observations in real financial markets. It is convenient to say that the Fokker–Planck equation (3.64) describes the dynamics of an ideal financial market. Although this equation yields possibly wrong results for short timescales, the solution approaches reality with increasing time difference $t - t_0$.

The Fokker–Planck equation corresponds to a set of Ito stochastic differential equations

$$d\xi_\alpha(t) = \mu_\alpha dt + \sum_{k=1}^{R} b_{\alpha,k} dW_k(t) \tag{3.65}$$

with R Wiener processes $W_k(t)$ and the $A \times R$ matrix b. This matrix is connected with the covariance rate via

$$\Phi_{\alpha\beta} = \sum_{k=1}^{R} b_{\alpha,k} b_{\beta,k} . \tag{3.66}$$

We must consider that the functions $W_k(t)$ show the characteristic properties of a Wiener process only above the Markov horizon. Roughly speaking, we have to replace the differential $dW_k(t)$ by the difference $\delta W_k(t) = W_k(t + \delta t_{\mathrm{Markov}}) - W_k(t)$ in order to keep the rules (2.137) of the Ito calculus.

We stress again that both the Ito stochastic differential equation and the corresponding Fokker–Planck equation describe an idealized substitute process that may be interpreted as a model of the financial market. This model approaches the real market only in a limited framework.

All R Wiener processes contributing to the stochastic differential equation can be interpreted as results of individual human decisions, while the coupling constants $b_{\alpha,k}$ represent more or less the action of external economic factors affecting the financial market [337]. In this theory, an economic factor is common to the set of stocks under consideration.

The matrix b may be calculated from the covariance rate Φ. To do this, we take into account that Φ is a symmetric positive-definite (or positive-semidefinite) matrix that can be written as

$$\Phi_{\alpha\beta} = \sum_{k=1}^{A} \varphi_{\alpha,k}\lambda_k\varphi_{\beta,k}, \tag{3.67}$$

where $\varphi_{\alpha,k}$ is the αth component of the kth normalized eigenvector and λ_k is the corresponding eigenvalue. If we compare (3.66) and (3.67), we get $b_{\alpha,k} = \sqrt{\lambda_k}\varphi_{\alpha,k}$. Thus, we may identify the eigenvalues λ_k as possible economic control factors, while the components of the normalized eigenvectors $\varphi_{\alpha,k}$ define the economic weight of this factor with respect to the stock α. In other words, the determination of possible economic control mechanisms is equivalent to the determination of the eigenvalues and eigenvectors of the matrix Φ. Obviously, the total number of economic factors is limited by $R \leq A$ because the matrix Φ has A nonnegative eigenvalues $\lambda_k \geq 0$. In particular, if Φ is a positive-semidefinite matrix and the eigenvalue $\lambda = 0$ is m-fold degenerated, we get $R = A - m$ since the corresponding eigenvectors do not contribute to the coupling parameters $b_{\alpha,k}$. The strength of the kth economic factor is given by

$$\Xi_k = \|b_{\alpha,k}\|^2 = \sum_{\alpha} b_{\alpha,k}^2 = \sum_{\alpha} \varphi_{\alpha,k}^2\lambda_k = \lambda_k, \tag{3.68}$$

where we have used the fact that the eigenvectors φ_k are normalized.

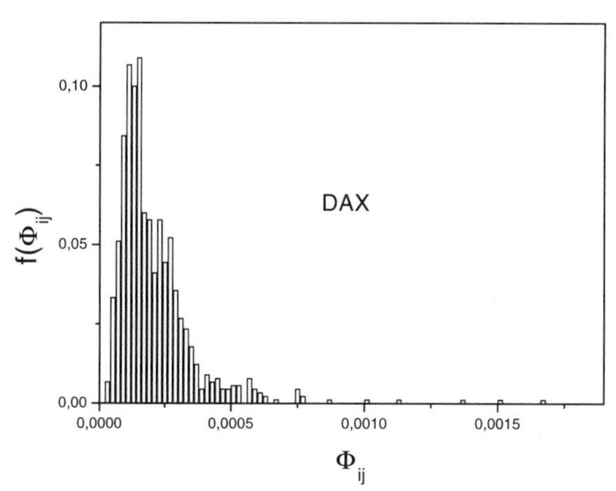

Fig. 3.12. The empirical frequency distribution function $f(\Phi_{\alpha\beta})$ of the matrix elements $\Phi_{\alpha\beta}$ corresponding to the German stock index DAX (reference time interval 12/00–12/01, daily observations).

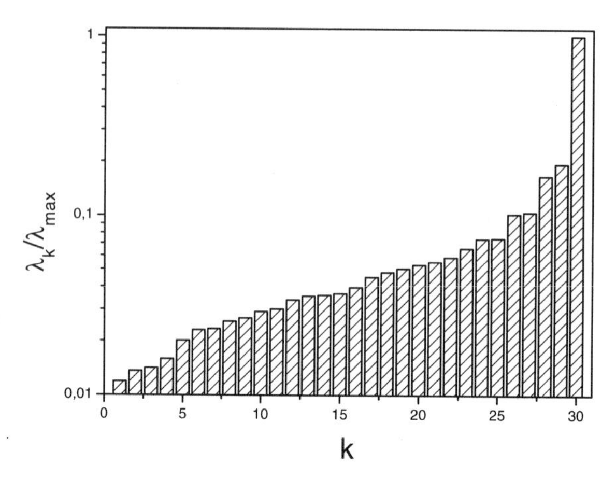

Fig. 3.13. The rank-ordered eigenvalues of the matrix Φ corresponding to the German stock index DAX (reference time interval 12/00–12/01, daily observations).

Within the framework of our theory, the existence of a few eigenvalues dominating the matrix Φ (or the corresponding covariance matrix $\tilde{\sigma}$) can be interpreted as evidence of a small number of economic factors that have an essential effect on the stochastic dynamics of the price fluctuations $\xi(t)$.

At first sight, the coefficients of the matrix Φ seem to be more or less randomly distributed variables (Figure 3.12). Thus, we should be able to use the experiences of physics and mathematics with respect to the random matrix theory [84, 163, 269]. But the empirical analyses (Figure 3.13) detect a prominent eigenvalue far larger than, and a small group of eigenvalues slightly larger than, what is expected from the random matrix theory [220, 311].

This result gives reason for two comments. On the one hand, the covariance matrix is obviously not a simple random matrix since it shows complicated correlations between the components. Therefore, we must be very carful when comparing the spectrum of eigenvalues of a symmetric positive-definite random matrix with the spectrum of the covariance matrix Φ.

On the other hand, we should extend the term "economic factors" to "economic-psychological effects." It is well-known and can be confirmed by empirical observations that the daily trading of all assets or at least of large groups of assets depends strongly on the actual mental situation of the traders. Bad or positive news, even if it concerns only a company or a small industrial sector, often has effects on the valuation of all other stocks or of an essential portion of the other stocks. These psychological effects lead to a strong common behavior that essentially influences the correlation between the share prices. It can be expected that these effects especially contribute mainly to the largest eigenvalues of the covariance matrix Φ.

3.3.4 The Log-Normal Law

Another quantity frequently used in finance is the return (3.4). The returns and the logarithmic price changes are connected via the relation $\xi_\alpha(t, \delta t) = \ln(R_\alpha(t, \delta t) + 1)$. Thus, the components $r_\alpha(t, \delta t) = R_\alpha(t, \delta t) + 1$ are distributed according to a log-normal probability distribution function because the $\xi_\alpha(t, \delta t)$ are distributed according to a Gaussian law.

The log-normal distribution is a fixed point in the functional space, not under addition but under multiplication, which is equivalent to the addition of logarithms. Therefore, the log-normal law is not a true stable distribution; it can be simply mapped onto the Gaussian probability distribution function. Substituting ξ_α by $\ln r_\alpha$, the Gaussian law will be transformed into the expression

$$p(r, t \mid 1, t_0) = \frac{1}{(2\pi \Delta t)^{A/2} \sqrt{\det \Phi} \prod_\alpha r_\alpha} \times \exp\left\{ -\sum_{\alpha\beta} \ln\left(\frac{r_\alpha}{e^{\mu_\alpha \Delta t}}\right) \frac{\left[\Phi^{-1}\right]_{\alpha\beta}}{2\Delta t} \ln\left(\frac{r_\beta}{e^{\mu_\beta \Delta t}}\right) \right\} \tag{3.69}$$

with $\Delta t = t - t_0$. Let us briefly discuss the properties of this probability distribution function for the one-dimensional case (i.e., for $A = 1$). Then, $e^{\mu \Delta t}$ is the measure of central tendency corresponding to the median $r_{1/2}(t)$ as defined in (2.8). Another measure of central tendency is the most probable value $r_{\max}(t)$, defined in (2.9). We obtain

$$r_{\max}(t) = \exp\left\{(\mu - \Phi)\, \Delta t\right\}, \tag{3.70}$$

while the mean is equal to

$$\overline{r}(t) = \exp\left\{(\mu + \Phi/2)\, \Delta t\right\}, \tag{3.71}$$

so that we get the relation $\overline{r}(t)^2 r_{\max}(t) = r^3_{1/2}(t)$. Obviously, the mean can be much larger than $r_{\max}(t)$. These differences are important for the interpretation of financial data on the basis of the returns. Suppose that we have only a small number of observations. Then, the most probable value will be sampled first, and the empirical average will not be far from $r_{\max}(t)$.

In contrast, the empirical average determined from a large number of observations approaches progressively the true average $\overline{r}(t)$. However, this value is eventually much larger than the true trend $\exp\{\mu \Delta t\}$. We remark that the corresponding Gaussian law satisfies $\overline{\xi}(t) = \xi_{1/2}(t) = \xi_{\max}(t)$.

The log-normal probability distribution function can be mistaken locally for a power law. To see this, we write

$$p(r, t \mid 1, t_0) = \frac{r_{1/2}(t)}{\sqrt{2\pi \Delta t \Phi}} \left(\frac{r}{r_{1/2}(t)}\right)^{-\varsigma(r/r_{1/2}(t))} \tag{3.72}$$

with

$$\varsigma(x) = 1 + \frac{\ln x}{2\Delta t\Phi}. \tag{3.73}$$

Since $\varsigma(r/r_{1/2}(t))$ is a slowly varying function of r, (3.72) shows that the log-normal law distribution becomes indistinguishable from the r^{-1} distribution at least for fluctuations in a moderate range around the median $r_{1/2}(t)$ and for sufficiently large values $\Delta t\Phi$. This is an important remark for the analysis of empirical frequency distribution functions of the returns using double-logarithmic plots. Especially for large $\Delta t\Phi$, the log-normal distribution mimics a power law very well over a large range.

The conditional probability distribution function of the returns (3.69) can be rewritten as a conditional probability density of the prices

$$p(X,t \mid X_0,t_0) = \frac{1}{(2\pi\Delta t)^{A/2}\sqrt{\det\Phi}\prod\limits_{\alpha} X_\alpha}$$

$$\times \exp\left\{-\sum_{\alpha\beta} \ln\left(\frac{X_\alpha}{X_{0,\alpha}e^{\mu_\alpha \Delta t}}\right)\frac{\left[\Phi^{-1}\right]_{\alpha\beta}}{2\Delta t}\ln\left(\frac{X_\beta}{X_{0,\beta}e^{\mu_\beta \Delta t}}\right)\right\}. \tag{3.74}$$

Thus, for sufficiently long time differences $\Delta t = t - t_0 \gg \delta t_{\text{Markov}}$, the prices of assets are log-normally distributed.

3.4 Standard Problems of Finance

3.4.1 The Escape Problem

It is often of interest to know how long an asset price whose dynamics are described by a Fokker–Planck equation (3.64) remains in a certain interval. In the following, we restrict ourselves to the case of one stock, $A = 1$. Furthermore, we assume that at $t_0 = 0$ the initial price $X_0 = X(0)$ has a value between $X_{\min}$ and $X_{\max}$ (i.e., $X_{\min} < X_0 < X_{\max}$). Then, the first passage time T is the time at which the price $X(t)$ first leaves the interval $[X_{\min}, X_{\max}]$. If we are interested only in the passage of an upper or lower boundary, we have to take into account $X_{\min} = 0$ or $X_{\max} = \infty$.

For the sake of simplicity, we translate the problem into a representation using logarithmic price changes (see Figure 3.14). Then, we have to deal with the boundaries $\xi_{\min}$ and $\xi_{\max}$ and the initial condition $\xi(0) = \eta$, which satisfies $\xi_{\min} < \eta < \xi_{\max}$.

The first-passage-time problem implies that a certain price evolution $\xi(t)$ is removed from the set of events (i.e., of all allowed price trajectories under consideration) if it reaches the upper or lower boundary. This means that the first-passage-time problem (or the escape problem) corresponds to a Fokker–Planck equation with absorbing conditions [143, 330]. The definition of the first-passage time requires that the probability

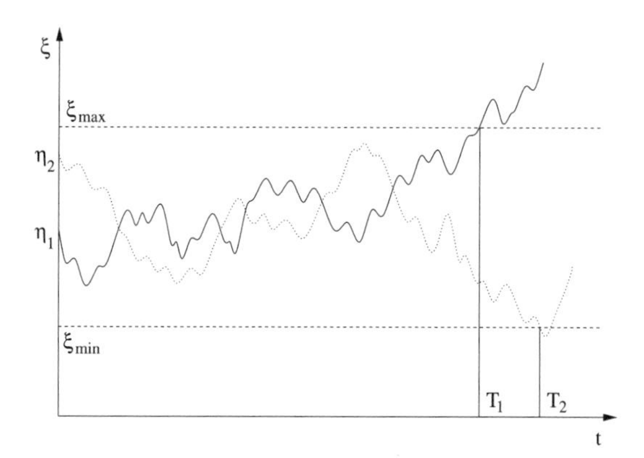

Fig. 3.14. Two price trajectories and the corresponding first-passage (escape) times.

$$P(T \geq t \mid \eta) = P_{\geq}(t, \eta) = \int d\xi p(\xi, t \mid \eta, 0) \tag{3.75}$$

define the fraction of "nonadsorbed" trajectories. We remark that the right-hand side is no longer the normalization condition (2.54) because there is a permanent probability current through the upper and lower boundaries. This current is implicitly determined [330] by the absorbing conditions $p(\xi, t \mid \xi_{min}, 0) = p(\xi, t \mid \xi_{max}, 0) = 0$. Due to these boundary conditions, the probability density $p(\xi, t \mid \eta, 0)$ considers only the fraction of price trajectories that remain in the "nonadsorbed" state during the whole evolution time t. Obviously, the boundary conditions are equivalent to

$$P_{\geq}(t, \xi_{min}) = P_{\geq}(t, \xi_{max}) = 0. \tag{3.76}$$

The initial condition $p(\xi, 0 \mid \eta, 0) = \delta(\xi - \eta)$ completes the escape problem. We get the corresponding relation

$$P_{\geq}(0, \eta) = 1 \quad \text{for} \quad \xi_{min} < \eta < \xi_{max}. \tag{3.77}$$

We are especially interested in the moments of the first-passage time because these moments contain the same information as the corresponding probability distribution. Since $P_{\geq}(t, \eta)$ is the probability that $T \geq t$, the corresponding probability density is given by

$$\hat{p}(t, \eta) = -\frac{dP_{\geq}(t, \eta)}{dt} \tag{3.78}$$

so that the nth moment is

$$T_n(\eta) = \int_0^\infty t^n \hat{p}(t, \eta) dt = -n \int_0^\infty t^{n-1} P_{\geq}(t, \eta) dt. \tag{3.79}$$

The solution of the first passage-time-problem can be achieved by use of the backward Fokker–Planck equation, which may be obtained directly from

(3.64) considering the relations between backward (2.89) and forward (2.85) equations

$$\frac{\partial}{\partial t'} p(\xi, t \mid \eta, t') = -\left[\frac{1}{2}\frac{\partial^2}{\partial \eta^2}\Phi + \frac{\partial}{\partial \eta}\mu\right] p(\xi, t \mid \eta, t'). \tag{3.80}$$

In order to connect (3.75) with the time development (3.80), we set $t = 0$ and $t' = -t$ and obtain

$$\frac{\partial}{\partial t} p(\xi, 0 \mid \eta, -t) = \left[\frac{1}{2}\frac{\partial^2}{\partial \eta^2}\Phi + \frac{\partial}{\partial \eta}\mu\right] p(\xi, 0 \mid \eta, -t). \tag{3.81}$$

Now, we consider that the financial market is in a steady state. Then, we get $p(\xi, t \mid \eta, 0) = p(\xi, 0 \mid \eta, -t)$ and therefore

$$\frac{\partial}{\partial t} P_{\geq}(t, \eta) = \left[\frac{1}{2}\frac{\partial^2}{\partial \eta^2}\Phi + \frac{\partial}{\partial \eta}\mu\right] P_{\geq}(t, \eta). \tag{3.82}$$

Now, we can derive a simple ordinary differential equation for the first moment $T_1(x)$ by using (3.79):

$$-1 = \left[\frac{1}{2}\frac{\partial^2}{\partial \eta^2}\Phi + \frac{\partial}{\partial \eta}\mu\right] T_1(\eta). \tag{3.83}$$

Similarly, we obtain equations for the higher moments $(n > 1)$

$$-nT_{n-1}(\eta) = \left[\frac{1}{2}\frac{\partial^2}{\partial \eta^2}\Phi + \frac{\partial}{\partial \eta}\mu\right] T_n(\eta), \tag{3.84}$$

which means that all of the moments of the first passage time can be found by a repeated integration. All of the ordinary differential equations have to be solved under the boundary condition

$$T_n(\xi_{\min}) = T_n(\xi_{\max}) = 0, \tag{3.85}$$

where we have considered the initial condition (3.77) and the expected final condition $P_{\geq}(\infty, \eta) = 0$. The solution of these equations is a standard procedure [194, 330].

3.4.2 The Portfolio Problem

On the condition of stationary financial markets, we are able to give information that guarantees an optimal investment. We assume that the relevant timescale Δt of the investment is far above the Markov horizon so that we can apply the central limit theorem. Thus, we expect a Gaussian law for the logarithmic price fluctuations ξ. In this case, all moments can be calculated from the knowledge of the trend μ and the covariance rate Φ.

Furthermore, we assume that A different stocks are traded in the financial market. An investor is usually not interested in the purchase of only one sort of stock but invests in stocks of different industrial sectors. The set of all stocks in the possession of the investor is called the portfolio. The fraction of shares α in the portfolio may be w_α. For the sake of simplicity, we focus here on the traditional budget condition

$$\sum_{\alpha=1}^{A} w_\alpha = 1. \tag{3.86}$$

We have not imposed any positivity constraints of the form $w_\alpha \geq 0$, so unrestricted short sales are permitted. Now, we want to determine the mean return of the entire portfolio. The actual return of the portfolio is the weighted return of the single assets contained in the portfolio

$$\mathcal{R} = \sum_{\alpha=1}^{A} w_\alpha R_\alpha = \sum_{\alpha=1}^{A} w_\alpha \left(e^{\xi_\alpha} - 1 \right), \tag{3.87}$$

where we have used (3.4) and (3.7). The mean return of the portfolio over a given time interval $\Delta t = t - t_0$ is then given by

$$\overline{\mathcal{R}} = \sum_{\alpha=1}^{A} w_\alpha \left[\int d\xi \exp\{\xi_\alpha\} p(\xi, t \mid 0, t_0) - 1 \right] = \sum_{\alpha=1}^{A} w_\alpha \overline{R}_\alpha(\Delta t) \tag{3.88}$$

with

$$\overline{R}_\alpha(\Delta t) = \exp\{\hat{\mu}_\alpha \Delta t\} - 1 \tag{3.89}$$

and $\hat{\mu}_\alpha = \mu_\alpha + \Phi_{\alpha\alpha}/2$. The standard deviation of the portfolio,

$$\sigma_P^2 = \overline{(\Delta \mathcal{R})^2} = \sum_{\alpha,\beta}^{A} S_{\alpha\beta}(\Delta t) w_\alpha w_\beta, \tag{3.90}$$

with

$$S_{\alpha\beta}(\Delta t) = \exp\{\hat{\mu}_\alpha \Delta t + \hat{\mu}_\beta \Delta t\} \left[\exp\{\Delta t \Phi_{\alpha\beta}\} - 1 \right], \tag{3.91}$$

is also called the volatility of the portfolio. The mean return is a financial measure for the expected yield of a portfolio, while the volatility σ_P characterizes the risk of the investor. An important task of financial mathematics is the determination of an optimal portfolio considering certain boundary conditions and other financial constraints [108, 256, 417]. We refer the reader to the comprehensive specialized literature [123, 178, 196, 226, 229, 255, 340, 401] for further details.

Let us explain briefly some main ideas. There are different starting points that would lead to specific classes of solutions. A very simple goal is the minimization of the risk in the case of a fixed return $\mathcal{R}^\star$ considering the constraint (3.86). This problem is called the minimum variance portfolio model. Forming the "Lagrangian"

$$F = \frac{1}{2} \sum_{\alpha,\beta}^{A} S_{\alpha\beta}(\Delta t) w_\alpha w_\beta$$

$$-\lambda_1 \left[\sum_{\alpha} w_\alpha - 1\right] - \lambda_2 \left[\sum_{\alpha=1}^{A} w_\alpha \overline{R}_\alpha(\Delta t) - \mathcal{R}^\star\right] \tag{3.92}$$

with the Lagrange multipliers λ_1 and λ_2 and differentiating gives the first-order conditions

$$\frac{\partial F}{\partial w_\alpha} = \sum_{\beta=1}^{A} S_{\alpha\beta}(\Delta t)\, w_\beta - \lambda_1 - \lambda_2 \overline{R}_\alpha(\Delta t) = 0. \tag{3.93}$$

The solution set is

$$w_\alpha = \sum_{\beta=1}^{A} \left[S^{-1}(\Delta t)\right]_{\alpha\beta} \left\{\lambda_1 + \lambda_2 \overline{R}_\beta(\Delta t)\right\}. \tag{3.94}$$

The multipliers can be determined from the constraints. We obtain

$$\lambda_1 = \frac{c - b\mathcal{R}^\star}{\Delta} \quad \text{and} \quad \lambda_2 = \frac{a\mathcal{R}^\star - b}{\Delta} \tag{3.95}$$

with $\Delta = ac - b^2 > 0$ and

$$a = \sum_{\alpha,\beta}^{A} \left[S^{-1}(\Delta t)\right]_{\alpha\beta} > 0, \quad b = \sum_{\alpha,\beta}^{A} \left[S^{-1}(\Delta t)\right]_{\alpha\beta} \overline{R}_\beta(\Delta t), \tag{3.96}$$

and

$$c = \sum_{\alpha,\beta}^{A} \left[S^{-1}(\Delta t)\right]_{\alpha\beta} \overline{R}_\alpha(\Delta t) \overline{R}_\beta(\Delta t) > 0. \tag{3.97}$$

Note that we have $\Delta > 0$ by the Cauchy–Schwarz inequality considering the assumptions that S is nonsingular and that all assets do not have the same mean. If all means were the same, then we have $\Delta = 0$, and this problem has no solution except when $\overline{R}_\alpha = \mathcal{R}^\star$ for all shares α. If we substitute the solution (3.94) into the portfolio volatility, we get a relation between the minimum variance and the return $\mathcal{R}^\star$:

$$\sigma_P^2 = \frac{a\mathcal{R}^{\star 2} - 2b\mathcal{R}^\star + c}{\Delta}. \tag{3.98}$$

This is the equation of a parabola. The global minimum variance portfolio is located by

$$\frac{d\sigma_P^2}{d\mathcal{R}^\star} = 2\frac{a\mathcal{R}^\star - b}{\Delta}; \tag{3.99}$$

that is, the minimum variance portfolio has a mean return of b/a and from (3.98) a volatility a^{-1}. Substituting for $\mathcal{R}^\star$ in (3.95) gives $\lambda_1 = a^{-1}$ and $\lambda_2 = 0$, so the global minimum variance portfolio is given by

$$w_\alpha = \frac{\sum_{\beta=1} \left[S^{-1}(\Delta t)\right]_{\alpha\beta}}{\sum_{\alpha,\beta} \left[S^{-1}(\Delta t)\right]_{\alpha\beta}}. \tag{3.100}$$

The personal risk behavior of investors differs considerably. Only a few investors are careful to make their financial decisions on the basis of the minimum variance model. Every investor pursues a certain aim with his financial decisions. Experience shows that investor preferences can be represented by a utility function V defined over the mean and the volatility of a portfolio. From a psychological point of view, the utility function characterizes the mental behavior of an investor. This function decreases monotonously with increasing interest in a portfolio. A simple example is Freund's function [127]

$$V = \sigma_P - \gamma \overline{\mathcal{R}}, \tag{3.101}$$

which is also known as the capital market line. The risk aversion coefficient γ classifies the investors. The larger this parameter, the larger the risk aversion of a given investor. The coefficient always has a positive value, and empirical investigations [117] suggest additionally $\gamma < 1$. Freund's function (3.101) reflects the main properties of utility functions. The standard assumptions are that all investors favor higher means and smaller variances, and a higher degree of preference corresponds to a smaller value of the utility function. This means

$$\frac{\partial V\left(\sigma_P, \overline{\mathcal{R}}\right)}{\partial \overline{\mathcal{R}}} < 0 \quad \text{and} \quad \frac{\partial V\left(\sigma_P, \overline{\mathcal{R}}\right)}{\partial \sigma_P} > 0. \tag{3.102}$$

Each curve $V = \text{const.}$ connecting portfolios with the same degree of preference is called an indifference curve. In other words, no point that an investor would favor exists along this curve.

Unfortunately, Freund's function is a very vague approximation of the psychological behavior of investors. In order to obtain a more appropriate representation, we use a psychometric construction . To do this, we take into account that a typical investor will neither exceed an upper risk nor remain under a lower return. These are very natural constraints that can be observed in practice. The upper risk is a native measure for the degree of preference V, while the lower return is usually fixed by the customary level of interest rates $\mathcal{R}_{\min}$. Mathematically, we have to deal with $\sigma_P \leq V$ and $\overline{\mathcal{R}} \geq \mathcal{R}_{\min}$.

To derive the utility function, we analyze the indifference curves of investors with equal degrees of preference but with different risk aversions. All investors are individual elements of a hierarchically ordered social system. In particular, each investor is anxious to conserve the distance to other investors.

We have two fundamental measures of distance in portfolio theory, namely the difference of the risk $\delta\sigma_P$ at constant mean return and the difference of the mean return $\delta\overline{\mathcal{R}}$ at constant risk (Figure 3.15). The product $\delta\sigma_P\delta\overline{\mathcal{R}}$ is an obvious local measure characterizing the mental difference between investors with respect to their financial behavior. The representation as a product and not as a sum seems to be plausible because distances appear to be logarithmic from a psychological point of view.

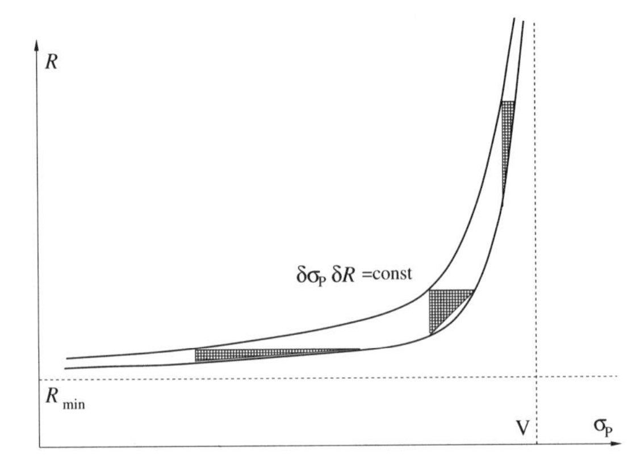

Fig. 3.15. Hyperbolic character of psychometric indifference curves. The area of the triangle $\delta\sigma_P\mathcal{R}$ is constant along two curves with equal asymptotics.

The psychometric construction of the utility function requires that $\delta\sigma_P\delta\overline{\mathcal{R}}$ be an invariant quantity along each indifference curve. The hyperbolas

$$(V - \sigma_P)\left(\overline{\mathcal{R}} - \mathcal{R}_{\min}\right) = \gamma \tag{3.103}$$

are the only family of curves satisfying these conditions. Here, γ plays the role of the risk aversion coefficient. The utility function now reads

$$V = \sigma_P + \frac{\gamma}{\left(\overline{\mathcal{R}} - \mathcal{R}_{\min}\right)}. \tag{3.104}$$

The construction of an optimal portfolio using utility functions is the standard procedure introduced above. We have to minimize V with respect to the fractions w_α considering the condition (3.86) and the relations (3.88) and (3.90).

3.4.3 Option Pricing Theory

Options and Trading Strategies. The actual value of a portfolio is not stable. An investor will sell or buy specific stocks in order to increase the value of the portfolio. The sale or purchase of stocks is controlled via financial contracts. The structure of the whole financial trading system has a large degree of complexity [58, 179, 268]. An important source of this complexity comes from the issuing of financial contracts on the fluctuating financial securities [102, 182, 276]. Examples of financial contracts are forward contracts, futures, and options.

The simplest financial contract is a forward contract . If such a contract is signed at time t_0, one of the parties agrees to buy a given amount of an asset at a given forward price or delivery price K at a specified future delivery date $t_0 + T$. The time T is called the maturity time. The other party agrees to sell the amount at the specified delivery price on the delivery date. The buyer of the contract is said to have a long position or to hold the contract long. The seller of the option contract is said to be in a short position or to have sold the option short. The actual price $X(t)$ of the underlying asset fluctuates, and the price $X(t_0+T)$ at the delivery date usually differs from the specified forward price K. The payoff at the short position is positive if $X(t_0+T) < K$, while the payoff at the long position is positive for $X(t_0+T) > K$. As a consequence of the forward contract, a positive payoff at the long position corresponds to a negative payoff at the short position, and vice versa.

A futures contract is a forward contract traded on an exchange of securities. Usually, the futures contracts are standardized. The two parties interact through an exchange institution, the so-called clearing house. The clearing house writes forward contracts with both parties and guarantees that the contracts will be executed at the delivery date.

Options are the most frequently used forms of financial contracts [81, 180]. An option is a contract between two people that conveys the right to buy or sell specified property at a specified price K for a designated time period T. The price K is now called the strike price or the exercise price, while the maturity time T is sometimes also called the expiration time or exercise time. The party who creates and offers the contract for sale at time t_0 is called the writer or seller. The other person is called the owner or buyer.

When the contract is made, the buyer pays cash to the writer for the right to buy or sell at a known price, which removes some risk in a future transaction. The owner of the option contract has the right to buy or sell the underlying asset but has no obligation to do so. The latter is the distinguishing characteristic of an option contract as opposed to forward and futures contracts, for which there is an obligation to execute or, using options terminology, exercise the contract.

The stockbroker knows two forms of option contracts: calls and puts. A call option is a contract that permits the holder to buy an asset during a specified time interval for a designated price and requires the seller to sell. At the time the contract is written, the parties agree to both the strike price and the maturity time. A put option differs from a call option in that it allows the holder to sell rather than purchase the asset.

Option contracts are also described by the type of restriction placed on the exercise period. A European option contract can be exercised only upon its termination at $t_0 + T$, whereas an American option can be exercised at any time up to and including the exercise date. The American option contract obviously offers the holder greater flexibility, which apparently makes its valuation more difficult. For further information, we refer the reader to the specialized literature. Here, we will focus our discussion on the European option.

It should be remarked that financial theorists have found that a number of investments or financial contracts other than stock options can also be interpreted as options. For instance, a firm's common stock can be considered a call option on its productive assets, with the exercise price equal to its debt obligations. Furthermore, a firm's research and development expenditures can be viewed as call options on the values of the productive ideas they create, with the exercise price equal to the investment outlay required to put these ideas into operation.

We now want to find the value of an option. This is important if one wants to buy or sell options. We assume that the price of the option depends on the current price $X(t)$ of the underlying asset, the strike price K, and the time difference $t_0 + T - t$ between the maturity date $t_0 + T$ and the actual time $t > t_0$. For our discussion, we choose $t_0 = 0$.

First, let us discuss a call option. Immediately following the completion of the contract at $t = 0$, the price of the option is the sale price at which the buyer pays cash to the seller. The problem of the definition of a fair purchase price is open and will be solved below. At the maturity time T, we have two possibilities. If $X(T) > K$, the contract is executed and the buyer gets assets of the value $X(T)$ for a price K. Thus, the option has the value $X(T) - K$. If $X(T) < K$, the contract is not executed, and the value of the option is 0. We will express the value of the option before maturity by the price function $C\left(X(t), T - t, K\right)$. This function has the final condition

$$C(X(T), 0, K) = \max(0, X(T) - K), \tag{3.105}$$

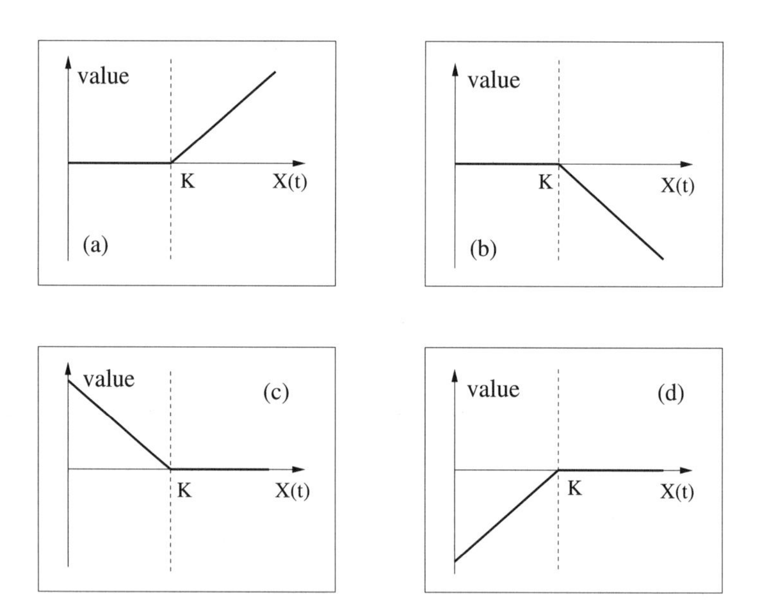

Fig. 3.16. The value of an option as a function of the price $X(t)$ at maturity time for the buyer of a call option (a), the seller of a call option (b), the buyer of a put option (c), and the seller of a put option (d). The strike price is K.

while the initial value $C(X(0),T,K)$ (i.e., the sale price of the option) is still open. The difference $\Delta C = C(X(T),0,K) - C(X(0),T,K)$ is the payoff for a buyer, while $-\Delta C$ is the payoff for the seller. In the case of a put option, we have the inverse situation; see Figure 3.16. Obviously, in call and put options, there is no symmetry between the two parties of the contract.

Usually, an investor may use different trading strategies in financial markets. A very risky strategy is the speculation. A special case of speculation is a naked position; that is, the investor holds only a single call option or a single put option. There is a huge risk for the speculator to lose his gamble. On the other hand, there is also the possibility for a large payoff.

Another strategy is a hedged position. A hedge is any combination of the underlying asset and call or put options. Writing both the call and the put creates a written or short straddle. Buying both produces a long straddle. The values of these hedges at maturity are shown in Figure 3.17.

Black and Scholes Equation. Hedging is the standard financial technique allowing the generation of portfolios with minimum risk. In order to examine more closely the procedure of hedging, we consider a simplified version of a hedging strategy with a portfolio of only one type of stocks. The generalization to a multicomponent portfolio is always possible. We suppose further

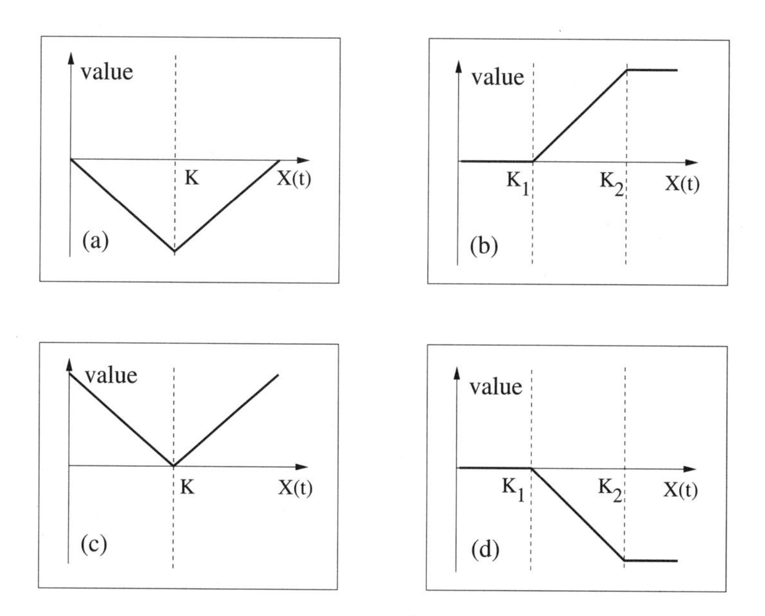

Fig. 3.17. Hedges: (a) long straddle obtained by buying a call and a put, both with strike price K; (b) bullish spread obtained by buying a call with strike price K_1 and selling a call with strike price $K_2 > K_1$; (c) short straddle by selling a call and a put, both with strike price K; (d) bearish spread obtained by selling a call with strike price K_1 and buying a call with strike price $K_2 > K_1$.

that the portfolio is constructed by buying n shares of price $X(t)$ and selling an option of the price $C(X(t), T-t, K)$. Thus, the value of the portfolio is

$$\mathcal{P}(t) = nX(t) - C(X(t), T-t, K). \tag{3.106}$$

The fluctuation of the share prices leads to fluctuations of the portfolio

$$\Delta\mathcal{P}(t) \approx n\Delta X(t) - \left[\frac{\partial C(X, T-t, K)}{\partial X}\right]_{X=X(t)} \Delta X(t). \tag{3.107}$$

A riskless investment requires

$$n = \left[\frac{\partial C(X, T-t, K)}{\partial X}\right]_{X=X(t)}. \tag{3.108}$$

Because $X(t)$ changes over time, the number n must also change over time in order to maximize the effectiveness of the hedging strategy and minimize the risk. However, it is important to realize that n is a slowly varying quantity following the trend of the stock prices, while the fluctuations $\Delta X(t)$ are very fast. The value of the portfolio is therefore

$$\mathcal{P}(t) = X(t)\left[\frac{\partial C(X, T-t, K)}{\partial X}\right]_{X=X(t)} - C(X(t), T-t, K). \tag{3.109}$$

For a financial market to function well, participants must thoroughly understand option pricing. The task is to find the rational and fair price $C(X, T-t, K)$.

At this point, we derive the most important result of option pricing theory: the Black and Scholes equation [50, 265]. There exist various other approaches using the binomial model [80] or more general concepts extending the Black and Scholes formula [265, 267]. In principle, the basic ideas used by Black and Scholes refer to a larger class of continuous-time self-financing strategies [167, 267]. The Black and Scholes equation is valid under the following conditions:

1. The market is assumed to be in a steady state at least over the time of the option contract.
2. There are no costs associated with exercising the option (i.e., the market is frictionless).
3. There are no riskless arbitrage opportunities (that is, there is no way to combine option contracts and the underlying stock into a portfolio that will produce a riskless profit).
4. The holder will exercise the option if it is profitable to do so.
5. There is no possibility of default on the contract.
6. Selling of securities is possible at any time.
7. The trading is continuous.
8. The stock pays no dividends during the option's life.

We start our derivation from the single-component representation of the Ito stochastic differential equation (3.65). Using (3.7) and (3.66), we obtain $d\xi = dX/X$ and therefore

$$dX = X\mu dt + \sqrt{\Phi} X dW, \tag{3.110}$$

representing a one-dimensional geometric Brownian diffusion of the stock price. Note that the stochastic term can be sufficiently described by a single Wiener process $W(t)$ because we consider only one stock. The price function depends explicitly on the price fluctuations. We get up to the first order of dt

$$dC = \frac{\partial C}{\partial t} dt + \frac{\partial C}{\partial X} dX + \frac{1}{2} \frac{\partial^2 C}{\partial X^2} (dX)^2 + o(dt). \tag{3.111}$$

Substituting (3.110) into (3.111), we obtain

$$dC = \left[\frac{\partial C}{\partial t} + \frac{\partial C}{\partial X} X\mu \right] dt + \frac{1}{2} \frac{\partial^2 C}{\partial X^2} \Phi X^2 dW^2 + \frac{\partial C}{\partial X} \sqrt{\Phi} X dW, \tag{3.112}$$

where we have taken into account that $dW^2 \sim dt$ (see Chapter 2). The value of the hedge portfolio is given by (3.109). From here, we get the total differential

$$d\mathcal{P}(t) = -dC + \frac{\partial C}{\partial X} dX. \tag{3.113}$$

Using (3.110) and (3.112), we obtain

$$\begin{aligned} d\mathcal{P}(t) = &- \left[\frac{\partial C}{\partial t} + \frac{\partial C}{\partial X} X\mu \right] dt - \frac{1}{2} \frac{\partial^2 C}{\partial X^2} \Phi X^2 dW^2 - \frac{\partial C}{\partial X} \sqrt{\Phi} X dW \\ &+ \frac{\partial C}{\partial X} \left(X\mu dt + \sqrt{\Phi} X dW \right) \end{aligned} \tag{3.114}$$

and therefore

$$d\overline{\mathcal{P}}(t) = -\frac{\partial C}{\partial t} dt - \frac{1}{2} \frac{\partial^2 C}{\partial X^2} \Phi X^2 dW^2. \tag{3.115}$$

Note that all contributions proportional to dW are cancelled mutually. This is a consequence of the riskless structure of the portfolio introduced above. Formally, we would have been able to start without the knowledge of (3.109). In this case, we require that the sum of the leading fluctuations (i.e., the sum of all terms with $dW \sim \sqrt{dt}$) vanish in order to make the portfolio riskless. This implies that we arrive again at (3.109).

The third key assumption in the list above concerns the absence of arbitrage opportunities. This means that the change in the value of the portfolio must equal the gain obtained by investing the same amount of money in the corresponding security providing an average return u per unit of time, namely

$$\frac{d\mathcal{P}(t)}{dt} = u\mathcal{P}(t). \tag{3.116}$$

This definition of the risk-free interest rate u allows us to connect this quantity with the parameters μ and Φ of the log-normal probability distribution function (3.74) for a single asset. To do this, we must simply replace $\mathcal{P}(t)$ by the mean price of the available stock $\overline{X}(t) \sim \overline{r}(t)$. Because of (3.71), we obtain from (3.116) the relation $u = \mu + \Phi/2$.

By equating (3.86), (3.115), and (3.116) and setting $dW^2 = dt$, we obtain the famous Black and Scholes equation

$$\frac{\partial C}{\partial t} + \frac{1}{2}\frac{\partial^2 C}{\partial X^2}\Phi X^2 + uX\frac{\partial C}{\partial X} = uC. \tag{3.117}$$

It should be considered that the relation $dW^2 = dt$ mathematically is not very precise with respect to its representation but is correct in the sense of the limit in the mean [143].

No assumption about the specific kind of option has been made except that we deal with European options. The appropriate price function $C(X, T-t, K)$ for the type of option chosen will be obtained by selecting the appropriate boundary conditions. For call options, we have to use the final condition (3.105). To solve (3.117) for a call option, we make the substitution

$$C(X, T-t, K) = \exp\{u(t-T)\}\, c(\eta, \tau) \tag{3.118}$$

with

$$\eta = \left(\frac{2u}{\Phi} - 1\right)\left\{\ln\left(\frac{X}{K}\right) + \left(u - \frac{\Phi}{2}\right)(T-t)\right\} \tag{3.119}$$

and

$$\tau = \left(1 - \frac{2u}{\Phi}\right)(t-T). \tag{3.120}$$

These transformations can be obtained, for instance, by inspecting the symmetry of the Black and Scholes differential equation. With this substitution, the Black and Scholes equation becomes equivalent to the heat-transfer equation of physics, which is the standard form of a parabolic partial differential equation

$$\frac{\partial c(\eta, \tau)}{\partial \tau} = \frac{\partial^2 c(\eta, \tau)}{\partial^2 \eta^2}, \tag{3.121}$$

which can be solved exactly for the boundary condition (3.105). We get

$$C(X, T-t, K) = X\phi(\eta_1) - K\exp\{u(t-T)\}\,\phi(\eta_0) \tag{3.122}$$

with

$$\eta_0 = \frac{\ln X - \ln K + (2u - \Phi)(T-t)}{2\sqrt{\Phi(T-t)}} \tag{3.123}$$

and

$$\eta_1 = \frac{\ln X - \ln K + (2u + \Phi)(T-t)}{2\sqrt{\Phi(T-t)}}, \tag{3.124}$$

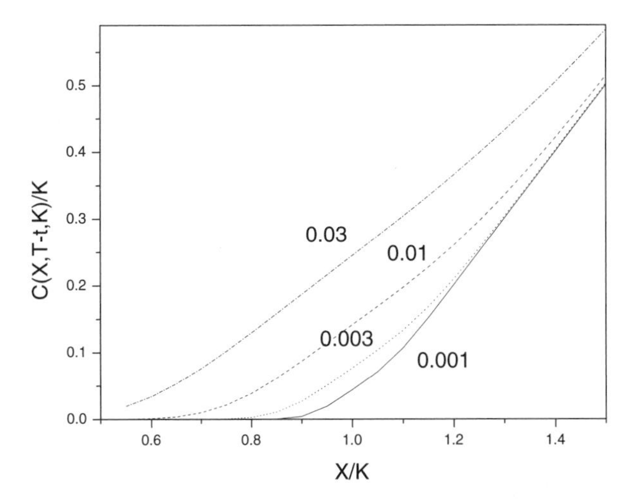

Fig. 3.18. The price function $C(X, T-t, K)$ (call option) for $\Phi = 3u$ and various time differences $u(T-t) = 0.03, 0.01, 0.003$, and 0.001. For $t \to T$, the price function approaches the values $C(X, T-t, K) = 0$ for $X < K$ and $C(X, T-t, K) = X - K$ for $X > K$.

where $\phi(x)$ is the cumulative density function for a Gaussian variable with zero mean and unit standard deviation (see Figure 3.18). For a European put option, we get a similar price function

$$C(X, T-t, K) = K \exp\{u(t-T)\} \phi(-\eta_0) - X\phi(-\eta_1), \qquad (3.125)$$

which satisfies the boundary condition $C(X(T), 0, K) = \max(0, K - X(T))$. Equations (3.122) and (3.125) allow the determination of the value of a European option for all times before the maturity time in order to provide a fair price only from the knowledge of the strike price, the actual price of the underlying stock, and the time remaining to maturity.

Generalized Option Pricing. Option pricing models such as the Black and Scholes model premised upon the underlying asset price following geometric Brownian motion have been found to exhibit serious specification errors when fitted to market data. For example, remarkable deviations have been found in American calls and puts on S&P 500 futures traded on the Chicago Mercantile Exchange [38, 423], in American foreign currency call options traded on the Philadelphia Stock Exchange [61], and in European Swiss franc-denominated call options on the dollar traded in Geneva [74]. Obviously, the Black and Scholes equation is the frame for nice models to understand and model option pricing in an ideal financial market.

To demonstrate more closely the differences between ideal and real markets, let us discuss the variance obtained from a time series analysis of various European options and the underlying assets by inversion of the Black and Scholes formula (3.122) or (3.125). The estimated variance rate Φ_{imp} is called the implied volatility rate. In a Black and Scholes market, the determination of the implied volatility would give a constant value for options with different strike prices and at different maturity times.

Furthermore, the value of the implied volatility should coincide with the variance rate Φ obtained from the price fluctuations of the assets. In real markets, we get, in general, $\Phi \neq \Phi_{\text{imp}}$. Various empirical investigations demonstrate that Φ_{imp} is a function of the strike price as well as the maturity time. In particular, the implied volatility Φ_{imp} increases the more the option price is in the money or out of the money. Note that an option is in the money if exercising it would produce a gain. For instance, a call contract is in the money if the stock price X is greater than the strike price. In contrast, an option is out of the money if exercising it would produce a loss.

Another problem concerns the apparently random character of the volatility of an asset price, especially for short timescales. This time dependence may have different causes.

On the one hand, the appropriate moving time window used for the empirical estimation of the volatility may be too small in comparison with the time difference δt between successive observations. Thus, only a few measurements contribute to the empirically determined actual volatility. However, one can argue that longer time intervals should provide better estimations. But the volatility estimated by using very long time periods may be quite different from the volatility observed in the lifetime of the option.

On the other hand, we may choose a very short time difference δt. Then, the danger exists that we are below the Markov horizon $\delta t < \delta t_{\text{Markov}}$ so that memory and correlation effects dominate the empirically determined volatility.

Short timescales become important for day-trading strategies. As we will see below, particularly the acceptance of the Gaussian law for the logarithmic price changes cannot be guaranteed for such short timescales. In fact, there is considerable time series evidence against the hypothesis that log-differenced asset prices are normally distributed at short timescales. Consequently, the main assumption of a geometric Brownian motion fails for short timescales so that the Black and Scholes equation loses its validity. In order to be able to calculate the value of European call and put options nevertheless, we have to generalize the definition of the price function $C(X(t), T-t, K)$.

Financial theorists believe that the value of a European option is given by the average expected payoff rescaled by the risk-free interest rate u of the market. Thus, we obtain for a call option

$$C(X, T-t, K) = \exp\{-u(T-t)\} \overline{(X(T)-K)^{+}} |_{X(t)=X} \tag{3.126}$$

with $(X-K)^{+}=0$ for $X<K$ and $(X-K)^{+}=X-K$ for $X\geq K$. The value of a European put option is defined by

$$C\left(X,T-t,K\right)=\exp\left\{-u\left(T-t\right)\right\}\overline{\left(K-X(T)\right)^{+}}\,|_{X(t)=X}\,. \tag{3.127}$$

In both formulas, we have used a conditioned average. This average is defined with respect to the conditional probability density

$$\overline{\left(X(T)-K\right)^{+}}\,|_{X(t)=X}=\int\left(X'-K\right)^{+}p\left(X',T\mid X,t\right)dX'. \tag{3.128}$$

If we apply the log-normal distribution (3.74) for a single asset in (3.128), we obtain again the Black and Scholes solutions (3.122) and (3.125) considering the relation $u=\mu+\Phi/2$ introduced above, which is necessary to eliminate the trend μ

The general concept of option pricing theory defined by (3.126) and (3.127) provides a flexible tool for the analytic or numerical determination of the price functions when the distribution of the asset prices is known. Especially the nonacceptance of the Gaussian law at short timescales has spurred the development of option pricing models for alternative probability distribution functions. Such functions may be obtainable from empirical observations [19, 59, 105, 259] or from the assumption of other stochastic processes considered in jump-diffusion models [266].

Other option pricing models consider stochastic interest rates [6, 265] or stochastic volatility rates [171, 180]. Stochastic volatility option pricing models [180, 368, 430] especially permit much more general patterns of volatility evolution.

However, almost all of these models are connected with the topics of financial mathematics. In contrast, the aim of the econophysical approach is the general understanding of the financial market considering universal laws that dominate the dynamics of this complex system.

3.5 Short-Time Regime

3.5.1 High-Frequency Observations

The most common stochastic model of stock price dynamics is the Gaussian behavior discussed above that assumes a geometric Brownian diffusion of the asset prices and a corresponding arithmetic Brownian motion of the logarithmic price differences. This model provides a first approximation of the behavior observed in empirical data. However, the Gaussian probability distribution function is a universal consequence of the central limit theorem in the limit of long times on the condition that the financial market is in a stationary state. Indeed, the Gaussian law can possibly deviate considerably from the probability distribution function determined empirically for short timescales.

Serious systematic deviations from the Gaussian model predictions are observed, which indicate that the empirically determined probability distributions exhibit a pronounced leptokurtic behavior. A highly leptokurtic function is characterized by a narrower and larger maximum and by fatter tails than in the Gaussian case.

Obviously, the degree of leptokurtosis increases with decreasing time difference δt between successive observations (Figure 3.19). We had already mentioned in the context of the option pricing theory that a lot of models had been developed in order to describe the short-time behavior of price fluctuations in terms of alternative probability distribution functions. Financial mathematicians would be able to develop hundreds of such models. Of course, all of these models have better qualities for the respective problem than the Gaussian distribution. But we should not forget that these models are only substitute processes approaching the complex dynamics of the financial market in an idealized sense within a more or less limited framework.

From a physical point of view, we have to search for universal principles that lead to leptokurtic probability distribution functions for price fluctuations. Let us determine the minimal conditions that the probability distribution of price fluctuations must satisfy for short timescales. Empirical studies [246, 249] suggest a pronounced form stability. We point out that it must be a metastable state since the expected asymptotic Gaussian behavior is not accessible otherwise.

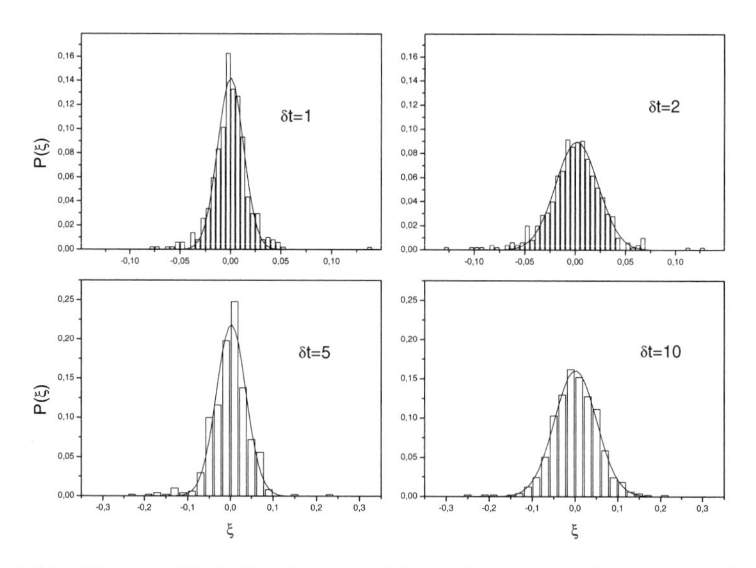

Fig. 3.19. The empirical distribution of logarithmic price fluctuations obtained for BASF stock from 11/00–07/02 for different time horizons, $\delta t = 1$, 2, 5, and 10 trading days. The curves are the corresponding fits with Gaussian law.

Additionally, we assume that the dynamics of the financial market are dominated by the Markov property (i.e., the time differences δt are still long compared with the Markov horizon). That may possibly be a dangerous approximation. In particular, we will demonstrate below that considerable memory effects can already occur on those timescales, which we take into account in the following. In order to overcome this problem, we assume that these effects eliminate themselves mutually over the entire observation period so that we can work with the Markov property as a sufficient approximation. Furthermore, we may use similar arguments supporting the stationarity hypothesis.

3.5.2 Lévy Distributions

While we were deriving the central limit theorem, we saw that the probability density function $p_{N\delta t}(\xi)$ of the logarithmic price changes with respect to the time interval $N\delta t$ can be expressed as a generalized convolution (3.29) of the elementary probability distribution functions $p_{\delta t}(\xi)$. This relation is equivalent to the algebraic equation

$$\hat{p}_{N\delta t}(k) = [\hat{p}_{\delta t}(k)]^N \tag{3.129}$$

connecting the characteristic functions $\hat{p}_{N\delta t}(k)$ and $\hat{p}_{\delta t}(k)$. We want to use this equation in order to determine the set of form-stable probability distribution functions that may be possible candidates for the probability distribution function of price fluctuations. A probability density $p_{N\delta t}(\xi)$ is called a form-stable function if it can be represented by a function g that is independent from the number N of convolutions,

$$p_{N\delta t}(\xi)d\xi = g(\xi')d\xi', \tag{3.130}$$

where the variables are connected by the linear relation $\xi' = \alpha_N \xi + \beta_N$. Because the vector ξ has the dimension A, the $A \times A$ matrix α_N describes an appropriate rotation and dilation of the coordinates, while the vector β_N corresponds to a global translation of the coordinate system. Within the formalism of the renormalization group, a form-stable probability density law corresponds to a fixed point of the convolution procedure. The Fourier transform of g is given by

$$\begin{aligned} \hat{g}(k) = \int g(\xi')\mathrm{e}^{ik\xi'}d\xi' &= \int p_{N\delta t}(\xi)\mathrm{e}^{ik(\alpha_N\xi+\beta_N)}d\xi \\ &= \mathrm{e}^{ik\beta_N}\hat{p}_{N\delta t}(\alpha_N k), \end{aligned} \tag{3.131}$$

where we have used the definition (2.14) of the characteristic function. The form stability requires that this relation must be fulfilled for all values of N. In particular, we obtain

$$\hat{p}_{N\delta t}(k) = \hat{g}(\alpha_N^{-1}k)\mathrm{e}^{-i\beta_N\alpha_N^{-1}k} \quad \text{and} \quad \hat{p}_{\delta t}(k) = \hat{g}(\alpha_1^{-1}k)\mathrm{e}^{-i\beta_1\alpha_1^{-1}k}. \tag{3.132}$$

Without any restriction, we can choose $\alpha_1 = 1$ and $\beta_1 = 0$. The substitution of (3.132) into the convolution formula (3.129) now yields

$$\hat{g}(\alpha_N^{-1}k)\mathrm{e}^{-i\beta_N\alpha_N^{-1}k} = \hat{g}^N(k). \tag{3.133}$$

Let us write

$$\hat{g}(k) = \exp\{\Phi(k)\}, \tag{3.134}$$

where $\Phi(k)$ is the cumulant generating function. Thus, (3.133) can be written as

$$\Phi(\alpha_N^{-1}k) - i\beta_N\alpha_N^{-1}k = N\Phi(k) \tag{3.135}$$

and after splitting off the contributions linearly in k,

$$\Phi(k) = iuk + \varphi(k), \tag{3.136}$$

we arrive at the two relations

$$\beta_N = \alpha_N u\left[\alpha_N^{-1} - N\right] \tag{3.137}$$

and

$$\varphi(\alpha_N^{-1}k) = N\varphi(k). \tag{3.138}$$

The first equation simply gives the total shift of the center of the probability distribution function resulting from N convolution steps. As discussed in the context of the central limit theorem, the drift term can be put to zero by a suitable linear change of the variables ξ. Thus, β_N is no object of further discussion. The second equation (3.138) is the true key for our analysis of the form stability. In the following investigation, we restrict ourselves to the one-variable case. The mathematical handling of the multidimensional case is similar, but the large number of possible degrees of freedom complicates the discussion.

The relation (3.138) requires that $\varphi(k)$ is a homogeneous function, $\varphi(\lambda k) = \lambda^\gamma\varphi(k)$, with the homogeneity coefficient γ. Considering that α_N must be a real quantity, we obtain $a_N = N^{-1/\gamma}$. Consequently, the function φ has the general structure

$$\varphi(k) = c_+|k|^\gamma + c_-k|k|^{\gamma-1} \tag{3.139}$$

with the three parameters c_+, c_-, and $\gamma \neq 1$.

A special solution occurs for $\gamma = 1$ because in this case $\varphi(k)$ merges with the separated linear contributions. Here, we obtain the special structure $\varphi(k) = c_+|k| + c_-k\ln|k|$. The rescaling $k \to \lambda k$ then leads to $\varphi(\lambda k) = \lambda\varphi(k) + c_-k\ln\lambda$, and the additional term $c_-\ln\lambda$ may be absorbed in the shift coefficient β_N.

It is convenient to use the more common representation [200, 232]

$$\hat{g}(k) = L_{a,b}^\gamma(k) = \exp\left\{-a|k|^\gamma\left[1 + ib\tan\left(\frac{\pi\gamma}{2}\right)\frac{k}{|k|}\right]\right\} \tag{3.140}$$

with $\gamma \neq 1$. For $\gamma = 1$, $\tan(\pi\gamma/2)$ must be replaced by $(2/\pi)\ln|k|$. A more detailed analysis [232, 345] shows that $\hat{g}(k)$ is a characteristic function of a probability distribution function if and only if a is a positive scale factor, γ is a positive exponent, and the asymmetry parameter satisfies $|b| \leq 1$.

Apart from the drift term, (3.140) is the representation of any characteristic function corresponding to a probability density that is form-invariant under the convolution procedure. The set of these functions is known as the class of Lévy functions. Obviously, the Gaussian law is a special subclass. The Lévy functions are fully characterized by the expression of their characteristic functions (3.140). Thus, the inverse Fourier transform of (3.140) should lead to the real Lévy functions $L_{a,b}^{\gamma}(\xi)$.

Unfortunately, there are no simple analytic expressions of the Lévy functions except for a few special cases, namely the Gaussian law ($\gamma = 2$), the Lévy–Smirnow law ($\gamma = 1/2$, $b = 1$)

$$L_{a,1}^{1/2}(\xi) = \frac{2a}{\sqrt{\pi}(2\xi)^{3/2}} \exp\left\{-\frac{a^2}{2\xi}\right\} \quad \text{for} \quad \xi > 0, \tag{3.141}$$

and the Cauchy law ($\gamma = 1$, $b = 0$)

$$L_{a,1}^{1/2}(\xi) = \frac{a}{\pi^2 a^2 + \xi^2}, \tag{3.142}$$

which is also known as Lorentzian. One of the most important properties of the Lévy functions is their asymptotic power law behavior. A symmetric Lévy function ($b = 0$) centered at zero is completely defined by the Fourier integral

$$L_{a,0}^{\gamma}(\xi) = \frac{1}{\pi} \int_0^{\infty} \exp\left\{-a|k|^{\gamma}\right\} \cos(k\xi) dk. \tag{3.143}$$

This integral can be written as a series expansion valid for $|\xi| \to \infty$,

$$L_{a,0}^{\gamma}(\xi) = -\frac{1}{\pi} \sum_{n=1}^{\infty} \frac{(-a)^n}{|\xi|^{\gamma n+1}} \frac{\Gamma(\gamma n + 1)}{\Gamma(n+1)} \sin\left(\frac{\pi\gamma n}{2}\right). \tag{3.144}$$

The leading term defines the asymptotic dependence

$$L_{a,0}^{\gamma}(\xi) \sim \frac{C}{|\xi|^{1+\gamma}}. \tag{3.145}$$

Here, $C = a\gamma\Gamma(\gamma)\sin(\pi\gamma/2)/\pi$ is a positive constant called the tail, and the exponent γ is between 0 and 2. The condition $\gamma < 2$ is necessary because a Lévy function with $\gamma > 2$ is unstable and converges to the Gaussian law. We will discuss this behavior below.

Lévy laws can also be asymmetric. Then, we have the asymptotic behavior $L_{a,0}^{\gamma}(\xi) \sim C_+/\xi^{1+\gamma}$ for $\xi \to \infty$ and $L_{a,0}^{\gamma}(\xi) \sim C_-/|\xi|^{1+\gamma}$ for $\xi \to -\infty$, and the asymmetry is quantified by the asymmetry parameter b via

$$b = \frac{C_+ - C_-}{C_+ + C_-}. \tag{3.146}$$

The completely antisymmetric cases correspond to $b = \pm 1$. For $b = +1$ and $\gamma < 1$, the variable ξ takes only positive values, while for $b = -1$ and $\gamma < 1$, the variable ξ is defined to be negative. For $1 < \gamma < 2$ and $b = 1$, the Lévy distribution is a power law $\xi^{-\gamma-1}$ for $\xi \to \infty$, while the function converges to zero for $\xi \to -\infty$ as $\exp\left(-|\xi|^{\gamma/(\gamma-1)}\right)$. The inverse situation occurs for $b = -1$.

All Lévy functions with the same exponent γ and the same asymmetry coefficient b are related by the scaling law

$$L_{a,b}^{\gamma}(\xi) = a^{-1/\gamma} L_{1,b}^{\gamma}\left(a^{-1/\gamma}\xi\right). \tag{3.147}$$

Therefore, we obtain

$$\overline{|\xi|^{\theta}} = \int |\xi|^{\theta} L_{a,b}^{\gamma}(\xi)\, d\xi = a^{\theta/\gamma} \int |\xi'|^{\theta} L_{1,b}^{\gamma}(\xi')\, d\xi' \tag{3.148}$$

if the integrals in (3.148) exist. An important property of all Lévy distributions is that the variance is infinite. This behavior follows directly from the substitution of (3.140) into (2.16). Roughly speaking, the Lévy law does not decay sufficiently rapidly at $|\xi| \to \infty$ for the integral (2.13) to converge. However, the absolute value of the spread (2.10) exists and suggests a characteristic scale of the fluctuations $D_{\mathrm{sp}}(t) \sim a^{1/\gamma}$. When $\gamma \leq 1$, even the mean and the average of the absolute value of the spread diverge. The characteristic scale of the fluctuations may be obtained from (3.148) via $\left[\overline{|\xi|^{\theta}}\right]^{1/\theta} \sim a^{1/\gamma}$ for a sufficiently small exponent θ. We remark that the median and the most probable value still exist also for $\gamma \leq 1$.

3.5.3 Convergence to Stable Lévy Distributions

The Gaussian probability distribution function is not only a form-stable distribution but also the fixed point of the classical central limit theorem. In particular, it is the attractor of all of the distribution functions having a finite variance. On the other hand, the Gaussian law is a special distribution of the form-stable class of Lévy distributions. It is then natural to ask wether all other Lévy distributions are also attractors in the functional space of probability distribution functions with respect to the convolution procedure.

There is a bipartite situation. Upon N convolutions, all probability distribution functions $p_{\delta t}(\xi)$ with an asymptotic behavior $p_{\delta t}(\xi) \sim C_{\pm} |\xi|^{-1-\gamma_{\pm}}$ and with $\gamma_{\pm} < 2$ are attracted to a stable Lévy distribution. In the case of asymptotically symmetric functions, $C_+ = C_- = C$ and $\gamma_+ = \gamma_- = \gamma$, the fixed point is the symmetric Lévy law with the exponent γ and the scale parameter $a \sim NC$.

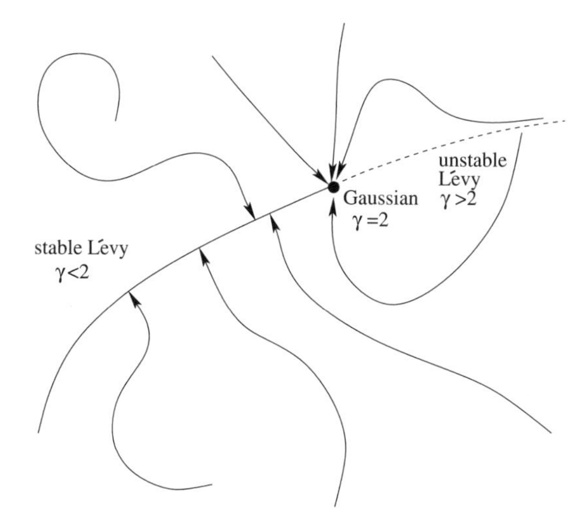

Fig. 3.20. The schematic convergence behavior of probability distribution functions in the functional space. The Gaussian law separates stable and unstable Lévy laws.

If the initial probability distribution functions have different tails, $C_+ \neq C_-$ but equal exponents, $p_{N\delta t}(\xi)$ converges to the asymmetric Lévy distribution with the exponent γ, the asymmetry parameter (3.146), and $a \sim N(C_+ + C_-)/2$.

If the asymptotic exponents $\gamma_\pm$ of the elementary probability density $p_{\delta t}(\xi)$ are different but $\min(\gamma_+, \gamma_-) < 2$, the convergence is to a completely asymmetric Lévy distribution with an exponent $\gamma = \min(\gamma_+, \gamma_-)$ and $b = 1$ for $\gamma_- < \gamma_+$ or $b = -1$ for $\gamma_- > \gamma_+$.

Finally, upon a sufficiently large number of convolutions, the Gaussian distribution also attracts all of the probability distribution functions decaying as or faster than $|\xi|^{-3}$ at large $|\xi|$. Therefore, Lévy laws with $\gamma < 2$ are sometimes called true Lévy laws (see Figure 3.20).

3.5.4 Scaling Behavior

Since Lévy distributions are form-stable functions having a nonvanishing basin of attraction in the functional space of the probability distributions, we may use these functions for a more precise representation of the probability distribution of price fluctuations.

Unfortunately, all Lévy distributions with $\gamma < 2$ have infinite variances. That limits their physical, but not their mathematical, meaning. Physically, Lévy distributions are meaningless with respect to finite systems. But in complex systems with an almost unlimited reservoir of hidden irrelevant degrees of freedom, such probability distribution functions are quite possible, at least over a wide range of stochastic variables. Well-known examples of such wild distributions [247, 248] have been found to quantify the velocity–length distribution of the fully developed turbulence (Kolmogorov law) [208, 209, 210], the size–frequency distribution of earthquakes (Gutenberg–Richter law) [385, 386], or the destructive losses due to storms [284]. Further examples related to social and economic problems are the distribution of wealth [437, 440], also known as Pareto law, the distribution of losses due to business interruption resulting from accidents [438, 439] in the insurance business, or the distribution of losses caused by floods worldwide [309], or the famous classical St. Petersburg paradox discussed by D. Bernoulli [119, 310].

A typical numerical realization of a cumulative Lévy process $S(t_{n+1}) = S(t_n) + \xi_n$, where ξ_n is a random Lévy distributed number, is shown in Figure 3.21. Intuitively, one has the feeling that such a process is similar to the development of a given asset or stock price. Note that the occurrence of large fluctuations in the time evolution of the share prices initiated the creation of several models of price dynamics (e.g., the jump-diffusion model [266]), but, as mentioned above, such models have to be understood as substitute processes approaching reality.

Now, we want to prove, whether the Lévy distribution is a suitable limit probability distribution function describing the frequency of logarithmic price changes. To do this, we use the scaling hypothesis. Let us demonstrate the concept of scaling first using the Gaussian law.

We assume that the elementary timescale δt of the underlying financial process is sufficiently short compared with the time difference Δt between

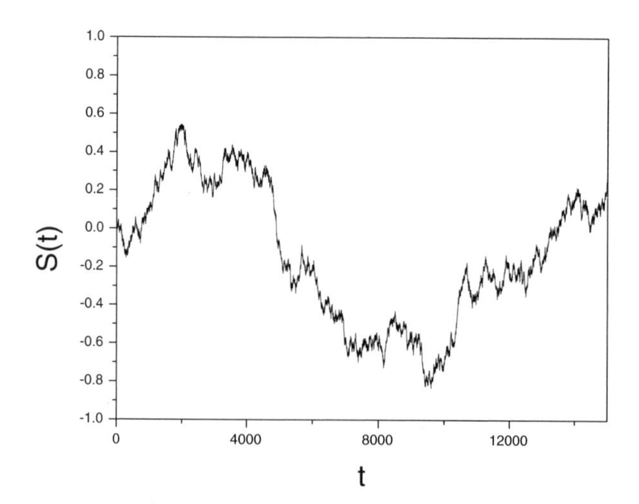

Fig. 3.21. A typical numerical realization of the cumulative Lévy process with the Lévy exponent $\gamma = 1.3$.

two successive observations of the logarithmic asset price. Furthermore, we make the hypothesis that the price changes are Gaussian-distributed also at the level of the elementary timescale. The probability distribution $p_{N\delta t}(\xi)$ of the sum of $N = \Delta t/\delta t$ random variables ξ obtained from the Gaussian distribution $p_{\delta t_0}(\xi)$ with the mean $\overline{\xi}(\delta t_0)$ and the variance σ^2 is again a Gaussian law with mean $\overline{\xi}(\Delta t) = N\overline{\xi}(\delta t)$ and variance $N\sigma^2$. Therefore, the rescaled variable

$$\hat{\xi} = \frac{\xi - N\overline{\xi}(\delta t)}{\sqrt{N}} = \frac{\xi - \overline{\xi}(\Delta t)}{\sqrt{N}} \tag{3.149}$$

has exactly the same probability density as the elementary variables. Therefore, the probability density of $\hat{\xi}$ is independent of N. This is the basic idea of the scaling procedure. In the first step, we have to prepare different time series of the same asset price but for different time differences Δt, $\Delta t'$, $\Delta t''$, ... between successive observations. Then, all of the rescaled variables

$$\frac{\xi - \overline{\xi}(\Delta t)}{\sqrt{\Delta t}}, \quad \frac{\xi - \overline{\xi}(\Delta t')}{\sqrt{\Delta t'}}, \quad \frac{\xi - \overline{\xi}(\Delta t'')}{\sqrt{\Delta t''}}, \quad \ldots \tag{3.150}$$

will exhibit the same probability distribution function and all of the data will collapse onto the same Gaussian law; that is, the probability distribution functions $p_{\Delta t}(\xi)$ are related to one universal master curve $f(x)$ via

$$p_{\Delta t}(\xi) = \frac{1}{\sqrt{\Delta t}} f\left(\frac{\xi - \overline{\xi}(\Delta t)}{\sqrt{\Delta t}}\right). \tag{3.151}$$

Unfortunately, this Gaussian concept does not work very well for short time differences Δt because the logarithmic price differences are just not Gaussian-distributed.

However, a similar procedure holds for Lévy distributions. For zero mean, the probability distribution function $p_{\Delta t}(\xi) = L^{\gamma}_{Na,b}(\xi)$ is a result of N convolution steps applied on the Lévy distribution $p_{\delta t}(\xi) = L^{\gamma}_{a,b}(\xi)$. Considering the scaling relation (3.147) as well as the probable translation of the center of the probability distribution function due to a nonzero mean, the distribution function

$$p_{\Delta t}(\xi) = \frac{1}{\Delta t^{1/\gamma}} f\left(\frac{\xi - \overline{\xi}(\Delta t)}{\Delta t^{1/\gamma}}\right) \tag{3.152}$$

is independent of Δt. In fact, this scaling function is valid for $1 < \gamma < 2$. For $\gamma < 1$, the mean is no longer defined, and we have to replace (3.152) by

$$p_{\Delta t}(\xi) = \frac{1}{\Delta t^{1/\gamma}} f\left(\frac{\xi}{\Delta t^{1/\gamma}}\right). \tag{3.153}$$

When the probability distribution functions for different time horizons satisfy (3.152) or (3.153), one says that the underlying process exhibits scaling properties. We remark that in the case of a Lévy distribution, the shape of the master curve $f(x)$ is controlled by the exponent γ and the asymmetry parameter b, while the third parameter a defines the scale of the fluctuations. A set of serious investigations of time series [250] support the existence of such a scaling behavior over a wide range.

In order to get a first approximation of the exponent γ, we should analyze the relative frequency of the special logarithmic price changes $\xi = 0$. Because $p_{\Delta t}(0) = \Delta t^{-1/\gamma} f(0) \sim \Delta t^{-1/\gamma}$, we expect a simple power law. The standard analysis in which the procedure is to qualify the existence of the power law using a double-logarithmic plot allows us to extract the exponent γ. An analysis of the short-time region $\Delta t \sim 1 - 10^3$ minutes suggests [250, 251] that the exponent γ has the value 1.4 ± 0.1. However, this power law is not a stable behavior. A crossover to the Gaussian law $p_{\Delta t}(0) \sim \Delta t^{-1/2}$ appears for sufficiently long time differences Δt.

An alternative way consists of a comparison of the empirically determined return probability $p_{\Delta t}(0)$ with the Gaussian return probability

$$p_{\Delta t}^{\mathrm{Gauss}}(0) \sim \frac{1}{\sqrt{2\pi}\sigma_{\mathrm{emp}}} \tag{3.154}$$

calculated from the empirically determined variance of the underlying asset price fluctuations. Note that the variance σ_{emp} is always finite because it is obtained from a finite set of finite values ξ. The difference between the two probabilities $p_{\Delta t}^{\mathrm{Gauss}}(0)$ and $p_{\Delta t}(0)$ decreases systematically for decreasing Δt above $\Delta t \approx 10^2$ minutes [251].

Both empirical analysis techniques suggest a characteristic crossover time of an order of magnitude 10^4 minutes (20 trading days). Above this time, we have a Gaussian behavior, while below this time an anomalous, Lévy-like regime appears. These results confirm our assumption that the probability distribution function of logarithmic price changes is never a true Lévy distribution or, more generally, a probability density of the basin of attraction of a Lévy law.

The empirically determined probability distribution functions $p_{\Delta t}(\xi)$ are rather a function of the basin of attraction of the Gaussian distribution. This conclusion corresponds to the expectation that the variance of the price changes is a finite quantity. On the other hand, we expect from the empirical analysis of high-frequency data that the probability distribution function of the logarithmic price changes is very close to a Lévy law over a wide range of Δt.

The data necessary for the characterization of the short-time behavior can be obtained from a relatively small time window (1 month) of high-frequency observations. This allows the computation of the time dependence of the Lévy exponent γ considering various subsets of the complete original data set [252]. The relatively small fluctuations of the numerically estimated exponents seem very likely that the Lévy exponent γ is a universal quantity characterizing the dynamics of financial markets.

Having obtained an estimation of the exponent using the simple techniques discussed above, we are able to construct the master curve for the

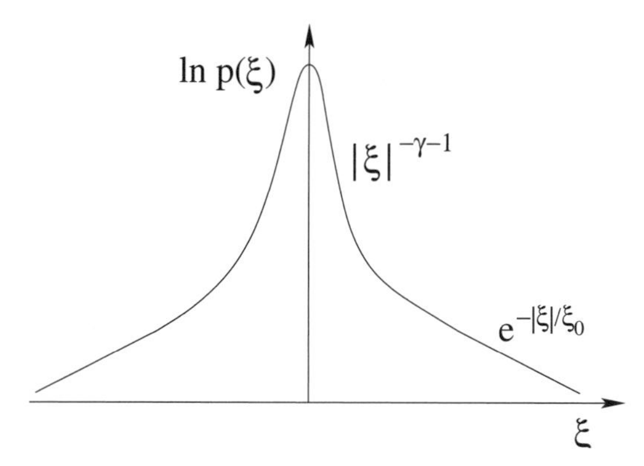

Fig. 3.22. The characteristic structure of the probability distribution function at short time horizons. The center can be described by a Lévy law with the characteristic scaling behavior. The tails show a behavior that can be fitted by an exponential decay.

probability distribution function. After the rescaling $p_{\Delta t}(\xi) \to \Delta t^{1/\gamma} p_{\Delta t}(\xi)$, and $\xi \to \left(\xi - \overline{\xi}\right)/\Delta t$, all of the empirically determined probability distribution functions with $\Delta t = 1 - 10^3$ minutes collapse approximately onto one curve [250].

We may reduce the mutual deviations of the rescaled curves by an additional adjustment of the exponent. The master curve is in almost every case an approximately symmetric function so that we obtain $b \approx 0$. We remark that the empirically determined probability distribution functions exhibit a slight skewness (left–right asymmetry), which is neglected in the following discussion. However, there are two essential problems suggesting that the scaling is only approximate. The first problem is that one can fit the master curve to a Lévy distribution only in a limited range. In particular, we find serious deviations for the asymptotic behavior. Whereas the tails of the Lévy distribution decay algebraically $|\xi|^{-1-\gamma}$, the master curve suggests an exponential decay (see Figure 3.22). But such a function is not form-stable, so the scaling procedure must break down, at least for long time intervals. Furthermore, the exponential decay corresponds to a finite variance, so the probability distribution $p_{\Delta t}(\xi)$ converges to a Gaussian law for $\Delta t \to \infty$.

The observation of this convergence in the present scaling procedure is very hard since the Gaussian law will be attracted onto the center of the master curve due to the use of the rescaling factor $\Delta t^{1/\gamma} \gg \Delta t^{1/2}$. The observation of the Gaussian law is possible by the application of the appropriate scaling factor $\Delta t^{1/2}$.

The range of intermediate and large fluctuations for relatively short time horizons may be fitted by the asymptotic law

$$p_{\Delta t}(\xi) \sim \frac{1}{|\xi|^{\gamma+1}} \exp\left\{-\frac{|\xi|}{\xi_0}\right\}, \tag{3.155}$$

with the characteristic scale ξ_0 defining the crossover between the algebraic and the exponential decays. The range $|\xi| < \xi_0$ can be fitted very well by a Lévy distribution with exponent γ and asymmetry coefficient $b = 0$. The Lévy-like behavior of the central region of the probability distribution function is also the cause for the apparent robustness of the probability against the scaling procedure. This supports our previous assumption that the probability distribution function $p_{\Delta t}(\xi)$ of the logarithmic price changes is closely neighbored to a stable Lévy law in the functional space, although $p_{\Delta t}(\xi)$ is always caught in the basin of attraction of the Gaussian law.

The second problem concerns the scale a of the Lévy distribution and the shift $\overline{\xi}$ of the center. The time evolution of these parameters obtained from various subsets of the complete original data set [252] shows strong fluctuations. These fluctuations indicate the more or less present nonstationarity of the financial market, but they have no essential influence on the stability of the Lévy exponent γ and the symmetry of the probability distribution.

3.5.5 Truncated Lévy Distributions

As we have seen, Lévy laws obey scaling relations but have an infinite variance. A real Lévy distribution is not observed in financial data. A stochastic process with finite variance and characterized by scaling relations in a large but finite region close to the center is the truncated Lévy distribution [249]. With respect to the observations in financial data, we have to ask for a distribution that in the tails is a power law multiplied by an exponential

$$p_{\delta t}(\xi) \sim \frac{C_{\pm}}{|\xi|^{\gamma+1}} \exp\left\{-\frac{|\xi|}{\xi_0}\right\}. \tag{3.156}$$

The characteristic function of a Lévy law truncated by an exponential as in (3.156) can be written explicitly as [211, 249]

$$\ln \hat{p}_{\delta t}(k) = a\frac{\left(1 + k^2\xi_0^2\right)^{\gamma/2}\cos\left(\gamma \arctan\left(k\xi_0\right)\right) - 1}{\xi_0^{\gamma}\cos\left(\pi\gamma/2\right)} \times \left[1 + ib\tan\left(\gamma\arctan\left(|k|\,\xi_0\right)\right)\frac{k}{|k|}\right]. \tag{3.157}$$

After N convolutions, we get the characteristic distribution function for a logarithmic price change with respect to the time interval $\Delta t = N\delta t$:

$$\ln \hat{p}_{\Delta t}(k) = -Na\frac{\left(1 + k^2\xi_0^2\right)^{\gamma/2}\cos\left(\gamma \arctan\left(k\xi_0\right)\right) - 1}{\xi_0^{\gamma}\cos\left(\pi\gamma/2\right)} \times \left[1 + ib\tan\left(\gamma\arctan\left(|k|\,\xi_0\right)\right)\frac{k}{|k|}\right]. \tag{3.158}$$

It can be checked that (3.157) recovers (3.140) for $\xi_0 \to \infty$. The behavior of $p_{\Delta t}(\xi)$ can be obtained from an inverse Fourier transform (2.15). In order to determine the characteristic scale of the probability distribution $p_{\Delta t}(\xi)$, we have to consider the main contributions to the inverse Fourier transform. This condition requires that the characteristic wave number k_{char} be of an order of magnitude satisfying $\ln \hat{p}_{\Delta t}(k_{\text{char}}) \simeq 1$. This relation is equivalent to

$$Na\left[\left(k^2+\xi_0^{-2}\right)^{\gamma/2}-\xi_0^{-\gamma}\right]\simeq 1. \tag{3.159}$$

For $N \ll \xi_0^\gamma$, (3.159) is satisfied if $k_{\text{char}}^2\xi_0^2 \gg 1$. Thus, we obtain immediately $k_{\text{char}} \sim (Na)^{-1/\gamma}$, and therefore the characteristic scale $\xi_{\text{char}} \sim (Na)^{1/\gamma}$, which characterizes an ideal Lévy distribution.

When on the contrary $N \gg \xi_0^\gamma$, the characteristic value of k_{char} becomes much smaller than ξ_0^{-1}, and we now find the relation $k_{\text{char}} \sim (Na)^{-1/2}\,\xi_0^{\gamma/2-1}$. The characteristic scale $\xi_{\text{char}} \sim (Na)^{1/2}\,\xi_0^{1-\gamma/2}$, corresponding to what we expect from the Gaussian behavior.

Hence, as expected, a truncated Lévy distribution is not stable. It flows to an ideal Lévy probability distribution function for small N and then to the Gaussian distribution for large N. The crossover from the initial Lévy-like regime to the final Gaussian regime occurs if the characteristic scale of the Lévy distribution reaches the truncation scale $\xi_{\text{char}} \sim \xi_0$ (i.e., if $Na \sim \xi_0^\gamma$).

This is exactly the behavior observed for the probability distribution of price changes. Here, we have to deal with a symmetric truncated Lévy distribution. The corresponding characteristic function is given by

$$\hat{p}_{\Delta t}(k) = \exp\left\{-Na\frac{\left(1+k^2\xi_0^2\right)^{\gamma/2}\cos\left(\gamma\arctan\left(k\xi_0\right)\right)-1}{\xi_0^\gamma\cos\left(\pi\gamma/2\right)}\right\}. \tag{3.160}$$

In particular, the variance obtained from (2.19) is given by

$$\sigma^2 = \frac{\Delta t a\gamma(1-\gamma)\xi_0^{2-\gamma}}{\delta t\cos(\pi\gamma/2)}, \tag{3.161}$$

which is in agreement with several high-frequency observations [251] for time differences $\Delta t > 10$ minutes.

In summary, the truncated Lévy distribution well describes the probability distribution functions of the logarithmic price differences at different timescales.

However, we remark again that the scale a shows partially strong fluctuations over long timescales, indicating that financial markets are possibly not in a complete steady state. Another more serious problem is that the assumption of the empirically motivated function (3.160) is at the moment only a model that was not derived from basic principles such as the Gaussian law and the Lévy distributions. We will discuss this problem below.

3.6 Large Fluctuations

3.6.1 Extreme Value Theory

The Value of Risk. For speculation strategies, it is helpful to have an estimate of the largest expected payoff and loss during a given time interval. This problem is equivalent to the determination of the expected largest and smallest values of a future time series $\{\xi(t_1), \xi(t_2), ... \xi(t_N)\}$ of logarithmic price fluctuations. The task is related to the extreme value theory [109, 138, 165]. This theory has many applications in the natural sciences, engineering, and social sciences [71, 107, 139, 164, 202].

In order to simplify the analysis, we focus on one certain asset price evolution observed in a stationary financial market. Furthermore, we assume that the time difference $\Delta t = t_{n+1} - t_n$ between successive observations is a constant value much larger than the Markov horizon. To proceed, we introduce the values of risk $\xi_{\max}$ and $\xi_{\min}$ that define the cumulative probability that a payoff $\xi > \xi_{\max}$ or a loss $\xi < \xi_{\min}$ is exceeded in any one time period Δt. Furthermore, we define the cumulative probabilities

$$P_{>}(\eta) = \int_{\eta}^{\infty} d\xi p_{\Delta t}(\xi) \quad \text{and} \quad P_{<}(\eta) = \int_{-\infty}^{\eta} d\xi p_{\Delta t}(\xi) . \tag{3.162}$$

The probability that the value of risk $\xi_{\max}$ is larger than the maximum of the time series, $\xi_{\max} > \xi_{\max}^{(N)} = \max(\xi(t_1), \xi(t_2), ... \xi(t_N))$, is given by the integral over the joint probability (3.10),

$$\Pi_{<}(\xi_{\max}) = \int_{-\infty}^{\xi_{\max}} d\xi^{(1)} \int_{-\infty}^{\xi_{\max}} d\xi^{(2)} \\ ... \int_{-\infty}^{\xi_{\max}} d\xi^{(N)} p_{\Delta t}\left(\xi^{(N)}, t_N; ...; \xi^{(1)}, t_1\right), \tag{3.163}$$

so that the independence condition (3.20) and the assumed stationarity lead to

$$\Pi_{<}(\xi_{\max}) = \left[\int_{-\infty}^{\xi_{\max}} d\xi p_{\Delta t}(\xi)\right]^N = \left[P_{<}(\xi_{\max})\right]^N . \tag{3.164}$$

Because $P_{<}(\xi_{\max}) + P_{>}(\xi_{\max}) = 1$, we arrive at

$$\Pi_{<}(\xi_{\max}) = \left[1 - P_{>}(\xi_{\max})\right]^N . \tag{3.165}$$

The function $\Pi_{<}(\xi_{\max})$ is the probability that all values of the time series are less than $\xi_{\max}$. This probability is a measure of risk aversion. Let us assume that we expect with a probability $p^{\star}$ that all price fluctuations are below an upper limit $\xi_{\max}$. Then, this boundary is given by the solution of $\Pi_{<}(\xi_{\max}) = p^{\star}$. We find

$$P_{>}(\xi_{\max}) = 1 - (p^{\star})^{1/N} \approx -\frac{\ln p^{\star}}{N}. \tag{3.166}$$

For typical applications, the probability $p^{\star}$ is chosen to have the value e^{-1} so that we obtain the simple relation $P_{>}(\xi_{\max}) \approx N^{-1}$. The probability $P_{>}(\xi_{\max})$ decreases with increasing N so that, because of (3.162), the value of risk increases with increasing N.

Furthermore, we conclude from (3.162) that the value of $\xi_{\max}$ is controlled by the asymptotic behavior of the probability distribution function $p_{\Delta t}(\xi)$ for $\xi \to \infty$. An analogous statement follows for the lower value of risk $\xi_{\min}$.

For example, let us assume that the Gaussian distribution function (3.63) holds for the logarithmic price changes. Then, we get for the vanishing trend $\mu = 0$ and the variance $\sigma^2 = \Phi \Delta t$

$$P_{>}(\xi_{\max}) = 1 - \phi\left(\frac{\xi_{\max}}{\sigma}\right), \tag{3.167}$$

where $\phi(x)$ is the cumulative standard unit normal distribution. For $N \to \infty$, (3.166) leads to the estimation:

$$\xi_{\max} \sim \sigma\sqrt{\ln N}. \tag{3.168}$$

On the other hand, if the Lévy law holds, we have to deal with the asymptotic behavior (3.145). This leads to another estimation

$$\xi_{\max} \sim \left(\frac{CN}{\gamma}\right)^{1/\gamma}. \tag{3.169}$$

Finally, the truncated Lévy law with the asymptotic behavior (3.156) leads to

$$\xi_{\max} \sim \xi_0 \ln \frac{CN}{\xi_0^{\gamma}\left(\ln CN/\xi_0^{\gamma}\right)^{1+\gamma}} \tag{3.170}$$

for $\xi_{\max}/\xi_0 \gg 1$, while (3.169) occurs for $\xi_{\max}/\xi_0 \ll 1$. These examples demonstrate in a simple manner that the random variables described by a probability distribution with tails decaying faster than an algebraic law show only mild fluctuations. An exponentially large number of observations is necessary to find a sufficiently large fluctuation.

Another situation occurs for algebraic probability distribution functions. Here, we find wild fluctuations leading to a large value of risk in financial data also for a relatively small set of observations. We remark that in the case of symmetric probability distribution functions, the lower value of risk is given by $\xi_{\min} = -\xi_{\max}$.

The results above imply an important remark concerning the empirical determination of the variance of possible Lévy processes[2]. We must realize

[2] For simplicity, we consider only symmetric Lévy distributions.

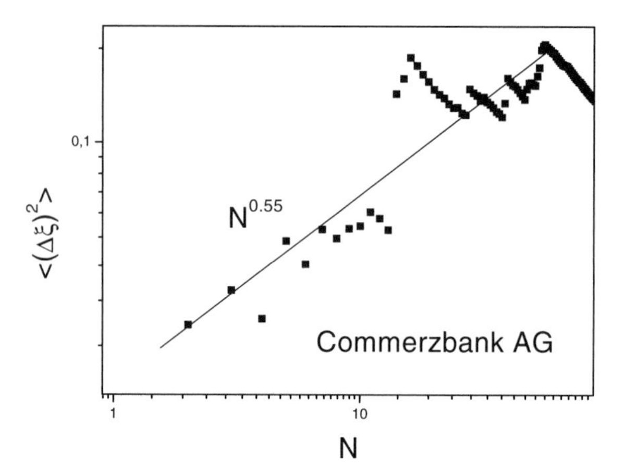

Fig. 3.23. The empirically determined second cumulant of the logarithmic daily share price fluctuations as a function of the number of observations N with first observation 09/01 for Commerzbank AG stock.

that, even if the variance of a Lévy probability distribution is infinite, one can always calculate the empirical variance from a finite set of observations. This fact comes about because a finite set of N measurements is usually restricted by an upper boundary $\xi_{\max}$ and a lower boundary $\xi_{\min} = -\xi_{\max}$, which are determined by the estimation (3.166). The probability of observing a value $|\xi| > \xi_{\max}$ in a typical series of N observations is small. Thus, the expected finite moments are defined by the truncation

$$m^{(n)} = \int_{-\xi_{\max}}^{\xi_{\max}} \xi^n p_{\Delta t}(\xi)\, d\xi, \tag{3.171}$$

where $p_{\Delta t}(\xi)$ may be a certain symmetric Lévy distribution. This integral can be approximated by using the asymptotic behavior (3.145). We obtain for $\gamma < n$ the result

$$m^{(2n)} \sim \xi_{\max}^{2n-\gamma} \sim N^{(2n-\gamma)/\gamma}, \tag{3.172}$$

while $m^{(2n+1)} = 0$ due to the symmetry of the probability distribution. Thus, the empirically determined variance typically increases with an increasing number of observations (Figure 3.23). In contrast to this behavior, a Gaussian law leads to

$$m^{(n)} \sim \Gamma\left(\frac{n+1}{2}\right) - \frac{(\ln N)^{(n-1)/2}}{N}. \tag{3.173}$$

This means that all empirically determined moments converge relatively fast to a fixed value. Therefore, sometimes the processes described by the Lévy law or another slowly decaying probability distribution function are erroneously denoted as nonstationary processes although the probability density is a time-independent quantity.

The Gumbel Distribution. Now, we ask for the asymptotic behavior for $N \to \infty$ of the cumulative probability $\Pi_< (\xi_{\max})$ defining the probability that the maximum observation is less than $\xi_{\max}$. To this end, we start from (3.165) and consider that the probability $P_> (\xi_{\max})$ converges monotonously to zero for $\xi_{\max} \to \infty$. Thus, we may write

$$\Pi_< (\xi_{\max}) = \exp\{-NP_> (\xi_{\max})\}, \tag{3.174}$$

where this expression becomes a better and better approximation as $\xi_{\max}$ increases so that $P_> (\xi_{\max})$ becomes smaller and smaller. The probability $\Pi_< (\xi_{\max})$ is very small for large N and small values of $\xi_{\max}$. But for sufficiently large $\xi_{\max}$, the probability $P_> (\xi_{\max})$ becomes an order of magnitude N^{-1} or lower and $\Pi_< (\xi_{\max})$ executes a transition from 0 to 1.

Let us assume that the probability distribution function has a tail falling faster than a power law. Then, the asymptotic behavior of $P_> (\xi_{\max})$ may be written as $\ln P_> (\xi_{\max}) \sim -c\xi_{\max}^a$ with $a > 0$. Now, we are able to conclude that the transition from $\Pi_< (\xi_{\max}) \approx 0$ to $\Pi_< (\xi_{\max}) \approx 1$ occurs at

$$NP_> (\xi_{\max}^\star) \sim 1. \tag{3.175}$$

Hence, we obtain the asymptotic relation $\xi_{\max}^\star \sim c^{-1/a} (\ln N)^{1/a}$. The expansion of $P_> (\xi_{\max})$ in powers of $\delta\xi_{\max} = \xi_{\max} - \xi_{\max}^\star$ around $\xi_{\max}^\star$ leads to

$$\begin{aligned} NP_> (\xi_{\max}) &\sim \exp\left\{-ca \left(\xi_{\max}^\star\right)^{a-1} \delta\xi_{\max}\right\} \\ &\quad \times \exp\left\{-\frac{ca(a-1)}{2} \left(\xi_{\max}^\star\right)^{a-2} \left(\delta\xi_{\max}\right)^2 + \ldots\right\}, \end{aligned} \tag{3.176}$$

where we have taken into account (3.175). The interval $\delta\xi_{\max}^\star$ over which the transition from 0 to 1 occurs is such that $\left(\xi_{\max}^\star\right)^{a-1} \delta\xi_{\max}^\star \sim 1$, leading to $\delta\xi_{\max}^\star \sim (\ln N)^{-1+1/a}$. Thus, the second-order term (and also all higher terms of this expansion) does not contribute essentially to the change of $NP_> (\xi_{\max})$ during the transition. In fact, we obtain $\left(\xi_{\max}^\star\right)^{a-2} \left(\delta\xi_{\max}^\star\right)^2 \sim (\ln N)^{-1}$ so that all higher terms disappear for $N \to \infty$. Finally, we collect all specific parameters in two nonuniversal, N-dependent numbers and arrive at

$$\Pi_< (\xi_{\max}) = \exp\left\{-\exp\left\{-\frac{\xi_{\max} - b_N}{a_N}\right\}\right\}. \tag{3.177}$$

This expression is the so-called Gumbel distribution [165], which determines the probability of finding the maximum value less than $\xi_{\max}$ in a set of N observations in the limit $N \to \infty$. Practically, this distribution holds very well for sufficiently large N.

The Gumbel distribution requires that the corresponding probability distribution have a tail with an asymptotic decay faster than a power law. We remark that in the case of a probability distribution with a tail falling as a power law $\xi^{-1-\gamma}$ for $\xi \to \infty$, the distribution of the extreme values converges to the Fréchet distribution

$$\Pi_{<}(\xi_{\max}) = \exp\left\{-\left[\max\left(0, 1 + \frac{\xi_{\max} - b_N}{\gamma a_N}\right)\right]^{-\gamma}\right\}. \tag{3.178}$$

Furthermore, any probability distribution with a finite right endpoint ξ_r and a functional behavior controlled by the leading term $(\xi_r - \xi)^\gamma$ $(\gamma > 0)$ close to this right endpoint offers an extreme-value distribution converging to the Weibull distribution

$$\Pi_{<}(\xi_{\max}) = \exp\left\{-\left[\max\left(0, \gamma + \frac{b_N - \xi_{\max}}{a_N}\right)\right]^{-\gamma}\right\}. \tag{3.179}$$

The remarkable result is that the maximum of any random series of N elements tends asymptotically for sufficiently large N to one of the three distribution functions introduced.

Rank-Ordering Statistics. Another possibility for dealing with extreme values is the rank-ordering technique [445]. In a certain sense, this method is a generalization of the extreme value theory. We consider again a time series of logarithmic price changes $\{\xi(t_1), \xi(t_2), ...\xi(t_N)\}$ and reorder them by increasing values

$$\xi^{(1)} \leq \xi^{(2)} \leq ... \leq \xi^{(n)} \leq ...\xi^{(N)}, \tag{3.180}$$

where $\xi^{(n)} \in \{\xi(t_1), \xi(t_2), ...\xi(t_N)\}$ for all $n = 1, ..., N$. The concept of the rank-ordering technique may be characterized as the determination of the nth value $\xi^{(n)}$ as a function of the rank n.

In order to quantify the rank-ordering problem, we start from the joint probability (3.10) and ask for the probability distribution function $\pi_n(\eta)$ that one variable has the value η while $n-1$ variables are less than η and $N-n$ variables are greater than η. Obviously, we have to consider the integral

$$\int_{-\infty}^{\eta} d\xi^{(1)} ... \int_{-\infty}^{\eta} d\xi^{(n-1)} \int_{\eta}^{\infty} d\xi^{(n+1)} ... \int_{\eta}^{\infty} d\xi^{(N)} p_{\Delta t}\left(\xi^{(N)}, t_N; ...; \xi^{(1)}, t_1\right), \tag{3.181}$$

which can also be written as

$$\left[\int_{-\infty}^{\eta} d\xi p_{\Delta t}(\xi)\right]^{n-1} p_{\Delta t}(\eta) \left[\int_{\eta}^{\infty} d\xi p_{\Delta t}(\xi)\right]^{N-n}. \tag{3.182}$$

Note that we have here applied the independence condition (3.20). There exist

$$\frac{N!}{(n-1)!(N-n)!} \tag{3.183}$$

various combinations of allowed rearrangements of the integral (3.181) leading to the same result (3.182). Thus, the total probability density to find at rank n the value η is given by

$$\pi_n(\eta) = \frac{N!}{(n-1)!(N-n)!} [P_<(\eta)]^{n-1} p_{\Delta t}(\eta) [P_>(\eta)]^{N-n}, \tag{3.184}$$

where we have used the cumulative probabilities (3.162). The expression (3.184) is valid for arbitrary probability distribution functions $p_{\Delta t}(\xi)$. We obtain an estimate of the nth value $\xi^{(n)}$ of the reordered time series if we determine the most probable value using the definition (2.9). Thus, the solution of

$$\left.\frac{\partial \pi_n(\eta)}{\partial \eta}\right|_{\eta=\xi^{(n)}} = 0 \tag{3.185}$$

yields the typical value $\xi^{(n)}$. Let us briefly demonstrate this technique. Differentiating (3.184) leads to

$$\frac{1}{\pi_n(\eta)} \frac{\partial \pi_n(\eta)}{\partial \eta} = (n-1) \frac{\partial \ln P_<(\eta)}{\partial \eta} + \frac{\partial \ln p_{\Delta t}(\eta)}{\partial \eta} + (N-n) \frac{\partial \ln P_>(\eta)}{\partial \eta}. \tag{3.186}$$

This equation can be written as

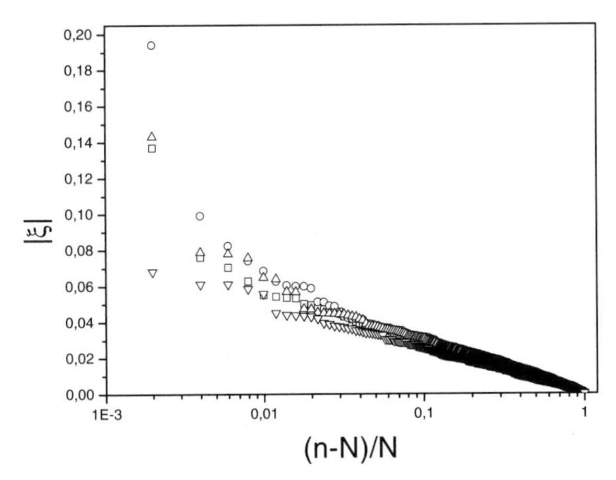

Fig. 3.24. The rank-ordered absolute values of the logarithmic daily price fluctuations for various companies of the chemical industry sector (squares: BASF; circles: Bayer, up triangles: Degussa; down triangles: Henkel; total time interval 11/00–07/02). The near equivalence of the curves suggests a widely common trading behavior with respect to this industrial sector.

$$\left[n - 1 - (N-1) P_<(\eta)\right] = P_<(\eta)\left[1 - P_<(\eta)\right] \frac{\partial 1/p_{\Delta t}(\eta)}{\partial \eta}. \tag{3.187}$$

In the case of an underlying Gaussian law, the absolute value of the right-hand side is always less than 1. Therefore, we get for large values of n and N,

$$\phi\left(\frac{\xi^{(n)}}{\sigma}\right) \approx \frac{n}{N}, \tag{3.188}$$

where $\phi(x)$ is again the standard error function. On the other hand, a Lévy distribution with the asymptotic behavior $p_{\Delta t}(\xi) \sim C|\xi|^{-1-\gamma}$ leads to the relation

$$\xi^{(n)} = \left[\frac{C}{\gamma} \frac{N\gamma + 1}{1 + (N-n)\gamma}\right]^{1/\gamma} \tag{3.189}$$

for $N - n \ll N$. This is illustrated in Figure 3.24 for the rank distribution of price variations obtained from various assets and a fixed time difference as well as in Figure 3.25 for the rank distribution of the logarithmic price changes obtained from the euro/US dollar exchange rate for various time intervals Δt. For sufficiently long time intervals, we get a Gaussian behavior aside some few very large fluctuations. The rank-ordered distribution for daily price changes is an intermediate regime, which demonstrates again the crossover from the Lévy law to the Gaussian behavior for this time horizon.

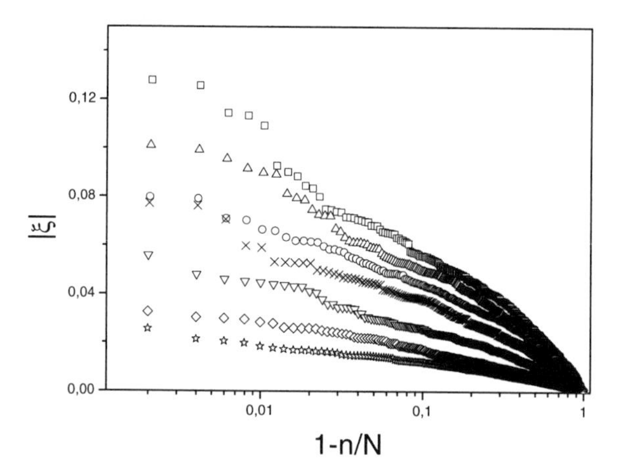

Fig. 3.25. The rank-ordered absolute values of the logarithmic fluctuations of the euro/US dollar exchange rate, for 11/00–07/02 for various time horizons (from bottom to top: $\delta t = 1, 2, 5, 10, 20, 50, 100$ trading days). The curves for $\delta t \geq 10$ can be fit very well by the inverse cumulative error function except for the region of large fluctuations $1 - n/N < 0.01$.

3.6.2 Partition Function Formalism

In the context of the rank-order statistics, we have formally introduced a separation of the allowed price changes in three parts. The three-piece scale consisted of the interval $[-\infty, \eta]$, $[\eta, \eta + d\eta]$, and $[\eta + d\eta, \infty]$. Afterwards, we determined the probability $\pi_n(\eta)$ that the first interval contains $n-1$ observations, the middle interval contains one observation, and the last interval contains the remaining $N-n$ observations.

Now, we investigate a generalization. We introduce M intervals $[\eta_{k-1}, \eta_k]$ with $k = 1, ..., M$ and $\eta_0 = -\infty$ and $\eta_M = \infty$. Then, we are able to calculate the probability $\pi(N_1, N_2, ..., N_M)$ that N_1 observations are located in the first interval, N_2 observations are located in the second interval, and so forth. A procedure similar to that used in the previous section leads to the formula

$$\pi(N_1, N_2, ..., N_M) = N! \prod_{k=1}^{M} \frac{(P_k)^{N_k}}{N_k!} \tag{3.190}$$

with

$$P_k = \int_{\eta_{k-1}}^{\eta_k} d\xi p_{\Delta t}(\xi) . \tag{3.191}$$

For a better representation, we introduce the frequencies $f_k = N_k/N$. Then, the probability $\pi(N_1, N_2, ..., N_M)$ may be written as $\pi_N(f_1, f_2, ..., f_M)$ and we obtain instead of (3.190)

$$\pi_N(f_1, f_2, ..., f_M) = N! \prod_{k=1}^{M} \frac{(P_k)^{N f_k}}{(N f_k)!} \tag{3.192}$$

with the constraint

$$\sum_{k=1}^{M} f_k = 1. \tag{3.193}$$

Considering that N is very large, the expression may be simplified. Using Stirlings formula $x! \approx x^x \exp\{-x\}$, we arrive at

$$\pi_N(f_1, f_2, ..., f_M) \approx \exp\{-NK[P, f]\} , \tag{3.194}$$

where $K[P, f]$ is the so-called information gain, or the Kullback information [218, 219]

$$K[P, f] = \sum_{k=1}^{M} f_k \ln\left(\frac{f_k}{P_k}\right) , \tag{3.195}$$

which is related to Shannon's information entropy [188, 189, 190, 371]. A more precise calculation shows that the exponential function in (3.194) must be multiplied with an extra factor, which leads to the correction

$$\frac{1}{N}\ln\pi_N(f_1,f_2,...,f_M) \approx -K[P,f]$$
$$-\frac{1}{2N}\sum_{k=1}^{m}\ln f_k + (1-m)\frac{\ln 2\pi N}{2N} + ...$$
$$= -K[P,f] + o\left(N^{-1}\ln N\right). \tag{3.196}$$

Therefore, the probability distribution $\pi_N(f_1,f_2,...,f_M)$ is completely dominated by the exponential for $N\to\infty$ so that the estimation (3.194) converges to the exact result, and the probability with which any given frequency distribution is realized is essentially determined by the Kullback information gain. The lower this quantity, the more likely is the frequency distribution. The information gain has the important property

$$K[P,f] \geq 0, \tag{3.197}$$

where the equality sign holds if and only if $f_k = P_k$ for all k. In the absence of constraints other than the normalization condition (3.193), the most probable frequencies are obtained from the maximum likelihood approach. This method is equivalent to the minimization of

$$F = \sum_{k=1}^{M} f_k \ln\left(\frac{f_k}{P_k}\right) + \lambda\left[\sum_{k=1}^{M} f_k - 1\right] \tag{3.198}$$

with respect to the frequencies f_k, where λ is the Lagrange multiplier fixing the normalization constraint. We obtain $f_k = P_k \exp\{-1-\lambda\}$ and $\lambda = -1$ due to (3.193). Hence, we recover the law of large numbers:

$$\lim_{N\to\infty} f_k = P_k. \tag{3.199}$$

The application of Lagrange multipliers is very useful in the improvement of estimated probability distribution functions in the presence of additional constraints. Let us assume S constraints that take the form

$$\sum_{k=1}^{M} c_k^{(\alpha)} f_k = C^{(\alpha)} \quad \text{for} \quad \alpha = 1,...,S \tag{3.200}$$

with arbitrary but fixed coefficients $c_k^{(\alpha)}$ and $C^{(\alpha)}$. The problem of finding the extremum of the probability distribution $\pi_N(f_1,f_2,...,f_M)$ under the $S+1$ constraints (3.200) and (3.193) can be solved by using instead of (3.198) the generalized Lagrangian

$$\sum_{k=1}^{M} f_k \ln\left(\frac{f_k}{P_k}\right) + \lambda\left[\sum_{k=1}^{M} f_k - 1\right] + \sum_{\alpha=1}^{S}\lambda^{(\alpha)}\left[\sum_{k=1}^{M} c_k^{(\alpha)} f_k - C^{(\alpha)}\right] \tag{3.201}$$

with S additional Lagrange multipliers $\lambda^{(\alpha)}$. The minimization of (3.201) now leads to M equations

$$\ln\left(\frac{f_k}{P_k}\right) + 1 + \lambda + \sum_{\alpha=1}^{S} \lambda^{(\alpha)} c_k^{(\alpha)} = 0 \tag{3.202}$$

that can be readily solved for the most probable frequencies

$$f_k^\star = P_k \exp\left\{-1 - \lambda - \sum_{\alpha=1}^{S} \lambda^{(\alpha)} c_k^{(\alpha)}\right\}. \tag{3.203}$$

In order to determine the Lagrange multipliers, we must substitute (3.203) into the constraints. In particular, the normalization condition yields

$$\exp\{1+\lambda\} = \sum_{k=1}^{M} P_k \exp\left\{-\sum_{\alpha=1}^{S} \lambda^{(\alpha)} c_k^{(\alpha)}\right\}. \tag{3.204}$$

It is now convenient to abbreviate the right-hand side by

$$Z\left(\lambda^{(1)}, \lambda^{(3)}, \ldots \lambda^{(S)}\right) = \sum_{k=1}^{M} P_k \exp\left\{-\sum_{\alpha=1}^{S} \lambda^{(\alpha)} c_k^{(\alpha)}\right\}, \tag{3.205}$$

which we will interpret as a partition function. In fact, many relations of classical thermodynamic equilibrium theory can be transferred onto the partition function formalism. For instance, the additional constraints may be written as

$$C^{(\alpha)} = -\frac{\partial \ln Z}{\partial \lambda^{(\alpha)}} \quad \text{for} \quad \alpha = 1, \ldots, S \tag{3.206}$$

corresponding to the thermodynamic equations of state. These nonlinear equations allow the determination of the Lagrange multipliers by numerical methods. The second-order derivatives form a positive-definite matrix. This follows immediately from

$$-\frac{\partial C^{(\alpha)}}{\partial \lambda^{(\beta)}} = \frac{\partial^2 \ln Z}{\partial \lambda^{(\alpha)} \partial \lambda^{(\beta)}} = \sum_{k=1}^{M} c_k^{(\alpha)} c_k^{(\beta)} f_k - \sum_{k=1}^{M} c_k^{(\alpha)} f_k \sum_{l=1}^{M} c_l^{(\beta)} f_l. \tag{3.207}$$

The sum may be interpreted as an average using the statistical weights f_k. Thus, the right-hand side of (3.207) is a component of the covariance matrix $\overline{c^{(\alpha)} c^{(\beta)}} - \overline{c^{(\alpha)}}\ \overline{c^{(\beta)}}$, which is always positive-definite. In this sense, each term $-\partial C^{(\alpha)}/\partial \lambda^{(\beta)}$ is also a component of a positive-definite matrix . This allows the construction of various helpful inequalities similar to the thermodynamic inequalities. For example, the diagonal elements of a positive-definite matrix are always positive, so $\partial C^{(\alpha)}/\partial \lambda^{(\alpha)} \leq 0$. Furthermore, (3.207) implies the symmetry relation

$$\frac{\partial C^{(\alpha)}}{\partial \lambda^{(\beta)}} = \frac{\partial C^{(\beta)}}{\partial \lambda^{(\alpha)}}. \tag{3.208}$$

We obtain the minimum Kullback information gain considering all given constraints if we replace f_k by (3.203):

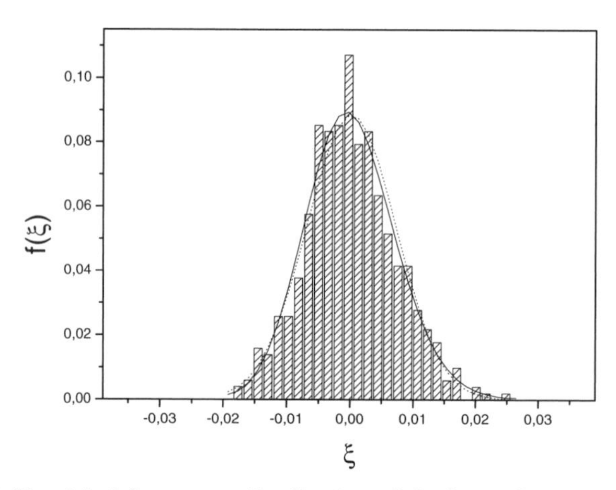

Fig. 3.26. Empirical frequency distribution of the logarithmic daily exchange rate for the euro/US dollar for 11/00–07/02). The dotted curve is the corresponding Gaussian distribution, and the straight line corresponds to the improved probability distribution function minimizing the Kullback information gain under certain constraints; see the text.

$$K_{\min} = K\left[P, f^{\star}\right] = -\ln Z - \sum_{\alpha=1}^{S} \lambda^{(\alpha)} \sum_{k=1}^{M} c_k^{(\alpha)} f_k^{\star}. \tag{3.209}$$

Let us use the concept of the partition function formalism for the computation of an appropriate distribution function. Suppose that we have a relatively short time series of daily price changes since the underlying asset is new to the market. However, we can determine an empirical probability distribution by constructing a histogram. This means that we divide the price scale in some intervals and collect all events falling in each of these intervals. The result is an empirical frequency distribution.

Furthermore, we can determine the first moments $m^{(n)}$ of this frequency distribution. Our hypothesis may be that the distribution function is an ideal Gaussian law with the trend $\tilde{\mu} = m^{(1)}$ and the variance $\sigma^2 = m^{(2)} - \tilde{\mu}^2$.

As an example, let us study the distribution of the logarithmic price changes of the euro/US dollar exchange rate of 700 trading days (Figure 3.26). The whole price scale is divided into 30 equal intervals except the two infinitely large boundary intervals, which are chosen to be empty. A simple numerical standard fit procedure has $\tilde{\mu} = 1.63 \times 10^{-4}$ and $\sigma = 7.08 \times 10^{-3}$, which allows us to calculate the ideal distribution function and afterwards the probabilities P_k.

The Kullback information gain is now a good measure for the distance between the observed frequency distribution f_k and the hypothetical proba-

Table 3.1. Moments and Lagrange multipliers corresponding to the probability distribution function of the daily euro/US dollar exchange rate for 11/00–07/02.

order n	1	2	3	4
moments $m^{(n)}$	1.63×10^{-4}	5.03×10^{-5}	1.08×10^{-7}	7.92×10^{-9}
Lagrange multipliers $\lambda^{(n)}$	14.74	7.66×10^{2}	-9.76×10^{4}	-1.35×10^{6}

bility distribution function given by the P_k. In our concrete case, we obtain $K[P, f] = 0.0256$.

In order to improve the hypothesis, we consider the constraints

$$\sum_{k=1}^{30} (\xi_k)^n f_k = m^{(n)} \quad \text{for} \quad \alpha = 1, ..., 4, \tag{3.210}$$

where ξ_k is the center of the kth interval and the $m^{(n)}$ are the empirically determined moments; see Table 3.1. The computation of the Lagrange multipliers using (3.205) and (3.206) is a numerically standard procedure.

The corresponding most likely distribution $f_k^\star$ (Figure 3.27) seems to be a better representation of the true probability distribution function than our original hypothesis. In fact, the Kullback information gain between the new hypothetical distribution function $f_k^\star$ and the empirical distribution function is now $K[f^\star, f] = 0.021$. Inspecting Figure 3.27, we find that the original

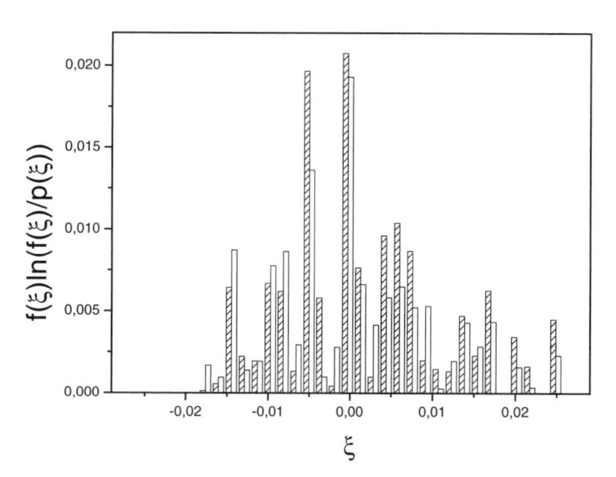

Fig. 3.27. Contributions to the Kullback information gain $f_k \ln(f_k/P_k)$ versus ξ_k (striped bars) and $f_k \ln(f_k/f_k^\star)$ versus ξ_k (empty bars). A noticeable reduction occurs mainly at the right tail.

hypothesis underestimates the tails, while the new hypothesis considers the effects of large fluctuations in an appropriate manner.

The maximum likelihood method provides also a simple framework to fit hypothetical distribution functions to empirically determined frequency distributions considering various constraints [82, 159, 242, 278, 408]. To this end, we have to consider a probability distribution function $p_{\Delta t}(\xi, \Theta)$, which is parametrized by a set $\Theta = \{\Theta_1, ...\Theta_L\}$ of L free parameters. Thus, the probabilities P_k depend also on these parameters, $P_k = P_k(\Theta)$, and the minimization of the Kullback information gain with respect to Θ,

$$\frac{\partial K[P(\Theta), f]}{\partial \Theta_i} = 0 \quad \text{for} \quad i = 1, ..., L, \tag{3.211}$$

leads to an optimum probability distribution. This method can be extended also onto probability distributions with constraints.

3.6.3 The Cramér Theorem

The central limit theorem states that the Gaussian law is a good description of the center of the probability distribution function $p_{\Delta t}(\xi)$ for sufficiently long time intervals Δt. We have demonstrated that the range of the center increases with increasing time intervals but is always limited for finite Δt. A similar statement is valid for the generalized version of the central limit theorem regarding the convergence behavior of Lévy laws. Fluctuations exceeding the range of the center are denoted as large fluctuations.

Of course, large fluctuations are rare events. This can be seen intuitively in Figure 3.25. The behavior of possible large fluctuations is affected only partially or not at all by the predictions of the central limit theorem, so we should ask for an alternative description. We start our investigation from the general formula (3.29) for a single asset price and assume that the market is in a steady state. The characteristic function can also be calculated for an imaginary $k \to iz$ so that the Fourier transform becomes a Laplace transform

$$\hat{p}_{\delta t}(z) = \int d\xi p_{\delta t}(\xi) \exp\{-z\xi\} \tag{3.212}$$

that holds under the assumption that the probability distribution function decays faster than an exponential for $|\xi| \to \infty$. We obtain again an algebraic relation for convolution of N elementary probability distribution functions $p_{\delta t}(\xi)$:

$$\hat{p}_{N\delta t}(z) = [\hat{p}_{\delta t}(z)]^N . \tag{3.213}$$

On the other hand, we assume that for sufficiently large N, the probability density $p_{N\delta t}(\xi)$ may be written as

$$p_{N\delta t}(\xi) = \exp\left\{-NC\left(\frac{\xi}{N}\right)\right\}, \tag{3.214}$$

where $\mathcal{C}(x)$ is the Cramér function [133, 225]. We will check by a construction principle whether such a function exists for the limit $N \to \infty$. To this end, we calculate the corresponding Laplace transform

$$\hat{p}_{N\delta t}(z) = N \int dx \exp\{-N[\mathcal{C}(x) + zx]\} \tag{3.215}$$

by using the method of steepest descent. This method approximates the integral by the value of the integrand in a small neighborhood around its maximum $\tilde{x}$. The value of $\tilde{x}$ depends not on N and is a solution of

$$\frac{\partial}{\partial x}\mathcal{C}(x) + z = 0. \tag{3.216}$$

With the knowledge of $\tilde{x}$, we can expand the Cramér function in powers of x around $\tilde{x}$,

$$\mathcal{C}(x) + zx = \mathcal{C}(\tilde{x}) + z\tilde{x} + \frac{1}{2}\frac{\partial^2}{\partial \tilde{x}^2}\mathcal{C}(\tilde{x})[x - \tilde{x}]^2 + \ldots . \tag{3.217}$$

Note that the first-order term vanishes due to (3.216). Substituting (3.217) into (3.215), we obtain the integral

$$\begin{aligned}\hat{p}_{N\delta t}(z) &= N \exp\{-N[\mathcal{C}(\tilde{x}) + z\tilde{x}]\} \\ &\quad \times \int dy \exp\left\{-N\left[\frac{1}{2}\frac{\partial^2 \mathcal{C}(\tilde{x})}{\partial \tilde{x}^2} y^2 + \ldots\right]\right\}\end{aligned} \tag{3.218}$$

with $y = x - \tilde{x}$. The leading term in the remaining integral is a Gaussian law of width $\delta y \sim N^{-1/2}$. With respect to this width, all other contributions of the series expansion can be neglected for $N \to \infty$. Therefore, we focus here on the second-order term. The corresponding Gaussian integral exists if $\partial^2\mathcal{C}/\partial x^2 > 0$. In this case, we obtain

$$\hat{p}_{N\delta t}(z) \sim \sqrt{N/\left(\partial^2\mathcal{C}(\tilde{x})/\partial\tilde{x}^2\right)} \exp\{-N[\mathcal{C}(\tilde{x}) + z\tilde{x}]\} . \tag{3.219}$$

For $N \to \infty$, the leading term of the characteristic function is given by

$$\hat{p}_{N\delta t}(z) \sim \exp\{-N[\mathcal{C}(\tilde{x}) + z\tilde{x}]\} . \tag{3.220}$$

Combining (3.213), (3.220), and (3.216) we obtain the equations

$$\mathcal{C}(x) + zx + \ln \hat{p}_{\delta t}(z) = 0 \quad \text{and} \quad \frac{\partial}{\partial x}\mathcal{C}(x) + z = 0, \tag{3.221}$$

which allow the determination of $\mathcal{C}(x)$. These two equations indicate that the Cramér function is the Legendre transform of $\ln \hat{p}_{\delta t}(z)$. Hence, in order to determine $\mathcal{C}(x)$ we must find the value of z that corresponds to a given $\tilde{x}$. The differentiation of (3.221) with respect to x leads to

$$\frac{\partial}{\partial x}\mathcal{C}(x) + z + x\frac{\partial z}{\partial x} + \frac{\partial \ln \hat{p}_{\delta t}(z)}{\partial z}\frac{\partial z}{\partial x} = \left[x + \frac{\partial \ln \hat{p}_{\delta t}(z)}{\partial z}\right]\frac{\partial z}{\partial x} = 0. \tag{3.222}$$

Because $\partial z/\partial x = -\partial^2\mathcal{C}(x)/\partial x^2 < 0$ (see above), we find the relation

$$x = -\frac{\partial \ln \hat{p}_{\delta t}(z)}{\partial z}, \tag{3.223}$$

from which we can calculate $z = z(x)$. Having $\mathcal{C}(x)$, the Cramér theorem reads

$$p_{N\delta t}(\xi) = \exp\left\{-N\mathcal{C}\left(\frac{\xi}{N}\right)\right\} \quad \text{for} \quad N \to \infty. \tag{3.224}$$

This theorem describes large fluctuations outside of the central region of $\hat{p}_{N\delta t}(\xi)$. The central region is defined by the central limit theorem, which requires $\xi \sim N^{\alpha}$ with $\alpha < 1$; see (3.54) and (3.55). Thus, the central region collapses to the origin in the Cramér theorem.

But outside of the center, we have $|\xi|/N > 0$. Obviously, the scaling of the variables differs between the central limit theorem and the Cramér theorem. While the rescaling $\xi \to \xi/\sqrt{N}$ leads to the form-stable Gaussian behavior of $p_{N\delta t}(\xi)$ in the limit $N \to \infty$, the rescaling ξ/N yields another kind of form stability concerning the expression $N^{-1} \ln p_{N\delta t}(\xi)$.

Furthermore, the properties of the initial elementary probability distribution disappear close to the center for $N \to \infty$. Therefore, the central limit theorem describes a universal phenomenon. The Cramér function conserves the properties of the elementary probability distribution functions due to (3.221) so that the large fluctuations show no universal behavior.

3.6.4 Extreme Fluctuations

Large fluctuations are a key point in the description of statistical properties of stock prices. However, the quantitative analysis is very difficult and requires extremely large databases [154, 155, 243]. The Cramér theorem provides a concept for the treatment of large fluctuations as a sum of an infinite number of successive price changes. This limit corresponds to large time intervals $\Delta t = N\delta t$, while the rescaled price fluctuations $\xi/N \sim \xi/\Delta t$ remain finite.

Another important regime that is of interest for the analysis of high-frequency data is the extreme fluctuation regime [136]. Here, we have to deal with finite N but $\xi/N \to \infty$.

In order to quantify this class of fluctuations, we start again from (3.29) and assume one asset and a stationary market. Because the regime of extreme fluctuations is characterized by small N and therefore by short timescales, the assumption of stationarity is justifiable. But the second necessary condition, namely the independence of successive observations, requires special attention.

We use the representation $p_{\delta t}(\xi) = \exp\{-f(\xi)\}$ and obtain

$$p_{N\delta t}(\xi) = \int \prod_{j=1}^{N} d\xi^{(j)} \delta\left(\xi - \sum_{j=1}^{N} \xi^{(j)}\right) \exp\left\{-\sum_{j=1}^{N} f\left(\xi^{(j)}\right)\right\}. \tag{3.225}$$

For simplicity, we restrict ourselves to the case of an extreme positive fluctuation $\xi \to +\infty$. We now have now two possibilities. On the one hand, the asymptotic behavior of the function $f(\xi)$ can be concave. Then, we have $f(x) + f(y) > f(x+y)$ so that the dominant contributions to (3.225) are obtained from configurations with all fluctuations very small except one extreme fluctuation almost equal to ξ. Therefore, we get

$$\ln p_{N\delta t}(\xi) \sim \ln p_{\delta t}(\xi) \sim -f(\xi). \tag{3.226}$$

On the other hand, if the asymptotic behavior of $f(\xi)$ is convex, $f(x) + f(y) < f(x+y)$, the minimum of the exponentials is given by the symmetric configuration $\xi^{(k)} = \xi/N$ for all $k = 1, ..., N$. The convexity condition requires a global minimum of the sum of all exponentials in (3.225) so that

$$\sum_{j=1}^{N} f\left(\xi^{(j)}\right) \geq N f\left(\frac{\xi}{N}\right). \tag{3.227}$$

We apply again the method of the steepest descent. To do this, we introduce the deviations $\delta\xi^{(k)} = \xi^{(k)} - \xi/N$ and expand the sum in (3.227) around its minimum,

$$\sum_{j=1}^{N} f\left(\xi^{(j)}\right) = N f\left(\frac{\xi}{N}\right) + \frac{1}{2} f''\left(\frac{\xi}{N}\right) \sum_{j=1}^{N} \left(\delta\xi^{(j)}\right)^2 + o\left(|\delta\xi|^3\right), \tag{3.228}$$

where we have used the constraint $\delta\xi^{(1)} + \delta\xi^{(2)} + ... + \delta\xi^{(N)} = 0$. We substitute this expression into (3.225). Then, with the assumption of convexity, $f''(\xi/N) > 0$, the integral (3.225) can be estimated. We get the leading term

$$p_{N\delta t}(\xi) \sim \exp\left\{-N f\left(\frac{\xi}{N}\right)\right\}. \tag{3.229}$$

This approximative result approaches the true value for $\xi/N \to \infty$. Apparently, (3.229) and (3.214) are identical expressions. But we should remember that (3.214) holds for $N \to \infty$ but finite ξ/N, while (3.229) requires $\xi/N \to \infty$. However, the Cramér function $\mathcal{C}(x)$ becomes equal to $f(x)$ for $x \to \infty$.

In summary, the knowledge of the tails of an elementary probability distribution $p_{\delta t}(\xi)$ of logarithmic price changes allows the determination of the tails of the probability distribution function $p_{N\delta t}(\xi)$ via

$$p_{N\delta t}(\xi) \sim \left[p_{\delta t}\left(\frac{\xi}{N}\right)\right]^N. \tag{3.230}$$

Therefore, we are also able to determine the probability distribution of extreme returns that are connected with the logarithmic price changes via $\xi(t) = \ln(R(t)+1) = \ln r(t)$.

The sum of N successive price changes $\xi(t_1) + \xi(t_2) + ... + \xi(t_N)$ with $t_{n+1} = t_n + \delta t$ is equivalent to the product $r(t_1) r(t_2) ... r(t_N)$. Hence, the probability distribution of extreme returns is given by

$$p_{N\delta t}(r) \sim \left[p_{\delta t}\left(r^{1/N}\right)\right]^N . \tag{3.231}$$

This expression has a very natural interpretation. The tail of the probability density $p_{N\delta t}(r)$ is controlled by the realizations where all terms in the product are of the same order. Therefore, the probability for an extremely large return over the time interval $N\delta t$ is just the product of the N elementary distribution functions, with each of their arguments being equal to the common value $r^{1/N}$.

3.6.5 A Mechanism for Extreme Price Changes

As mentioned above, the observation of the very rare extreme fluctuations requires large databases. Therefore, the time distance Δt between successive observations is necessarily small. Usually, the price fluctuations are analyzed over fixed but relatively short time intervals Δt. But also on these short timescales, each fluctuation is made up of more elementary fluctuations. Suppose that the elementary events of asset prices are single transactions. This assumption is correct with respect to time series obtained on a tick-by-tick frequency, including every quote or transaction price of the market.

A further divisibility is impossible without an essential extension of the set of relevant degrees of freedom. We should not expect that the short time differences between neighboring data points are still above the Markov horizon. Thus, memory effects may become important. Usually, these effects dominate the center of the probability distribution function of price changes. In the context of extreme fluctuations, we may neglect the memory. In a similar way, we can argue that the market is in a steady state.

Let us assume that the logarithmic price changes of single transactions are randomly distributed with a certain probability distribution function $p_0(\xi)$ while the random price changes $\xi_n = \xi^{(n)} + \xi^{(2)} + ... + \xi^{(n)}$ after n transactions are defined by the probability distribution function $p_n(\xi)$. In particular, the extreme fluctuations describing the tails of the probability density are given by $p_n(\xi) \sim p_0(\xi/n)^n$.

The tick-by-tick series are irregularly spaced because the market ticks arrive at random times. Thus, the number of ticks per time interval is Δt, a random number. Since the individual transactions occur randomly, this number is Poisson distributed,

$$\pi_n = \frac{\lambda^n}{n!} \mathrm{e}^{-\lambda}, \tag{3.232}$$

with λ the average number of ticks per time interval Δt. It then follows that the unconditional probability distribution function of extreme logarithmic price changes per time interval Δt is given by

$$p_{\Delta t}(\xi) \sim \sum_{n=0}^{\infty} p_n(\xi_n) \frac{1}{n!} \lambda^n \mathrm{e}^{-\lambda} \sim \sum_{n=0}^{\infty} \left[\lambda p_0 \left(\frac{\xi}{n} \right) \right]^n \frac{1}{n!}. \tag{3.233}$$

We use $\lambda p_0(\xi) = \exp\{-f(\xi)\}$ and the Stirling formula in order to obtain

$$\begin{aligned} p_{\Delta t}(\xi) &\sim \sum_{n=0}^{\infty} \exp\left\{ -\left[nf\left(\frac{\xi}{n}\right) + n(\ln n - 1) \right] \right\} \\ &\sim \sum_{n=0}^{\infty} \exp\{-F(\xi, n)\}. \end{aligned} \tag{3.234}$$

The main problem is to calculate the sum for sufficiently large ξ. To do this, we apply again the method of steepest descent. The main contributions to the sum in (3.234) stem from the minimum of $F(\xi, n)$. The corresponding value $n^\star$ of n is defined by the equation

$$\left. \frac{\partial F}{\partial n} \right|_{n=n^\star} = f\left(\frac{\xi}{n^\star}\right) + \ln n^\star - \frac{\xi}{n^\star} f'\left(\frac{\xi}{n^\star}\right) = 0. \tag{3.235}$$

The second derivative at the minimum point is given by

$$\left. \frac{\partial^2 F}{\partial n^2} \right|_{n=n^\star} = \frac{1}{n^\star} \left[1 + \left(\frac{\xi}{n^\star}\right)^2 f''\left(\frac{\xi}{n^\star}\right) \right] > 0 \tag{3.236}$$

so that $n^\star$ is a real minimum if the convexity condition is satisfied. Thus, we obtain

$$p_{\Delta t}(\xi) \sim \exp\{-F(\xi, n^\star)\} = \exp\left\{ n^\star \left[1 - \frac{\xi}{n^\star} f'\left(\frac{\xi}{n^\star}\right) \right] \right\}. \tag{3.237}$$

The minimum value $n^\star$ follows from the equation (3.235). We introduce the variable $w = \xi / n^\star$. Thus, we get

$$f(w) - wf'(w) - \ln w = -\ln \xi. \tag{3.238}$$

This equation allows the determination of w as a function of ξ if the function $f(w)$ is known. Usually, that is not the case. Therefore, we solve (3.238) in terms of a perturbation theory. To do this, we focus on an analysis of the fluctuations ξ around a certain value, say ξ_0, by setting $\ln \xi = \ln \xi_0 + \varepsilon$ with $\varepsilon \ll \ln \xi_0$. Then, we may solve the unperturbed equation

$$f(w_0) - w_0 f'(w_0) - \ln w_0 = -\ln \xi_0 \tag{3.239}$$

from which we obtain the unperturbed minimum point $n_0^\star = \xi_0 / w_0(\xi_0)$. The first-order correction follows by substituting $w = w_0 + \delta w$ into (3.238) and expanding this equation up to first order in δw. Thus, we obtain the solution

$$\delta w = \frac{w_0 \varepsilon}{1 + w_0^2 f''(w_0)}. \tag{3.240}$$

Furthermore, we get for $F(\xi, n^\star)$ up to first order in $\delta w \sim \ln \varepsilon = \ln(\xi/\xi_0)$

$$F(\xi, n^{\star}) = \xi \left[\frac{w_0 f'(w_0) - 1}{w_0} + \frac{\ln(\xi/\xi_0)}{w_0} \right], \tag{3.241}$$

so

$$p_{\Delta t}(\xi) \sim \exp\left\{ -\xi \frac{w_0 f'(w_0) - 1 + \ln(\xi/\xi_0)}{w_0} \right\}. \tag{3.242}$$

This is a remarkable result: for sufficiently large ξ_0 and therefore large w_0, the probability distribution function can mimic an exponential decay very well over a wide region $\xi^- < \xi < \xi^+$ with $|\ln(\xi^{\pm}/\xi_0)| \sim w_0 f'(w_0)$. In other words, the extreme fluctuations in the tails of the probability distribution function of price changes may be approximately described by a leading term $p_{\Delta t}(\xi) \sim \exp\{-g\xi\}$. This behavior is precisely what is expected for the tails of a truncated probability distribution function. We remark that the occurrence of an exponential decay is independent of the concrete structure of the elementary probability distribution function $p_0(\xi)$.

In order to describe the tails of the distribution function of returns, we use the relation $\ln r = \xi$ with $r = R + 1$. Hence, we arrive at

$$p_{\Delta t}(r) \sim r^{-1-g-\beta(r)} \sim R^{-1-g-\beta(R)} \quad \text{for} \quad r, R \to \infty, \tag{3.243}$$

with the constant exponent $g = (w_0 f'(w_0) - 1)/w_0$ and the slowly varying quantity $\beta(r) = \ln(\ln r/\ln r_0)/w_0$. Obviously, the probability distribution functions of extreme returns approach a power law over a wide region, in agreement with various studies [154].

3.7 Memory Effects

3.7.1 Time Correlation in Financial Data

In the previous chapters, we assumed that the time difference between successive price observations is above the Markov horizon. Then, we were able to use the separation (3.20) representing the probability of the occurrence of a complete time series $\{\xi(t_1, \delta t), ..., \xi(t_N, \delta t)\}$ as a product of single probabilities. Let us now analyze whether such an assumption is justifiable or whether we have to deal with the more general formulation (3.10).

To this end, we restrict ourselves to the evolution of a single stock. The extension onto a multivariable price vector is always possible. We expect that deviations from the Markov behavior occur at relatively short timescales. Since the empirical detection of possible correlation effects requires a relatively small time window of high-frequency data, we may assume a stationary financial market during the observation. This assumption is supported by the fact that the investigation of high-frequency data allows us to extend the analysis of correlation effects over a large number of points in time even if the total time interval over which the data are analyzed is not very long.

The independence of the logarithmic price changes of stocks or other financial assets is typically investigated by analyzing various autocorrelation functions (2.102). A standard quantity is the autocorrelation function (2.103), which reads for a single asset

$$C(t,t';\delta t) = C(t-t';\delta t) = \overline{\xi(t,\delta t)\xi(t',\delta t)} - \overline{\xi(t,\delta t)}\ \overline{\xi(t',\delta t)}, \tag{3.244}$$

where we have used explicitly the stationarity assumption. In this expression, the effects of a possible trend are eliminated. If we consider relatively short timescales δt, the trend $\overline{\xi(t,\delta t)} = \overline{\xi}\,(\delta t)$ is also of an order of magnitude δt (see (3.63)) so that the second term in (3.244) is proportional to δt^2. Therefore, this term may be neglected in the case of high-frequency data (i.e., small δt), and (3.244) reduces to the simpler autocorrelation function

$$F(t,t';\delta t) = F(t-t';\delta t) = \overline{\xi(t,\delta t)\xi(t',\delta t)}. \tag{3.245}$$

Another important remark is that the autocorrelation function depends on two timescales, δt and $t-t'$. Therefore, one must be very careful in comparing correlation functions that are obtained from various observations and in extracting general conclusions. Formally, we obtain from (2.102)

$$F(t-t';\delta t) = \int d\xi d\xi' p_{\delta t}\,(\xi,t;\xi',t')\,\xi\xi' \tag{3.246}$$

so that a theoretical determination of the autocorrelation requires the knowledge of the joint probability $p_{\delta t}\,(\xi,t;\xi',t')$. This probability distribution function is obtainable by integrating the general joint probability (3.10) over all variables not of interest.

If the price changes at different times t and t' are independent, the joint probability separates $p_{\delta t}\,(\xi,t;\xi',t') = p_{\delta t}\,(\xi,t)\,p_{\delta t}\,(\xi',t')$ and the autocorrelation function is of an order of magnitude $F(t-t';\delta t) \sim \delta t^2 \to 0$. On the other hand, we find that for $t=t'$ the autocorrelation function is equivalent to the variance $F(0;\delta t) = \sigma$. Such a peak structure indicates independence between successive observations of price changes.

Empirical investigations show that the correlation functions decay relatively fast. The decay is characterized by a correlation time much shorter than a trading day. Typically, one observes a correlation function that may be fitted by an exponential decay [253]

$$F(t-t';\delta t) \sim \exp\left\{-\frac{|t-t'|}{\tau_{\text{char}}}\right\}, \tag{3.247}$$

where the characteristic time τ_{char} is of an order of magnitude $10^0 - 10^2$ minutes. We conclude from (3.247) that the financial data have a Markov horizon of $\delta t_{\text{Markov}} \sim \tau_{\text{char}}$.

The Fourier transform of the autocorrelation function is the spectral function (2.106). In particular, we obtain from (3.247)

$$S(\omega) \sim \frac{\tau_{\text{char}}}{1+\omega^2\tau_{\text{char}}^2} \tag{3.248}$$

with the asymptotic behavior $S(\omega) \sim \omega^{-2}$ for large ω. That is the typical behavior that one observes for the price changes of stocks [251]. Various empirical observations [253] support the ω^{-2} decay, which is in agreement with the hypothesis that the stochastic dynamics of the logarithmic price changes may be described by a Wiener process corresponding to a random walk in the price space.

Another powerful test in detecting the presence of correlations is the investigation of the variance σ of an individual asset price as a function of the time Δt between successive price observations. For small time intervals Δt, the variance can be estimated by the second moment

$$\sigma^2(\Delta t) \approx \overline{\xi(t, \Delta t)^2} \tag{3.249}$$

of the logarithmic price fluctuations. This quantity is directly connected to the autocorrelation function. We demonstrate this connection using the divisibility of the logarithmic price changes,

$$\xi(t, \Delta t) = \sum_{n=0}^{N-1} \xi(t + n\delta t, \delta t), \tag{3.250}$$

with $\Delta t = N\delta t$. Thus, we obtain

$$\begin{aligned} \sigma^2(\Delta t) &= \sum_{n,m=0}^{N-1} \overline{\xi(t + n\delta t, \delta t)\xi(t + m\delta t, \delta t)} \\ &= \sum_{n,m=0}^{N-1} F(\delta t\,(n - m)\,; \delta t) \approx \frac{2}{\delta t^2} \int_0^{\Delta t} dt \int_0^t dt' F(t - t'; \delta t). \end{aligned} \tag{3.251}$$

For $\Delta t \gg \delta t_{\text{Markov}}$, the main contributions of the autocorrelation function $F(t - t'; \delta t)$ to this integral come from a small stripe $|t - t'| \leq \delta t_{\text{Markov}}$ so that $\sigma^2 \sim \Delta t$. In fact, the empirical behavior detected in financial data is well-described by a power law $\sigma \sim \Delta t^{1/2}$ in the time window from approximately $10^0 - 10^2$ trading minutes to 10^2 trading days [88, 251].

However, we observe a superdiffusive regime $\sigma \sim \Delta t^{\beta}$ with $\beta \approx 0.8$ below the Markov horizon, $\Delta t \leq \delta t_{\text{Markov}}$. This anomalous behavior may be explained by the fact that financial time series have a memory of only a few minutes.

We should remember that the majority of our considerations were based on the assumption that the financial market is in a steady state. As discussed above, the existence of stationarity plays no essential role at very short time intervals. A similar statement holds also for very long time intervals due to the validity of the central limit theorem.

However, the type of stationarity that would be possible for this regime is at best asymptotic stationarity. But because of the small number of available data with respect to the price changes $\xi(t, \delta t)$ over long time horizons δt, it cannot be decided whether these price changes are already controlled by an asymptotic stationary process.

In the intermediate range between these two regimes, the assumption of stationarity can lead to real problems. These problems become recognizable by inspecting the empirical moving volatility $\sigma(t, T, \delta t)$ obtained from (3.15) or (3.16). In spite of the use of an apparently suitable moving time window T, these quantities show considerable variations.

As discussed above, the trend $\overline{\xi(t, \delta t)}$, which may be estimated by the moving mean value of logarithmic price fluctuations (3.14), is strongly affected by underlying economic and social processes. But this quantity may be eliminated from the dynamics of the financial market by using the reduced fluctuations $\hat{\xi}(t, \delta t) = \xi(t, \delta t) - \overline{\xi(t, \delta t)}$. This means that we can speak about an ideal financial market in a wider sense if all moments constructed from these quantities are time-independent.

The obvious fluctuations of the volatility lead to a further step generalizing the term stationary. This step consists of the introduction of additional random processes. For instance, a typical model is given by the Ito stochastic differential equations

$$d\hat{\xi} = \sigma dW_1 \quad \text{and} \quad d\sigma = g(\sigma, \hat{\xi})dt + h(\sigma, \hat{\xi})dW_2 \tag{3.252}$$

with two independent Wiener processes dW_1 and dW_2 and with two functions $g(\sigma, \hat{\xi})$ and $h(\sigma, \hat{\xi})$ describing possible drift and diffusion effects related to the volatility σ.

Such models are frequently used in financial mathematics [180, 241, 368]. The merit of these models is that the stationarity may be conserved in a wider sense although the actual volatility is a fluctuating quantity. The problem is that we have to deal with two independent degrees of freedom, $\hat{\xi}$ and σ. We point out that (3.252) describes a pure model on the basis of empirical experience.

The stochastic behavior of the volatility and therefore the correlation function $\overline{\sigma(t)\, \sigma(t')}$ depend on the functions $g(\sigma, \hat{\xi})$ and $h(\sigma, \hat{\xi})$. As mentioned in subsection 3.3.3, the Wiener processes are connected to the price fluctuations by various economic factors [337], or rather economic-psychological factors. Therefore, we may interpret (3.252) in such a manner that the presence of volatility fluctuations in real markets suggests that there might be some other fundamental economic and financial processes in addition to those controlling the price changes directly. In order to estimate the randomness of these additional processes, we analyze the empirical correlation function of the moving volatility

$$\begin{aligned} C_{\mathrm{vola}}(t - t'; T, \delta t) = {} & \langle \sigma(t, T, \delta t)\, \sigma(t, T, \delta t) \rangle \\ & - \langle \sigma(t, T, \delta t) \rangle \langle \sigma(t, T, \delta t) \rangle, \end{aligned} \tag{3.253}$$

where we must consider three different timescales: the time difference δt between successive observations, the width of the time window T, and the difference $t - t'$. Besides (3.253), other characteristic quantities are also used for an estimation of the volatility fluctuations, such as the average over the absolute values of the price changes in an appropriate time window, and quantities obtained from various kinds of maximum likelihood methods [296].

Typically, the autocorrelation function of the volatility shows a pronounced power law decay [76, 237, 238, 303, 323]. For instance, the autocorrelation function for the fluctuations of the absolute values of the S&P 500 price changes can be fitted by a power law decay $|t-t'|^{-\nu}$with a characteristic exponent $\nu \sim 1/3$ in the time interval from approximately 10^0 to 10^2 trading days [238]. Other studies [237, 238, 252] on the spectral function are consistent with the results obtained from the analysis of the autocorrelation functions. All of these observations support the hypothesis that the fundamental processes mentioned above that controll the volatility are long-range-correlated.

It should be remarked that the values of the autocorrelation function of the volatility fluctuations $C_{\text{vola}}(t-t'; T, \delta t)$ for $|t-t'| > 0$ are relatively small in comparison with the standard deviation of the volatility $C_{\text{vola}}(0; T, \delta t)$. This and the fact that the definition of the volatility correlation function usually requires higher joint probability distribution functions, such as the two-point distribution used in (3.246), indicate that the long-time correlations are not necessarily in contradiction to the pairwise independence of the logarithmic price changes discussed above. Rather, these fluctuations contain information about the character of the stationarity of the market.

We remark that our discussion suggests that the price changes cannot be described completely by a stationary stochastic process in a strict sense since the volatility is a time-dependent quantity in real financial markets. Under certain conditions, the strict steady state concept is a reasonable approximation that may be helpful for the discussion of many financial problems. But, in general, one should expect deviations from this ideal behavior.

3.7.2 Ultrashort Timescales

Nonlinear Fokker–Planck Equation. The time evolution of an arbitrary asset price may be described by the use of the conditional probability $p(x, t \mid x_0, t_0)$ with $x = \ln X$. As we have already discussed, a possible trend can be neglected on sufficiently short timescales. A general description of the probability distribution function of the asset price is formally possible in terms of the Nakajima–Zwanzig equation (2.50). This equation reads in our special case

$$\frac{\partial p\,(x, t \mid x_0, t_0)}{\partial t} = -\hat{M} p\,(x, t \mid x_0, t_0) + \int_{t_0}^{t} dt' \hat{K}\,(t - t')\, p\,(x, t' \mid x_0, t_0)\,. \tag{3.254}$$

This expression is an exact formulation of the time evolution of the probability $p\,(x, t \mid x_0, t_0)$. The general problem is, however, that the frequency operator $\hat{M}$ and the memory kernel $\hat{K}$ are indefinable because they contain the dynamics of all other degrees of freedom. At least, we want to attempt to estimate these quantities using our knowledge of financial systems and the dynamics of complex systems.

The memory term describes the feedback of the price evolution with its own history. In order to derive a possible functional structure of the memory kernel, we formally introduce the price density $g(x,t) = \delta(x - \ln X(t))$, where $X(t)$ is the asset price as a function of the time t. Obviously, the joint probability density $p(x,t;x_0,t_0)$ is equivalent to the correlation function $\overline{g(x,t)g(x_0,t_0)}$, where the average procedure includes all allowed price trajectories. On the other hand, the time-dependent field $g(x,t)$ may be interpreted as a set of relevant variables $G_x(t)$ labeled by the logarithmic price x. The evolution of these quantities is described by a system of Mori–Zwanzig equations (2.121),

$$\frac{\partial G_x(t)}{\partial t} = \sum_{x'} \Omega_{xx'} G_{x'}(t) + \sum_{x'} \int_{t_0}^{t} K_{xx'}(t-t') G_{x'}(t') dt' + f_x(t), \quad (3.255)$$

where we have assumed that the market is stationary. As discussed in subsection 2.7.1, these equations are also exact relations connecting the dynamics of the relevant quantities $G_{x'}(t)$ to the dynamics of all other degrees of freedom, which are collected in the time-dependent residual forces $f_x(t)$. The frequency matrix $\Omega_{xx'}$ is given by (2.115), while the memory term is defined as a normalized correlation matrix (2.123) of the residual forces $f_x(t)$,

$$K_{xx'}(t-t') = -\sum_{x''} H_{xx''} \overline{f_{x''}(t) f_{x'}(t')}; \quad (3.256)$$

see (2.123). Note that the matrix H is defined by

$$\sum_{x''} H_{xx''} \overline{G_{x''}(t) G_{x'}(t)} = \delta_{xx'}. \quad (3.257)$$

An important relation between the relevant variables and the residual forces is the orthogonality relation (2.119) ,

$$\overline{G_x(t_0) f_{x'}(t)} = 0, \quad (3.258)$$

where t_0 is the initial time with respect to (3.255). This property allows the determination of the evolution equation for the correlation functions $\overline{G_x(t)G_{x_0}(t_0)}$ from (3.255) as demonstrated in (2.127). Because

$$p(x,t;x_0,t_0) = \overline{g(x,t)g(x_0,t_0)} = \overline{G_x(t)G_{x_0}(t_0)} \quad (3.259)$$

and $p(x,t;x_0,t_0) = p(x,t \mid x_0,t_0)\, p(x_0,t_0)$ (see (2.56)), we reproduce the Nakajima–Zwanzig equation introduced above:

$$\frac{\partial p(x,t \mid x_0,t_0)}{\partial t} = \sum_{x'} \Omega_{xx'} p(x,t \mid x_0,t_0)$$

$$+\sum_{x'}\int_{t_0}^{t} K_{xx'}(t-t')p(x,t \mid x_0,t_0)\,dt'. \tag{3.260}$$

Now, we are able to estimate the memory kernel $K_{xx'}(t-t')$. Obviously, the quantities $G_x(t)$ form a set of orthogonal quantities due to

$$\begin{aligned}\overline{G_x(t)G_{x'}(t)} &= \overline{g(x,t)g(x',t)} = \overline{\delta\left(x-\ln X(t)\right)\delta\left(x'-\ln X(t)\right)} \\ &\sim \delta\left(x-x'\right) \sim \delta_{xx'}.\end{aligned} \tag{3.261}$$

Considering (3.257), we get immediately $H_{xx'} \sim \delta_{xx'}$. Furthermore, we separate the residual forces into a fast part $f_x^{\text{fast}}(t)$ and a slow part $f_x^{\text{slow}}(t)$.

The fast part represents, for example, the contributions of random buying decisions while the collective dynamics of the whole financial market are hidden behind the slowly varying forces. For instance, the permanent occurrence of combined buy and sell orders leads to a feedback between individual stock prices and therefore also to a feedback of the specific asset price $X(t)$ with its own history.

The dynamics of both contributions to the residual forces may be uncorrelated, $\left\langle f_x^{\text{fast}}(t) f_{x'}^{\text{slow}}(t')\right\rangle = 0$, so that the memory term splits into $K_{xx'}(t-t') = K_{xx'}^{\text{fast}}(t-t') + K_{xx'}^{\text{slow}}(t-t')$. The fast term can be approximated very well by a Markov ansatz $K_{xx'}^{\text{fast}}(t-t') = \Xi_{xx'}\delta\left(t-t'\right)$. Thus, (3.260) may be written as

$$\begin{aligned}\frac{\partial p\left(x,t \mid x_0,t_0\right)}{\partial t} &= \sum_{x'}\left[\Omega_{xx'} + \Xi_{xx'}\right] p\left(x',t \mid x_0,t_0\right) \\ &\quad +\sum_{x'}\int_{t_0}^{t} K_{xx'}^{\text{slow}}(t-t')p\left(x',t' \mid x_0,t_0\right)dt'.\end{aligned} \tag{3.262}$$

This equation is similar to equation (3.254) if we identify the matrix $\Omega_{xx'}+\Xi_{xx'}$ with the operator $-\hat{M}$ and the slow memory $K_{xx'}^{\text{slow}}$ with $\hat{K}\left(t-t'\right)$. Without any memory, $\hat{K}=0$, all irrelevant degrees of freedom are apparently external stochastic variables.

This situation is comparable with the behavior of asset prices at very long timescales well above the Markov horizon. Thus, equation (3.262) reduces to a simple Markov equation so that the first term of (3.262) can be interpreted as the right-hand side of a Fokker–Planck equation (3.64); that is,

$$\hat{M} \rightarrow -\frac{\Phi_0}{2}\frac{\partial^2}{\partial x^2}. \tag{3.263}$$

To construct the structure of the slow-memory part, we use an idea of Kawasaki [197] and present the slow residual forces in terms of polynomials of the relevant variables

$$f_x^{\text{slow}}(t) = \sum_{x'} A_{xx'}G_{x'}(t) + \sum_{x'x''} B_{xx'x''}G_{x'}(t)G_{x''}(t)$$

$$+ \sum_{x'x''x'''} C_{xx'x''x'''} G_{x'}(t) G_{x''}(t) G_{x'''}(t) + \dots . \tag{3.264}$$

This general expansion can be specified under consideration of the special structure of the relevant quantities $G_x(t)$. First, we notice that all summations refer to different logarithmic prices. Otherwise, a term that contains $G_x^2(t)$ reduces to the next lower order due to $G_x^2(t) = \delta^2 (x - \ln X(t)) \sim \delta(0) G_x(t)$. The initial correlation between residual forces and the relevant quantities is given by

$$\begin{aligned} \overline{G_y(t_0) f_x^{\text{slow}}(t_0)} &= \sum_{x'} A_{xx'} \overline{G_y(t_0) G_{x'}(t_0)} \\ &\quad + \sum_{x'x''} B_{xx'x''} \overline{G_y(t_0) G_{x'}(t_0) G_{x''}(t_0)} + \dots . \end{aligned} \tag{3.265}$$

All correlations higher than second order vanish because the corresponding prices are not completely identical; for instance,

$$\begin{aligned} \overline{G_y(t) G_{x'}(t) G_{x''}(t)} &= \overline{\delta(y - X(t))\, \delta(x' - X(t))\, \delta(x'' - X(t))} \\ &\sim \delta(x' - y) \delta(x'' - y) = 0. \end{aligned} \tag{3.266}$$

Note that the last step follows from the condition $x' \neq x''$ discussed above. The orthogonality relation $\overline{G_y(t_0) f_x(t)} = \overline{G_y(t_0) f_x^{\text{slow}}(t)} = 0$ requires $A_{xx'} = 0$ due to

$$\begin{aligned} 0 = \overline{G_y(t_0) f_x^{\text{slow}}(t_0)} &= \sum_{x'} A_{xx'} \overline{G_y(t_0) G_{x'}(t_0)} \\ &= \sum_{x'} A_{xx'} \delta(x' - y) \overline{G_y(t_0)} \sim A_{xy}. \end{aligned} \tag{3.267}$$

Thus, the expansion (3.264) starts with the second-order term

$$\begin{aligned} f_{\mathbf{r}}^{\text{slow}}(t) &= \sum_{\mathbf{r}'\mathbf{r}''} B_{xx'x''} G_{x'}(t) G_{x''}(t) \\ &\quad + \sum_{x'x''x'''} C_{xx'x''x'''} G_{x'}(t) G_{x''}(t) G_{x'''}(t) + \dots . \end{aligned} \tag{3.268}$$

Now, we can insert this relation into the definition of the memory kernel. We obtain from (3.256) because of $H_{xx'} \sim \delta_{xx'}$

$$\begin{aligned} K_{xy}^{\text{slow}}(t - t') &\sim \overline{f_x^{\text{slow}}(t) f_y^{\text{slow}}(t')} \\ &= \sum_{x'x''y'y''} B_{xx'x''} B_{yy'y''} \overline{G_{x'}(t) G_{x''}(t) G_{y'}(t') G_{y''}(t')} \\ &\quad + \sum_{x'x''x'''y'y''} C_{xx'x''x'''} B_{yy'y''} \times \\ &\qquad \overline{G_{x'}(t) G_{x''}(t) G_{x'''}(t) G_{y'}(t') G_{y''}(t')} \\ &\quad + \dots . \end{aligned} \tag{3.269}$$

The correlation functions can be approximated by using the standard decoupling procedure

$$\overline{G_{x'}(t)G_{x''}(t)G_{y'}(t')G_{y''}(t')} \approx \overline{G_{x'}(t)G_{y'}(t')}\;\overline{G_{x''}(t)G_{y''}(t')} + \overline{G_{x'}(t)G_{y''}(t')}\;\overline{G_{x''}(t)G_{y'}(t')} . \qquad (3.270)$$

Formally, the correlation function $\overline{G_x(t)G_{x'}(t')}$ is equivalent to the joint probability $p(x,t;x',t') = p(x,t \mid x',t')p(x',t')$. In a stationary market, the unconditional probability distribution function $p(x',t')$ is a slowly varying quantity[3] compared with the conditional probability $p(x,t \mid x',t')$ that depends strongly on the logarithmic price difference $\xi = x - x'$. Therefore, the probability density $p(x',t')$ is assumed to be a constant that may be incorporated into the coefficients of the expansion.

We therefore obtain a series expansion of the slow memory in terms of the conditional probability density $p(x,t \mid x',t')$ starting with the second order. The absence of linear contributions is a consequence of the orthogonality relation (3.258). We focus in the following discussion on the second-order term of (3.269), which seems to be the leading term of a schematic expansion of $K_{xy}^{\text{slow}}(t-t')$. Higher contributions can be treated in the same way. Due to the decoupling, we get

$$\begin{aligned} K_{xy}^{\text{slow}}(t-t') &= \sum_{x'x''y'y''} B_{xx'x''}B_{yy'y''}p(x',t \mid y',t')p(x'',t \mid y'',t') \\ &= \sum_{x'x''y'y''} B_{xx'x''}B_{yy'y''}p(x',t \mid y'',t')p(x'',t \mid y',t'). \qquad (3.271) \end{aligned}$$

This memory can be simplified further by considering some reasonable assumptions about the structure of $B_{xx'x''}$. In particular, we expect (3.264) to be dominated by coefficients $B_{xx'x''}$ referring to infinitesimally neighboring logarithmic prices x, x', and x''. If we take into account the expected symmetry of $p(x,t \mid x',t') = p(-x,t \mid x',t')$, we arrive at the general representation

$$K_{xx'}^{\text{slow}}(t-t') = \sum_{n=0}^{\infty} \gamma_n \left(\frac{\partial}{\partial x'}\right)^{2n} p(x,t \mid x',t')^2. \qquad (3.272)$$

Note that because of the symmetry mentioned odd powers of $\partial/\partial x$ must vanish identically. The strength parameters γ_n of the memory are determined by the underlying hidden dynamics of the irrelevant degrees of freedom. Thus, after some partial integrations, the evolution equation (3.262) can be written as

$$\frac{\partial p(x,t \mid x_0,t_0)}{\partial t} = \frac{\Phi_0}{2}\frac{\partial^2}{\partial x^2}\Delta p(x,t \mid x_0,t_0) \qquad (3.273)$$

[3] The probability distribution $p(x',t')$ follows directly from $p_{\Delta t}(\xi) = p_{\Delta t}(x - x_0)$, where $\Delta t = t - t_0$ is the lifetime of the specific stock and $x_0 = \ln X(t_0)$ is the corresponding logarithmic price observed at the point in time of the introduction of the share on the market, which passed long ago.

$$+\sum_{n=0}^{\infty}\gamma_n \int dx' \int_{t_0}^{t} dt' p(x,t \mid x',t')^2 \left(\frac{\partial}{\partial x'}\right)^{2n} p(x',t' \mid x_0,t_0). \tag{3.274}$$

The normalization condition

$$\int dx' p(x',t' \mid x_0,t_0) = 1 \tag{3.275}$$

requires $\gamma_0 = 0$. Considering only the leading order of the series in (3.274), we get

$$\frac{\partial p(x,t \mid x_0,t_0)}{\partial t} = \frac{\Phi_0}{2}\frac{\partial^2}{\partial x^2} p(x,t \mid x_0,t_0)$$
$$+\gamma_1 \int dx' \int_{t_0}^{t} dt' p(x,t \mid x',t')^2 \left(\frac{\partial}{\partial x'}\right)^{2} p(x',t' \mid x_0,t_0). \tag{3.276}$$

This nonlinear Fokker–Planck equation is the basis for the following investigation [358]. Similar equations are used for the description of the dynamics of other complex systems, such as catalytic reactions [361, 366], climate fluctuations [352], or several types of diffusion controlled by feedback mechanisms [353, 354].

We remark that (3.276) is only valid for a short time difference $t - t_0$. This is because we have considered only the leading terms of a schematic expansion of the memory kernel during the derivation of (3.276). The upper boundary of validity can be estimated to be the Markov horizon δt_{Markov}.

Naive Scaling Procedure. Considering a stationary market, equation (3.276) may also be written in terms of the logarithmic price changes $\xi = x - x_0 = \ln(X/X_0)$,

$$\frac{\partial p_t(\xi)}{\partial t} = \frac{\Phi_0}{2}\frac{\partial^2 p_t(\xi)}{\partial \xi^2} + \gamma_1 \int d\eta \int_0^t dt' p_{t-t'}(\xi-\eta)^2 \frac{\partial^2}{\partial \eta^2} p_{t'}(\eta), \tag{3.277}$$

where we have set $t_0 = 0$. The solution of this equation is characterized by two regimes.

On the one hand, the memory term can be neglected at short times $t \to 0$, and we get a simple Gaussian law considering the initial condition $p_0(\xi) = \delta(\xi)$. On the other hand, the nonlinearity dominates (3.277) at sufficiently long times. Therefore, we expect a crossover from the initially Gaussian solution into another functional structure.

We remark also that the long-time regime with respect to (3.277) is always below the Markov horizon δt_{Markov}. For $t \sim \delta t_{\text{Markov}}$, we expect that the memory kernel can no longer be expressed by the leading term of the schematic expansion discussed above.

In the next paragraph, we will now derive the asymptotic solution of (3.277) for long times (but below δt_{Markov}) using the technique of the dynamic renormalization group.

To this end, we use the following trick. We write (3.277) as an integro-differential equation in a multidimensional space of dimension d,

$$\frac{\partial p_t(\xi)}{\partial t} = \frac{\Phi_0}{2} \frac{\partial^2 p_t(\xi)}{\partial \xi^2} + \gamma_1 \int d^d\eta \int_0^t dt' p_{t-t'}(\xi - \eta)^2 \frac{\partial^2}{\partial \eta^2} p_{t'}(\eta), \tag{3.278}$$

and analyze the behavior of $p_t(\xi)$ in terms of the renormalization procedure [358, 362]. At the end of our calculations, we then carry out the limit $d \to 1$.

Such tricks are very popular in statistical physics. The original problem is extended onto a more general class, then it is solved, and finally the general solution obtained is reduced to the solution of the problem. Note also that some very exotic limits will be reasonable, such as $d \to 0$ for the replica method [93, 299, 300, 302] applied in the spin-glass theory or the transition to fields with a vanishing number of components in the theory of polymers [146, 297, 356].

Let us assume that the solution of (3.278) approaches a homogeneous function of the type

$$p_t(\xi) = \lambda^{-d} \phi\left(\lambda^{-z} t, \lambda^{-1} \xi\right) \tag{3.279}$$

in the long-time limit. Note that this suggestion is also supported by several numerical simulations [352, 353, 354]. The quantity λ is an arbitrary scaling parameter, whereas z is called the dynamic exponent. We remark that this behavior refers again to the center of the probability distribution function, similar to what we saw during the derivation of the central limit theorem.

Substituting (3.279) into (3.278), we obtain

$$\begin{aligned} \frac{\partial \phi(\tau, x)}{\partial \tau} &= \lambda^{z-2} \frac{\Phi_0}{2} \frac{\partial^2 \phi(\tau, x)}{\partial x^2} + \gamma_1 \lambda^{2z-2-d} \int d^d x' \int_0^\tau d\tau' \\ &\quad \times \phi^2(\tau - \tau', x - x') \frac{\partial^2}{\partial x'^2} \phi(\tau', x') \end{aligned} \tag{3.280}$$

with $\tau = \lambda^{-z} t$ and $x = \lambda^{-1} \xi$. Let us analyze the change of this equation under an increase of the scaling parameter. In the absence of the nonlinear term (i.e., $\gamma_1 = 0$), the equation is made scale-invariant upon the choice of $z = 2$.

The nonlinearity added to this scale-invariant equation then has a prefactor λ^{2-d}. The difference from the original equation for $\lambda = 1$ is the coupling parameter, which changes from γ_1 to $\gamma_1 \lambda^{2-d}$. Thus, for $d > 2$, a small nonlinearity scales to zero and becomes irrelevant. In other words, the behavior for large λ and $d > 2$ is well-described by the linear diffusion equation.

A nontrivial behavior occurs for $d < 2$. In this case, the nonlinearity wins more and more with increasing λ and dominates (3.280). Such a simple estimation is called a naive scaling procedure. This method allows the determination of the critical dimension $d_c = 2$, where the nonlinearity changes from an irrelevant term to a dominant contribution.

Furthermore, the equation (3.280) allows a first estimation of the asymptotic behavior at large scales. For $d > d_c$ and the choice $z = 2$, we arrive at a simple diffusion equation for $\lambda \to \infty$. Setting $\lambda = t^{1/2}$, the scaling relation (3.279) requires an asymptotic behavior of the solution

$$p_t(\xi) = \lambda^{-d}\phi\left(\lambda^{-2}t, \lambda^{-1}\xi\right) = \frac{1}{t^{d/2}}\phi\left(1, \frac{\xi}{t^{1/2}}\right) = \frac{1}{t^{d/2}}\psi\left(\frac{\xi}{t^{1/2}}\right). \qquad (3.281)$$

The functional structure of ψ still remains open in the framework of our estimation, but it is clear that the solution $p_t(\xi)$ approaches the Gaussian law for $t \to \infty$. The characteristic scale of the asymptotic probability distribution function is $t^{1/2}$. Below the critical dimension, the nonlinearity dominates (3.280), and we should now use the choice $z = (2+d)/2$. Then, the left-hand side and the memory term of (3.280) are in balance, while the diffusion term disappears for $\lambda \to \infty$. Setting $\lambda = t^{2/(2+d)}$, we expect an asymptotic solution

$$\begin{aligned} p_t(\xi) &= \lambda^{-d}\phi\left(\lambda^{-z}t, \lambda^{-1}\xi\right) \\ &= \frac{1}{t^{2d/(2+d)}}\phi\left(1, \frac{\xi}{t^{2/(2+d)}}\right) = \frac{1}{t^{2d/(2+d)}}\psi\left(\frac{\xi}{t^{2/(2+d)}}\right), \end{aligned} \qquad (3.282)$$

and the characteristic scale of the probability distribution function is now $t^{2/(2+d)}$. Consequently, the corresponding variance is $\sigma^2 \sim t^{4/(2+d)}$, so that in particular for $d = 1$, the time dependence of variance of the logarithmic price fluctuations (3.249) is given by $\sigma(\Delta t) \sim \Delta t^{2/3}$.

Finally, we still want to determine the functional structure of the probability distribution function $p_t(\xi)$ that can be expected at the end of the ultrashort-time regime. We start from equation (3.278) and consider that this regime is dominated by the nonlinear memory term. Therefore, we may neglect the linear diffusion term. The Fourier transform of this equation with respect to the price changes ξ leads to the time evolution of the characteristic function. Considering the scaling function (3.282), we obtain

$$\hat{p}_t(k) = \int d^d\xi e^{ik\xi} p_t(\xi) = \widehat{\psi}\left(kt^{1/z}\right) \qquad (3.283)$$

with $\widehat{\psi}(k) = \int d^d x e^{ikx}\psi(x)$. Hence, the evolution equation (3.278) now reads

$$\frac{\partial\widehat{\psi}\left(kt^{1/z}\right)}{\partial t} \sim -k^2\gamma_1 \int_0^t dt' \frac{\widehat{\psi}^{(2)}\left(k(t-t')^{1/z}\right)}{(t-t')^{d/z}} \widehat{\psi}\left(kt'^{1/z}\right) \qquad (3.284)$$

with $\widehat{\psi}^{(2)}(k) = \int d^d x e^{ikx}\psi^2(x)$. The transformations $\widehat{\psi}\left(\tau^{1/z}\right) = \widetilde{\psi}(\tau)$ and $\widehat{\psi}^{(2)}\left(\tau^{1/z}\right) = \widetilde{\psi}^{(2)}(\tau)$ with $\tau = k^z t$ now lead to

$$\frac{\partial\widetilde{\psi}(\tau)}{\partial\tau} \sim -k^{2+d-2z}\gamma_1 \int_0^\tau d\tau' \frac{\widetilde{\psi}^{(2)}(\tau-\tau')}{(\tau-\tau')^{1/z}} \widetilde{\psi}(\tau'). \qquad (3.285)$$

The dependence on k disappears due to the dynamic exponent $z = (2+d)/2$ estimated above. Then, the Laplace transform with respect to τ allows us to write

$$\overline{\psi}(q) \sim \frac{1}{q + \gamma_1 F(q)} \tag{3.286}$$

with $\overline{\psi}(q) = \int d\tau e^{-q\tau} \widetilde{\psi}(\tau)$ and $F(q) = \int dx e^{-q\tau} \tau^{-1/z} \widetilde{\psi}^{(2)}(\tau)$. Here, we have considered the normalization condition of the probability distribution function, which is equivalent to $\widehat{\psi}(0) = \widetilde{\psi}(0) = 1$; see equation (3.283). Assuming that the coupling constant γ_1 is sufficiently small, (3.286) may be approximated by $\overline{\psi}(q) \sim [q + \gamma_1 F(0)]^{-1}$. Thus, the inverse Laplace transform leads to

$$\widetilde{\psi}(\tau) \sim \exp\{-\gamma_1 F(0)\tau\} \tag{3.287}$$

and therefore to the characteristic function

$$\hat{p}_t(k) = \exp\{-\gamma_1 F(0) k^z t\} \tag{3.288}$$

corresponding to a symmetric Lévy distribution (3.140). Such a probability distribution function is also observed for the logarithmic price change at and above the Markov horizon. We conclude that the relative stability of the Lévy distribution above $\delta t_{\mathrm{Markov}}$ is a consequence of the generalized central limit theorem, whereas the generation of the Lévy law is due to the memory effects below the Markov horizon. We identify the Lévy exponent γ with the dynamic exponent $z = 3/2$ for $d = 1$. This naive scaling estimation is in good agreement with the observations of financial markets, $\gamma = 1.4 \pm 0.1$ [250, 251].

Dynamic Renormalization Group. The concept of the naive scaling procedure requires the complete neglect either of the linear diffusion term or of the nonlinearity in (3.278). Therefore, it cannot be decided within the framework of this method whether the true solution of (3.278) converges to a scaling function of type (3.279) and whether the interplay between linear and nonlinear terms influences the dynamic exponent z.

To answer this question, we have to derive the asymptotic solution of (3.278) using the dynamic renormalization group. It is reasonable to start from the Fourier transform of (3.278) with respect to ξ. Considering $\overline{p}(k,t) = \int p_t(\xi) \exp\{ik\xi\} d\xi$, we obtain the formal representation

$$\frac{\partial \overline{p}(k,t)}{\partial t} = -\frac{\Phi_0 k^2}{2} \overline{p}(k,t) - \gamma_1 k^2 \Sigma(\overline{p}, \overline{p}; k, t) * \overline{p}(k,t), \tag{3.289}$$

where the symbol $*$ defines the convolution procedure with respect to the time; that is,

$$\Sigma(\overline{p}, \overline{p}; k, t) * \overline{p}(k,t) = \int_0^t dt' \Sigma(\overline{p}, \overline{p}; k, t - t') \overline{p}(k, t'), \tag{3.290}$$

and the memory kernel is given by

$$\Sigma\left(\overline{p},\overline{p};k,t\right)=-\gamma_1 k^2\int\frac{d^dk'}{\left(2\pi\right)^d}\overline{p}(k-k',t)\overline{p}(k',t). \tag{3.291}$$

Formally, (3.289) can be rewritten as

$$\overline{p}(k,t)=\overline{p}_0(k,t)+\overline{p}_0(k,t)*\Sigma\left(\overline{p},\overline{p};k,t\right)*\overline{p}(k,t) \tag{3.292}$$

with the bare propagator $\overline{p}_0(k,t)=\exp\left\{-\Phi_0k^2t/2\right\}$. The repeated substitution of (3.292) into itself leads to a perturbation series. The first terms of this expansion are given by

$$\begin{aligned}\overline{p}&=\overline{p}_0+\overline{p}_0*\Sigma\left(\overline{p}_0,\overline{p}_0\right)*\overline{p}_0\\&\quad+2\overline{p}_0*\Sigma\left(\overline{p}_0,\overline{p}_0*\Sigma\left(\overline{p}_0,\overline{p}_0\right)*\overline{p}_0\right)*\overline{p}_0\\&\quad+\overline{p}_0*\Sigma\left(\overline{p}_0,\overline{p}_0\right)*\overline{p}_0*\Sigma\left(\overline{p}_0,\overline{p}_0\right)*\overline{p}_0+\ldots\,.\end{aligned} \tag{3.293}$$

This series may be rearranged. To do this, we take into account that the expansion

$$\begin{aligned}\overline{p}&=\overline{p}_0+\overline{p}_0*\Sigma\left(\overline{p}_0,\overline{p}_0\right)*\overline{p}\\&\quad+2\overline{p}_0*\Sigma\left(\overline{p}_0,\overline{p}_0*\Sigma\left(\overline{p}_0,\overline{p}_0\right)*\overline{p}_0\right)*\overline{p}+\ldots\end{aligned} \tag{3.294}$$

again produces (3.293) after repeated substitutions of (3.294) into itself. The main difference between (3.293) and (3.294) is that all reducible terms appearing in (3.293) are hidden in the algebraic structure of (3.294). For instance, the last represented term of (3.293) occurs after the first substitution of (3.294) into itself.

The Laplace transform of (3.294) with respect to the time allows a further simplification. We get

$$\overline{p}(k,\omega)=\left[\omega+\frac{1}{2}\Phi_0k^2-\Xi(k,\omega)\right]^{-1} \tag{3.295}$$

with

$$\Xi\left(k,\omega\right)=\mathcal{L}\left[\Sigma\left(\overline{p}_0,\overline{p}_0\right)\right]+2\mathcal{L}\left[\Sigma\left(\overline{p}_0,\overline{p}_0*\Sigma\left(\overline{p}_0,\overline{p}_0\right)*\overline{p}_0\right)\right]+\ldots\,, \tag{3.296}$$

where $\mathcal{L}$ indicates the Laplace transform and ω the corresponding Laplace variable. The correction $\Xi\left(k,\omega\right)$ to the bare propagator is also denoted as self-energy [153, 191]. Various diagrammatic representations of this perturbation theory [7, 53, 183, 184, 228, 444] exist in quantum field theory and statistical mechanics. We refrain from such a representation because we want to focus only on the lowest order of the present perturbation theory.

In the context of the renormalization group approach, it is convenient to introduce an upper limit Λ. This border restricts the k-integration to a large but finite sphere $|k|\leq\Lambda$. This allows us to apply the concept of the renormalization group technique introduced above by repeated applications of decimation steps and rescaling steps.

During the decimation step, we carry out the integration over an infinitely thin shell $[\Lambda',\Lambda]$ with $\Lambda'=\Lambda-\delta\Lambda$ considering that Λ is a sufficiently large scale. Furthermore, we write $\Phi_0=\Phi_0'+\delta\Phi_0$ and $\gamma_1=\gamma_1'+\delta\gamma_1$. The aim is now to compensate the changes due to the integration over the shell $[\Lambda', \Lambda]$ by the changes $\delta\Phi_0$ and $\delta\gamma_1$ of the constants Φ_0 and γ_1. Considering (3.295), this requires

$$\left(\frac{k^2}{2} - \frac{\partial \Xi'}{\partial \Phi'_0}\right) \delta\Phi_0 = \frac{\partial \Xi'}{\partial \Lambda'} \delta\Lambda + \frac{\partial \Xi'}{\partial \gamma'_1} \delta\gamma_1, \tag{3.297}$$

where all changes are assumed to be infinitely small. First, we discuss the meaning of the left-hand side of (3.297).

It is simple to check that $\Xi' \sim k^2$. Because of the fact that we work with a perturbation theory with a sufficiently small coupling parameter γ_1, the self-energy Ξ' should be small compared with $k^2\Phi_0$. Hence, the left-hand side may be estimated as $k^2\delta\Phi_0/2$. In order to determine the corrections $\delta\Phi_0$ and $\delta\gamma_1$ as a function of $\delta\Lambda$, we ask only for the leading orders in k and ω, which completely determines the behavior of the large-scale regime at sufficiently long times.

Furthermore, we consider only the leading order of Λ^{-1}. The lowest-order correction to $\delta\Phi_0$ arises from $\mathcal{L}[\Sigma(\overline{p}_0, \overline{p}_0)]$. Considering the explicit functional structure of $\mathcal{L}[\Sigma(\overline{p}_0, \overline{p}_0)]$,

$$\mathcal{L}[\Sigma(\overline{p}_0, \overline{p}_0)] = -\gamma'_1 k^2 \int_\Lambda \frac{d^d k'}{(2\pi)^d} \frac{1}{\omega + \frac{1}{2}\Phi'_0 \left[(k-k')^2 + k'^2\right]}, \tag{3.298}$$

we obtain

$$\left.\frac{\partial \Xi'}{\partial \Lambda'}\right|_{1.\text{order}} = \frac{\partial \mathcal{L}[\Sigma(\overline{p}_0, \overline{p}_0)]}{\partial \Lambda'} = -\gamma'_1 k^2 \frac{S_d \Lambda'^{d-3}}{(2\pi)^d \Phi'_0} \tag{3.299}$$

and

$$\left.\frac{\partial \Xi'}{\partial \gamma'_1}\right|_{1.\text{order}} = \frac{\partial \mathcal{L}[\Sigma(\overline{p}_0, \overline{p}_0)]}{\partial \gamma'_1} = \frac{1}{\gamma'_1} \mathcal{L}[\Sigma(\overline{p}_0, \overline{p}_0)], \tag{3.300}$$

where S_d is the surface area of a unit sphere in d dimensions. The contributions that are necessary to determine $\delta\gamma_1$ follow from the next higher order. If we exclude all of these terms for the moment, we receive the leading term of the infinitesimal change of Φ_0,

$$\delta\Phi_0 = -2\gamma'_1 \frac{S_d \Lambda'^{d-3}}{(2\pi)^d \Phi'_0} \delta\Lambda = -2\gamma_1 \frac{S_d \Lambda^{d-3}}{(2\pi)^d \Phi_0} \delta\Lambda. \tag{3.301}$$

The differentiation of the second-order terms of the series expansion (3.296) with respect to Λ yields two terms. The derivation of the outer integral gives the next higher contribution to $\delta\Phi_0$ and will be neglected in the further procedure. The inner integral produces the contribution

$$\left.\frac{\partial \Xi'}{\partial \Lambda'}\right|_{2.\text{order,inner}} = -\gamma'_1 \frac{2 S_d \Lambda'^{d-3}}{\Phi_0'^2 (2\pi)^d} \mathcal{L}[\Sigma(\overline{p}_0, \overline{p}_0)], \tag{3.302}$$

which allows the determination of the leading term of $\delta\gamma_1$. To do this, we insert (3.302) and (3.300) into (3.297) and obtain

$$\delta\gamma_1 = \gamma_1'^2 \frac{2S_{d-1}\Lambda'^{d-3}}{\Phi_0'^2 (2\pi)^d} \delta\Lambda = \gamma_1^2 \frac{2S_d\Lambda^{d-3}}{\Phi_0^2 (2\pi)^d} \delta\Lambda. \tag{3.303}$$

To summarize the previous calculations, we get the changes of γ_1 and Φ_0 as the result of one decimation step,

$$\gamma_1' = \gamma_1 - \delta\gamma_1 = \gamma_1 - 2\gamma_1^2 \frac{S_d\Lambda^{d-3}}{\Phi_0^2 (2\pi)^d} \delta\Lambda \tag{3.304}$$

and

$$\Phi_0' = \Phi_0 - \delta\Phi_0 = \Phi_0 + 2\gamma_1 \frac{S_d\Lambda^{d-3}}{(2\pi)^d \Phi_0} \delta\Lambda. \tag{3.305}$$

A more detailed analysis of the perturbation expansion [7, 444] shows that this approximation becomes exact up to logarithmic corrections at the critical dimension.

The second step of the renormalization procedure was the rescaling procedure. As discussed in the context of the naive scaling method, an infinitesimal change $t \to (1+\delta\lambda)^{-z} t$ and $\xi \to (1+\delta\lambda)^{-1}\xi$ (and consequently $k \to (1+\delta\lambda)k$ and $\Lambda' \to (1+\delta\lambda)\Lambda'$) requires $\Phi_0' \to (1+\delta\lambda)^{z-2}\Phi_0'$ and $\gamma_1' \to (1+\delta\lambda)^{2z-2-d}\gamma_1'$. In particular, the choice $(1+\delta\lambda)\Lambda' = \Lambda$, that is, $\delta\lambda = \delta\Lambda/\Lambda$ restores the initial equation (3.295). Thus, the relationships

$$\delta_{\text{total}}\Phi_0 = \Phi_0 \left[z - 2 + 2\gamma_1 \frac{2S_d\Lambda^{d-2}}{(2\pi)^d \Phi_0^2} \right] \delta\lambda \tag{3.306}$$

and

$$\delta_{\text{total}}\gamma_1 = \gamma_1 \left[2z - 2 - d - 2\gamma_1 \frac{S_{d-1}\Lambda^{d-2}}{\Phi_0^2 (2\pi)^d} \right] \delta\lambda \tag{3.307}$$

appear after a complete cycle from decimation and rescaling. Let us now repeat the application of decimation and rescaling steps N times. Then, we get the total scaling factor $\lambda = (1+\delta\lambda)^N = \exp(N\delta\lambda)$, and we can replace the discrete recursive relations (3.306) and (3.307) by corresponding differential equations considering the limit $N \to \infty$ and $\delta\lambda \to 0$,

$$\frac{\partial \ln \Phi_0}{\partial \ln \lambda} = z - 2 + \widetilde{\gamma}_1 \quad \text{and} \quad \frac{\partial \ln \widetilde{\gamma}_1}{\partial \ln \lambda} = 2 - d - 3\widetilde{\gamma}_1, \tag{3.308}$$

with

$$\widetilde{\gamma}_1 = 2\gamma_1 \frac{2S_d\Lambda^{d-2}}{(2\pi)^d \Phi_0^2} . \tag{3.309}$$

These equations are also called as the flow equations of the renormalization group. Finally, we want to fix the degree of freedom remaining from the rescaling procedure. During the derivation of the central limit theorem, we achieved this by the conservation of the second cumulant of the probability distribution function in the course of the renormalization procedure. In the present case, it is reasonable to fix the value of Φ_0. This requires $z = 2 - \widetilde{\gamma}_1$.

The second equation of (3.308) has two fixed points, $\widetilde{\gamma}_1^{(1)} = 0$ and $\widetilde{\gamma}_1^{(2)} = (2-d)/3$. The so-called Gaussian fixed point $\widetilde{\gamma}_1^{(1)}$ is stable above the critical dimension $d_c = 2$. This means that each coupling constant $\gamma_1 \sim \widetilde{\gamma}_1$ converges to zero with increasing λ. In other words, the nonlinearity becomes irrelevant at sufficiently large scales, and the solution of (3.278) approaches the Gaussian law. It is quite normal to say that the Gaussian fixed point is stable with respect to the addition of a certain nonlinearity. This result is in good agreement with the qualitative statements of the naive scaling procedure discussed above.

Conversely, for $d < 2$, the strength of a weak nonlinear term grows, indicating that another fixed point with nonzero γ_1 determines the behavior of the probability distribution function $p_t(\xi)$. At sufficiently large scales, the coupling constant arrives at the final value $\widetilde{\gamma}_1^{(2)}$ and the dynamic exponent z is given by

$$z = 2 - \frac{2-d}{3} \,. \tag{3.310}$$

The final value of $\widetilde{\gamma}_1$ is also a measure characterizing the quality of the underlying perturbation theory. Remember that we have pointed out above that the value of γ_1 must be a small parameter. Even if the initial value of $\gamma_1$appearing in the original equation (3.278) is sufficiently small, the rescaled value of γ_1 increases under the renormalization procedure and arrives at the finite value. This is the reason that the results presented above are only an estimation for $d < 2$. An exception is the case $d = 2$. Here, the coupling parameter always decreases: $\gamma_1 \sim 1/\ln\lambda$. If one considers higher orders of the perturbation expansion (3.296), the dynamic exponent can be expressed by a series expansion

$$z = 2 + \sum_{i=1}^{\infty} \alpha_i \epsilon^i \quad \text{with} \quad \epsilon = 2 - d \tag{3.311}$$

with the numerical coefficients α_i. Such a representation is called an ϵ-expansion. However, it is often very difficult to determine the higher coefficients of (3.311).

In order to come back to the original problem, we set $d = 1$. Thus, we expect that the existing feedback effects generate a probability distribution

$$p_{\delta t}(\xi) = \frac{1}{\delta t^{1/z}} \psi\left(\frac{\xi}{\delta t^{1/z}}\right) \tag{3.312}$$

of the logarithmic price fluctuations ξ with respect to a time horizon δt. The lowest order of an ϵ-expansion within the renormalization group approach suggests $z = 5/3$.

Therefore, the time dependence of the variance offers a superdiffusive behavior given by $\sigma \sim \delta t^{0.6}$. However, the value of this exponent is approximately 20 percent less than the exponents that are obtained from high-frequency market data.

This fact requires two comments. On the one hand, we should take into account that the superdiffusive regime in financial data is observed at a time interval of approximately 30 minutes. An appropriate fitting procedure using a power law over such a relatively short time interval can usually show an error up to 10 percent. On the other hand, the dynamic exponent z predicted above is only an estimation corresponding to the first order of an ϵ-expansion.

However, it remains a fact that the nonlinear Fokker–Planck equation (3.278) seems to be useful to describe the regime of price fluctuations below the Markov horizon $\delta t_{\mathrm{Markov}}$. Furthermore, as discussed above, the memory effects lead to the formation of the Lévy-like probability distribution, which becomes metastable over a long time period well above the Markov horizon due to the generalized central limit theorem.

3.7.3 Autoregressive Processes

Processes with Volatility Fluctuations. As mentioned above, the volatilities of asset price fluctuations are believed to change over time. The present fluctuations of the volatility have led to the development of various models considering additional stochastic processes [44, 91, 102, 180]. We have presented the relative general model (3.252), which allows various specifications. Many different models have been proposed in the academic literature. However, all of these special models reflect mostly only a part of our empirical experiences (i.e., these models are always very special substitute processes that explain specific phenomena of the evolution of the financial market).

Only a few of these models have found good practical or theoretical applications. In particular, some of the models considering a time-dependent volatility are quoted for the explanation of the slow decay of the volatility autocorrelation function.

We distinguish formally between two classes of models. The common property of all models is that the fluctuations of the logarithmic price $\hat{\xi}$ around its trend are expressed by the simple Ito stochastic differential equation $d\hat{\xi} = \sigma(t)\, dW_{\xi}(t)$, where W_{ξ} is a Wiener process and $\sigma(t)$ is the time-dependent volatility.

Stochastic volatility models contain further Wiener processes controlling the dynamics of the volatility. Very popular is the model of Hull and White [180, 368, 430] with $d\sigma = g(\sigma)dt + h(\sigma)dW_{\sigma}$. This theory allows the derivation of various specifications and generalizations [22, 230, 241]. Other types of stochastic models assume correlations between the stochastic processes [104, 186, 187, 224, 244, 283]. For example, this may be realized using the substitutions $W_{\xi} = W_1$ and $W_{\sigma} = W_1 \cos\theta + W_2 \sin\theta$ with the two independent Wiener processes W_1 and W_2 and an arbitrary fixed angle θ.

Conditional volatility models contain no further stochastic process. Here, (3.252) is reduced to

$$d\hat{\xi} = \sigma dW \quad \text{and} \quad d\sigma = g(\sigma, \hat{\xi})dt. \tag{3.313}$$

The current volatility is the result of an ordinary differential equation. Therefore, the volatility is also called the conditional volatility or conditional variance because it depends on the relative fluctuations. The random character of the volatility is a consequence of the coupling with the Ito stochastic differential equation for $\hat{\xi}$ due to the function $g(\sigma, \hat{\xi})$.

The choice of the function $g(\sigma, \hat{\xi})$ depends on the model and on the empirical experience. But it should be guaranteed that σ is a positive quantity independent from the trajectory of the price fluctuations. The formal solution of the second equation of (3.313) is a function depending on the whole history of $\hat{\xi}$. For example, the special function $g = \left(\sigma_0^2 - \sigma^2\right)/2\tau\sigma + \kappa\hat{\xi}^2(t)/2\sigma$ leads to the time-ependent variance

$$\sigma^2 = \sigma_0^2 + \kappa \int_{-\infty}^{t} \exp\left\{-\frac{t-t'}{\tau}\right\} \hat{\xi}^2(t')dt'. \tag{3.314}$$

Since the function $g(\sigma, \hat{\xi})$ is anyhow an empirical construction, we can refrain from a dubious construction of this function and the solution of the corresponding differential equation and instead immediately provide a possible structure of the solution of (3.313). A relatively general representation is the functional series expansion

$$\begin{aligned}\sigma^2(t) = \beta^{(0)} &+ \int_{-\infty}^{t} dt' \beta^{(1)}(t-t')\hat{\xi}^2(t') \\ &+ \frac{1}{2}\int_{-\infty}^{t} dt' \int_{-\infty}^{t} dt'' \beta^{(2)}(t-t', t-t'')\hat{\xi}^2(t')\hat{\xi}^2(t'') + \ldots .\end{aligned} \tag{3.315}$$

The appearance of the quadratic fluctuations and a suitable choice of the response functions $\beta^{(n)}$ $(n = 1, 2, ...)$ guarantee that the conditional variance is a positive-definite quantity. The specification of the functions $\beta^{(n)}$ is equivalent to the construction of the function $g(\sigma, \hat{\xi})$.

The representation (3.315) is similar to relations that are known from the nonlinear response theory, such as the nonlinear material equations connecting polarization and an electric field or magnetization and a magnetic field in classical electrodynamics. The description of the time-dependent volatility by a linear relation

$$\sigma^2(t) = \beta^{(0)} + \int_{-\infty}^{t} dt' \beta(t-t')\hat{\xi}^2(t') \tag{3.316}$$

is a probably sufficient approximation similar to the linear response theory of materials [215, 216]. While the physical response function describes the delay of the effect of an external field on the macroscopic state of a material, in the present case the response function $\beta(t-t')$ generates a delayed influence of the price fluctuations on the current volatility. But there is an essential difference: The external fields are usually independent from the response; however, there is a feedback in financial theory due to the first equation of (3.313).

The discrete version of (3.316) is given by

$$\sigma^2(t) = \beta^{(0)} + \sum_{n=1}^{\infty} \beta_n \hat{\xi}^2(t - n\delta t, \delta t) \tag{3.317}$$

with usually positive response coefficients β_n and $\beta^{(0)}$. The modeling of a time-varying volatility on the basis of (3.317), or more generally on the discrete version of (3.315), has been one of the most important research topics in various financial applications over the last twenty years.

The first development to capture such a volatility was the autoregressive conditional heteroskedasticity (ARCH) model [110]. Many different types of ARCH models have been proposed in the academic literature, such as ARMA–ARCH [421], CHARMA [405], threshold ARCH [441], or double-threshold ARCH [234].

A model with quite attractive features is the generalized ARCH (GARCH) model [54]. This model also has various extensions or specifications, such as IGARCH [4], AGARCH [55, 56], EGARCH [280], or the components GARCH model [111, 112].

The ARCH Process. The original model of autoregressive conditional heteroskedasticity [110] has the variance equation

$$\sigma_m^2 = \beta^{(0)} + \sum_{n=1}^{N} \beta_n \hat{\xi}_{m-n}^2 \tag{3.318}$$

with a memory of N time periods and the definitions $\hat{\xi}_n = \hat{\xi}(t_n, \delta t)$, $\sigma_n = \sigma(t_n)$, and $t_{n+1} = t_n + \delta t$. Furthermore, some constraints on the coefficients β_n are necessary to ensure that the conditional variance is always positive. Obviously, ARCH(N) is nothing but a weighted moving average (3.17) over the squared price fluctuations. For an ARCH(N) model, the series (3.317) is truncated after N steps.

The unfortunate acronym ARCH is nevertheless essential. Heteroskedasticity means changing variance, so conditional heteroskedasticity means changing conditional variance. This model captures the conditional heteroskedasticity of asset price fluctuations by using a moving average of past squared errors: if a major market movement in either direction occurred $m \leq N$ time periods δt ago, the error square will be large, and assuming

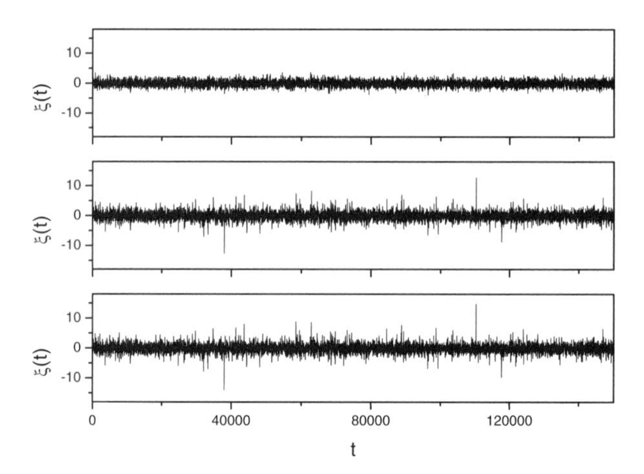

Fig. 3.28. Various realizations of ARCH(1) processes: $\beta_1 = 0$ (top), $\beta_1 = 0.447$ (middle), $\beta_1 = 0.488$ (bottom).

its coefficient is nonzero, the effect will be to increase the actual variance. In other words, large fluctuations tend to follow large fluctuations of either sign. Mandelbrot [246] had first described this phenomenon, which today we call conditional heteroskedasticity (Figure 3.28).

The unconditional variance of an ARCH process is simply the average over the conditional variance using the probability distribution function of the corresponding ARCH(N) model.

Because the discrete version of the price fluctuation equation of (3.313) reads $\hat{\xi}_n = \sigma_n \Delta W_n$ with $\Delta W_n = W(t_n) - W(t_{n-1})$ and $W(t)$ a Wiener process, we obtain from (2.134) $\overline{(\Delta W_n)^2} = \delta t$. The time period δt may be absorbed in the conditional variance so that we can write $\overline{(\Delta W_n)^2} = 1$. Thus, we get $\overline{\hat{\xi}_n^2} = \overline{\sigma_n^2}$, where we have considered the mutual independence of the differences ΔW_n and ΔW_m for $n \neq m$. Furthermore, the stationarity requires $\overline{\sigma_n^2} = \overline{\sigma^2}$ for all n. Hence, we obtain from (3.318)

$$\overline{\sigma^2} = \frac{\beta^{(0)}}{1 - \sum_{n=1}^{N} \beta_n} . \tag{3.319}$$

Therefore, the coefficients β_n must satisfy the constraint

$$\sum_{n=1}^{N} \beta_n < 1 . \tag{3.320}$$

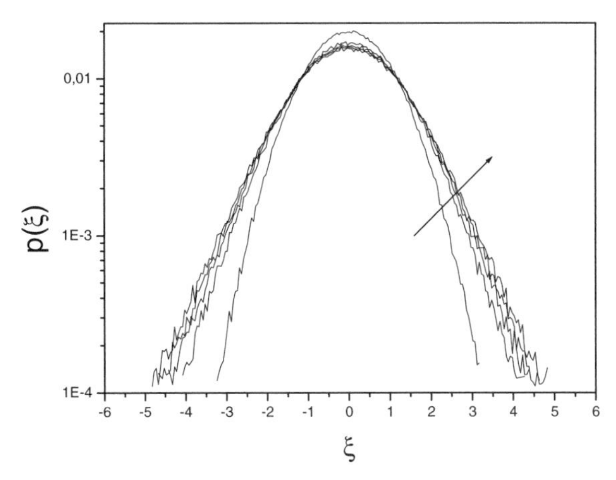

Fig. 3.29. The frequency distribution of an ARCH(1) process for $\beta_0 = 1$ and $\beta_1 = 0$ (Gaussian law), 0.333, 0.408, 0.447, 0.471, 0.488 (in the direction of the arrow).

For the sake of simplicity, we focus now on the ARCH(1) process. Here, we have the unconditional variance $\overline{\sigma^2} = \sigma_0^2/(1-\beta_1)$. Furthermore, we get from (2.132) the relation $\overline{(\Delta W)^4} = 3$ so that we find

$$\overline{\sigma^4} = \left(\beta^{(0)}\right)^2 + 2\beta_1 \overline{\sigma^2} \beta^{(0)} + 3\beta_1^2 \overline{\sigma^4}. \tag{3.321}$$

Then, the excess kurtosis (see Subsection 2.1.5) is

$$\lambda_4 = \frac{\overline{\hat{\xi}^4}}{\left(\overline{\hat{\xi}^2}\right)^2} - 3 = \frac{3\overline{\sigma^4}}{\left(\overline{\sigma^2}\right)^2} - 3 = \frac{6\beta_1^2}{1 - 3\beta_1^2}, \tag{3.322}$$

which is positive and finite for $\beta_1 < 1/\sqrt{3}$. The probability distribution function of the logarithmic price changes will then be leptokurtic because the fluctuating conditional variance allows for more outliers or unusually large observations (Figure 3.29). In order to determine the volatility autocorrelation function of the ARCH(1) process, we start from

$$\overline{\sigma_{n+m+1}^2 \sigma_n^2} = \overline{\left(\beta^{(0)} + \beta_1 \hat{\xi}_{n+m}^2\right) \sigma_n^2} = \beta^{(0)} \overline{\sigma^2} + \beta_1 \overline{\sigma_{n+m}^2 \sigma_n^2} \tag{3.323}$$

for $m > 0$. Considering (3.319), we arrive at

$$C_{\text{vola},m+1} = \overline{\sigma_{n+m+1}^2 \sigma_n^2} - \overline{\sigma^2}^2 = \beta_1 \left(\overline{\sigma_{n+m}^2 \sigma_n^2} - \overline{\sigma^2}^2\right) = \beta_1 C_{\text{vola},m}. \tag{3.324}$$

This recursion law is equivalent to

$$C_{\text{vola}}(t) \sim C_0 \exp\left\{-t/\tau_{\text{vola}}\right\} \tag{3.325}$$

with $\tau_{\text{vola}} = -\delta t / \ln \beta_1$. Thus, the autocorrelation of the ARCH(1) process shows an exponential decay in contradiction to various observations, suggesting a power law decay. This problem appears also for higher ARCH(N) processes [54], where the autocorrelation function is a weighted sum of various exponential decays.

The GARCH Process. In various applications using the ARCH(N) model, a large value of N is required. This usually poses some problems in the determination of the parameters $\beta^{(0)}$ and β_n describing the evolution of a given time series. Overcoming this problem and some other inadequacies of the ARCH(N) model leads to the introduction of a generalized ARCH process, the so-called GARCH(N, M) process [54]. This generalized model adds M autoregressive terms to the moving averages of squared errors. The variance equation takes the form

$$\sigma_m^2 = \beta^{(0)} + \sum_{n=1}^{N} \beta_n \hat{\xi}_{m-n}^2 + \sum_{n=1}^{M} \overline{\beta}_n \sigma_{m-n}^2. \tag{3.326}$$

From here, we obtain immediately that the unconditional variance of the GARCH(N, M) model is given by

$$\overline{\sigma^2} = \frac{\beta^{(0)}}{1 - \sum\limits_{n=1}^{N} \beta_n - \sum\limits_{m=1}^{M} \overline{\beta}_m}. \tag{3.327}$$

Obviously, the GARCH(N, M) process is a special case of (3.317) where the coefficients β_n are obtainable from generalized geometric series. We demonstrate this behavior for the GARCH(1, 1) process. Repeated substitution of (3.326) into itself yields

$$\begin{aligned}
\sigma_m &= \beta^{(0)} + \beta_1 \hat{\xi}_{m-1}^2 + \overline{\beta}_1 \sigma_{m-1}^2 \\
&= \left(1 + \overline{\beta}_1\right) \beta^{(0)} + \beta_1 \hat{\xi}_{m-1}^2 + \overline{\beta}_1 \beta_1 \hat{\xi}_{m-2}^2 + \overline{\beta}_1^2 \sigma_{m-2}^2 \\
&\ \vdots \\
&= \beta^{(0)} \sum_{n=1}^{\infty} \overline{\beta}_1^{n-1} + \beta_1 \sum_{n=1}^{\infty} \overline{\beta}_1^{n-1} \hat{\xi}_{m-n}^2;
\end{aligned} \tag{3.328}$$

that is, the GARCH(1, 1) process corresponds to an initial volatility $\sigma_0'^2 = \sigma_0^2 / (1 - \overline{\beta}_1)$ and a response function $\beta(t) \sim \exp\{-t/\tau_0\}$ with $\tau_0 = -\delta t / \ln \overline{\beta}_1$. Surprisingly, the parsimonious GARCH(1, 1) model, which has just one squared fluctuation and one autoregressive term, is in fact sufficient for most purposes since it has an infinite memory. The excess kurtosis of GARCH(1, 1) is given by the relation

$$\lambda_4 = \frac{6\beta_1^2}{1 - 3\beta_1^2 - 2\beta_1 \overline{\beta}_1 - \overline{\beta}_1^2}. \tag{3.329}$$

Thus, a finite unconditional variance requires $\beta_1 + \overline{\beta}_1 < 1$ while a finite positive excess kurtosis appears for $1 > 2\beta_1^2 + \left(\beta_1 + \overline{\beta}_1\right)^2$. In financial markets, it is common to find the so-called lag coefficient $\overline{\beta}_1$ in excess of 0.7, but the error coefficient β_1 tends to be smaller. The value of these parameters determines the shape of the resulting volatility time series. On the one hand, large lag coefficients mean that shocks to the conditional variance take a long time to die out, so volatility is persistent.

On the other hand, large error coefficients indicate that the volatility is quick to react to market changes and volatilities tend to be more spiky.

An analogous computation as used for the derivation of (3.324) yields the recursion law $C_{\text{vola},m+1} = \left(\beta_1 + \overline{\beta}_1\right) C_{\text{vola},m}$ for the volatility autocorrelation function. We obtain again an exponential decay

$$C_{\text{vola}}(t) \sim C_0 \exp\left\{-t/\tau_{\text{vola}}\right\} \tag{3.330}$$

with the correlation time $\tau_{\text{vola}} = -\delta t / \ln(\beta_1 + \overline{\beta}_1)$. However, an important difference between ARCH(1) and GARCH(1,1) is detected by comparing the characteristic timescales for these two processes.

Let us consider only processes with finite unconditional variance and finite excess kurtosis. The constraint of the finiteness of the kurtosis implies that the error coefficient β_1 of ARCH(1) must be lower than $1/\sqrt{3}$. Hence, this process may be characterized by a relatively short maximal correlation time in the volatility fluctuations of $\tau_{\text{vola}} \sim 2\delta t / \ln 3$. In GARCH(1,1), we can observe an arbitrarily long correlation time. To see this, let us write $\beta_1 + \overline{\beta}_1 = \varepsilon$ with $\varepsilon = \exp\left\{-\tau_{\text{vola}}/\delta t\right\}$.

We find that $\varepsilon < 1$ for all finite correlation times τ_{vola}. Hence, the corresponding GARCH(1,1) process has a finite unconditional volatility, and the excess kurtosis is finite if $1 - \varepsilon^2 > 2\beta_1^2$, which is always possible with a suitable choice of the error coefficient β_1. That is why the GARCH(1,1) process is widely used to model financial time series.

The empirical analysis [3, 4] of various volatility time series suggests a sum of lag and error coefficients close to 1 and relatively small error coefficients. For instance, the empirically determined coefficients of the US dollar rate GARCH(1,1) volatility for sterling and the Japanese yen, obtained in the time period 1983 to 1991 by analyzing the daily data [4], are $\beta_1 = 0.052$ and $\overline{\beta}_1 = 0.931$ (sterling) and $\beta_1 = 0.094$ and $\overline{\beta}_1 = 0.839$ (yen). Thus, we obtain a correlation time of $\tau_{\text{vola}}^{\text{sterling}} \approx 58$ trading days and $\tau_{\text{vola}}^{\text{yen}} \approx 15$ trading days.

There also remains the problem that GARCH(1,1) offers an exponential decay in contradiction to the empirical evidence that the volatility autocorrelation functions of asset and stock prices or exchange rates are characterized by a power law decay. It is also possible to determine the volatility autocorrelation function of a GARCH(N, M) process. Here, we obtain [54]

$$C_{\text{vola},m} = \sum_{i=1}^{\max(M,N)} \left(\beta_i + \overline{\beta}_i\right) C_{\text{vola},m-i} \tag{3.331}$$

so that the autocorrelation function can be expressed as a superposition of exponential decays. Such a combination may mimic in an approximate way the power law correlation of the volatility over a finite window.

The IGARCH Process. When $\beta_1 + \overline{\beta}_1 = 1$, we can rewrite the variance equation of the GARCH(1,1) process as

$$\sigma_m^2 = \beta^{(0)} + (1-\lambda)\,\hat{\xi}_{m-1}^2 + \lambda\sigma_{m-1}^2. \tag{3.332}$$

This model [4] is called the integrated GARCH process (IGARCH). Using (3.328) with $\overline{\beta}_1 = \lambda$ and $\beta_n = 1-\lambda$, we obtain the representation

$$\sigma_m^2 = \frac{\beta^{(0)}}{1-\lambda} + (1-\lambda)\sum_{n=1}^{\infty}\lambda^{n-1}\hat{\xi}_{m-n}^2. \tag{3.333}$$

Obviously, the IGARCH model with $\beta^{(0)} = 0$ is equivalent to an infinite exponential weighted moving average. To this end, we rewrite (3.17) as

$$\left\langle \hat{\xi}_m^2 \right\rangle_\omega = \frac{\sum\limits_{n=-\infty}^{m} \omega\left(t_m - t_n\right)\hat{\xi}_n^2}{\sum\limits_{n=-\infty}^{m} \omega\left(t_m - t_n\right)} = \frac{\sum\limits_{n=0}^{\infty} \omega\left(t_n\right)\hat{\xi}_{m-n}^2}{\sum\limits_{n=0}^{\infty} \omega\left(t_n\right)} \tag{3.334}$$

and use $\omega(t) \sim \exp(-t/\tau)$ with $\tau = \delta t/\ln\lambda$ in order to obtain

$$\sigma_m^2 = \left\langle \hat{\xi}_{m-1}^2 \right\rangle_\lambda. \tag{3.335}$$

We remark that the unconditional variance is undefined in the IGARCH model.

The AGARCH Process. As remarked in subsection 3.5.4, there is empirical evidence for a slight skewness in the probability distribution of the price fluctuations. The normal GARCH(1,1) model does not always fully account for this right–left asymmetry of empirical financial data. However, the skewness is easily accommodated by introducing just one additional parameter η in the conditional variance equation. This leads to the asymmetric GARCH(1,1) model (AGARCH), which has conditional variance

$$\sigma_m^2 = \beta^{(0)} + \beta_1\left(\hat{\xi}_{m-1} - \eta\right)^2 + \overline{\beta}_1\sigma_{m-1}^2. \tag{3.336}$$

For $\eta > 0$, negative shocks to the price fluctuations induce larger conditional variances than positive shocks. The AGARCH model is therefore appropriate when we expect more volatility following a market fall than following a market rise. This so-called leverage effect is a common feature of a large class of financial markets.

The EGARCH Process. The GARCH process has several drawbacks, including the inability to capture asymmetric volatility and to impose nonnegativity restrictions. While the skewness may be considered in the framework of an AGARCH model, the need constraint is eliminated in the exponential GARCH model [280]. This EGARCH model considers the conditional variance equation in logarithmic terms

$$\ln \sigma_m^2 = \ln \beta^{(0)} + H\left(Z_{m-1}\right) + \overline{\beta}_1 \ln \sigma_{m-1}^2 \tag{3.337}$$

with $Z_m = \hat{\xi}_m / \sigma_m$ and the function

$$H(Z) = \frac{1}{2}\left(\lambda_+ + \lambda_-\right) Z + \frac{1}{2}\left(\lambda_+ - \lambda_-\right) |Z| \,. \tag{3.338}$$

This asymmetric function has the slope λ_- for $Z < 0$, and λ_+ for $Z > 0$ provides the leverage effect just as in the AGARCH model.

The Components GARCH Process. A popular alternative specification of the GARCH process is the components GARCH process [111, 112]. Using the unconditional variance (3.327) of the GARCH(1,1) model, the variance equation may be written as

$$\begin{aligned} \sigma_m^2 &= \overline{\sigma^2}(1 - \beta_1 + \overline{\beta}_1) + \beta_1 \hat{\xi}_{m-1}^2 + \overline{\beta}_1 \sigma_{m-1}^2 \\ &= \overline{\sigma^2} + \beta_1 \left[\hat{\xi}_{m-1}^2 - \overline{\sigma^2}\right] + \overline{\beta}_1 \left[\sigma_{m-1}^2 - \overline{\sigma^2}\right], \end{aligned} \tag{3.339}$$

where $\overline{\sigma^2}$ may be interpreted as a certain baseline GARCH(1,1) model. The estimation of $\overline{\sigma^2}$ over a rolling time window suggests also that this quantity changes over time. The components model incorporates the varying volatility into the GARCH model. To this end, we replace (3.339) by the variance equation

$$\sigma_m^2 = q_m + \beta_1 \left[\hat{\xi}_{m-1}^2 - q_{m-1}\right] + \overline{\beta}_1 \left[\sigma_{m-1}^2 - q_{m-1}\right], \tag{3.340}$$

where the time-dependent baseline q_m is obtainable by an additional equation. A widely used possibility is given by

$$q_m = \overline{\sigma^2} + \chi\left(q_{m-1} - \overline{\sigma^2}\right) + \overline{\chi}\left(\hat{\xi}_{m-1}^2 - \sigma_{m-1}^2\right). \tag{3.341}$$

The equations (3.340) and (3.341) together define the components model. For $\chi = \overline{\chi} = 0$, we obtain the original GARCH model, while for $\chi = 1$, the baseline is just a random walk.

Convergence and Scaling. The ARCH and GARCH processes fit very well the empirically determined probability distribution functions of logarithmic price changes. The correspondence of the leptokurtic character is an especially remarkable feature. However, all of these more or less mathematical models fail to describe the behavior for different time horizons δt.

Above, we have mentioned one critical problem: the volatility autocorrelation correlation function of ARCH and GARCH models shows an exponential decay instead of the expected power law.

Another serious problem is related to the scaling properties of the probability distribution functions obtained from real financial data [253]. We know that, for processes with finite variance, the central limit theorem applies. Thus, the probability distribution function of the logarithmic price differences over a large time interval $\Delta t = N\delta t$ should approach a Gaussian law. In fact, the probability distribution function of the sum of N successive price changes

$$S(t, \Delta t) = \sum_{k=0}^{N-1} \xi(t - k\delta t, \delta t) \tag{3.342}$$

obtained from an ARCH or GARCH process implies a decrease in the excess kurtosis with increasing N [99]. In any case, the attractor of all possible ARCH and GARCH processes with finite variance is the Gaussian probability distribution function.

However, the fact that these processes describe well the probability distribution function for a given time horizon δt and that the probability distribution function of the sum (3.342) converges to a Gaussian law does not ensure that the same process describes well the stochastic dynamics of a financial market for any time horizon δt. There are strong indications obtained from numerical simulations [253] that the scaling properties of a probability distribution function for different time horizons cannot be captured by autoregressive processes.

In summary, ARCH and GARCH processes are only of moderate interest from a physical point of view because they fail to describe properly the time evolution of the probability distribution function of the price changes. The power of these models lies in their application within finance and financial mathematics. In particular, ARCH and GARCH processes are suitable classes of stochastic processes modeling the time evolution of the price for a given stock or asset at a fixed time horizon δt.

3.7.4 Time-Reversible Symmetry

Suppose that we have a sufficiently large time series of an arbitrary asset price with fixed time horizon

$$\{\xi_1, ..., \xi_{n-1}, \xi_n, \xi_{n+1}, ...\} \tag{3.343}$$

with $\xi_n = \xi(t_n, \delta t)$ and $t_n = t_{n-1} + \delta t$. Then, it may be of interest whether such a series shows a time-reversal symmetry in a statistical sense [312]. A stationary Markov process is always a time-reversal; that is, the probability for the realization of the time series (3.343) is equivalent to the probability for the realization of the reversed series

$$\{...., \xi_{n+1}, \xi_n, \xi_{n-1}, ..., \xi_1\} . \tag{3.344}$$

However, the existence of memory effects leads to a violation of the time-reversal symmetry due to the causal structure of the corresponding evolution equations. We will now answer the question of how the existence or the lack of such a symmetry may be tested. The two-point correlation function of a stationary process (2.104) is always an even function. If we use the relative price changes $\hat{\xi} = \xi - \overline{\xi}$, we obtain due to the stationarity

$$C(\tau;\delta t) = \overline{\hat{\xi}(t,\delta t)\hat{\xi}(t+\tau,\delta t)} = \overline{\hat{\xi}(t-\tau,\delta t)\hat{\xi}(t,\delta t)} = C(-\tau;\delta t). \tag{3.345}$$

This means that this function is not suitable for a test of statistical time-reversal symmetry. This statement holds also for higher correlation functions with regularly spaced points in time:

$$\overline{\hat{\xi}(t,\delta t)\hat{\xi}(t+\tau,\delta t)\hat{\xi}(t+2\tau,\delta t)...\hat{\xi}(t+n\tau,\delta t)}. \tag{3.346}$$

Limiting the analysis of the time series to the calculation of such intrinsically symmetric correlation functions, this corresponds to a loss of information since the price changes $\xi(t,\delta t)$ can be symmetric or asymmetric under time reversal. Another situation appears if the observation times are not equidistant. The simplest correlation function that is not necessarily an invariant quantity under a reversion of the time is

$$\widetilde{C}(\tau;\delta t) = \overline{\hat{\xi}(t,\delta t)\hat{\xi}(t+\tau,\delta t)\hat{\xi}(t+3\tau,\delta t)}. \tag{3.347}$$

The stationarity now leads to

$$\widetilde{C}(-\tau;\delta t) = \overline{\hat{\xi}(t,\delta t)\hat{\xi}(t+2\tau,\delta t)\hat{\xi}(t+3\tau,\delta t)}. \tag{3.348}$$

If the underlying time series is a time-reversal series in a statistical sense, we expect $\widetilde{C}(\tau;\delta t) = \widetilde{C}(-\tau;\delta t)$. Therefore, the correlation function should be a reasonable measure for a time-reversal asymmetry of the time series

$$\begin{aligned} C_{\text{asym}}(\tau;\delta t) &= \widetilde{C}(\tau;\delta t) - \widetilde{C}(-\tau;\delta t) \\ &= \overline{\hat{\xi}(t,\delta t)\left[\hat{\xi}(t+\tau,\delta t) - \hat{\xi}(t+2\tau,\delta t)\right]\hat{\xi}(t+3\tau,\delta t)}. \end{aligned} \tag{3.349}$$

The detection of serious deviations of $C_{\text{asym}}(\tau;\delta t)$ implies the existence of some statistical asymmetry in the time series.

Another suitable measure of statistical time-reversal symmetry is the two-point correlation function of different functions. For instance, all combinations

$$C_{\text{asym}}^{(n,m)}(\tau;\delta t) = \overline{\hat{\xi}^n(t,\delta t)\hat{\xi}^m(t+\tau,\delta t)} - \overline{\hat{\xi}(t,\delta t)^m\hat{\xi}(t+\tau,\delta t)^n} \tag{3.350}$$

with $n \neq m$ are possible measures for testing the time-reversal symmetry. A few applications [14, 325] of such methods have been performed in order to test the symmetry of financial time series and seem to indicate that statistical time asymmetry is present. We document this behavior by an analysis of the cubic correlation function

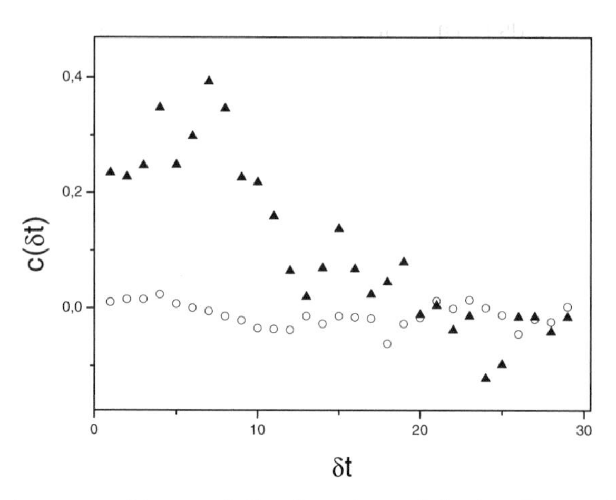

Fig. 3.30. Average cubic correlation functions of the locally trend-corrected logarithmic price fluctuations (circles) and the locally-trend corrected volatility (triangles). Each point is the average over all empirically determined asymmetric cubic correlation functions of all stocks of the German stock index (DAX) for 11/00–07/02.

$$c(\delta t) = \frac{\overline{\eta(t,\delta t)^3 \eta(t+\delta t,\delta t)} - \overline{\eta(t,\delta t)\eta(t+\delta t,\delta t)^3}}{\left(\overline{\eta(t,\delta t)^2}\right)^2}, \tag{3.351}$$

where we choose
(i) $\eta(t) = \xi(t,\delta t)$ (the logarithmic price fluctuation),
(ii) $\eta = |\xi(t,\delta t)|$ (a measure of the volatility),
(iii) $\eta(t) = [2\xi(t,\delta t) - \xi(t+\delta t,\delta t) - \xi(t-\delta t,\delta t)]/3$ (a measure of locally "trend-corrected" price fluctuation), and
(iv) $\eta(t) = [2|\xi(t,\delta t)| - |\xi(t+\delta t,\delta t)| - |\xi(t-\delta t,\delta t)|]/3$ (a measure of locally "trend-corrected" volatility).

In Figure 3.30, we compare the cubic correlation functions of (iii) and (iv) obtained from daily series averaged over all of the stock prices contained in the DAX.

Obviously, there is strong empirical evidence for the existence of a time-reversal asymmetry in the volatility fluctuations, whereas the trend-corrected fluctuations of the prices themselves are more or less symmetric in a statistical sense. This analysis supports the assumption that daily fluctuations of price changes of stocks or assets are approximately independent for a time horizon δt much larger then the Markov horizon $\delta t_{\text{Markov}} \sim 10^1$ trading minutes.

On the other hand, the absence of a time-reversal symmetry in the volatility fluctuations suggests that there might be other, probably long-range-correlated, fundamental economic and financial processes controlling the dynamics of a financial market.

A similar interpretation follows from an inspection of Figure 3.31, which considers possible trends in the price fluctuations. The asymmetry is detected for both the price fluctuations and the absolute price fluctuations. Here, the asymmetry is observed over a timescale of some trading month. Obviously, the fundamental economic processes behind this effect are slow and therefore probably offer a global character. In other words, the existence of a pronounced time arrow in the financial data indicates that the dynamics of financial markets are an irreversible process.

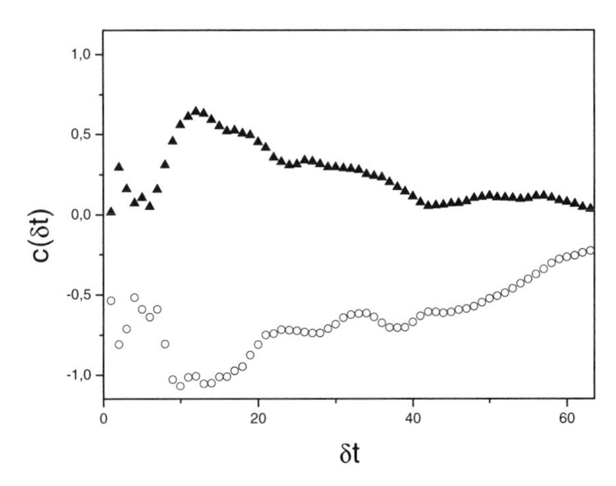

Fig. 3.31. Average cubic correlation functions of the logarithmic price fluctuations (circles) and the volatility (triangles). Each point is the average over all empirically determined asymmetric cubic correlation functions of all stocks of the German stock index (DAX) for 11/00–07/02.

4. Economic Systems

4.1 The Task of Economics

Economics helps us to understand the nature and organization of societies, the arguments underlying many of the great public issues of the day, and the operation and behavior of business firms and other economic decision-making units. To perform effectively and responsibly as citizens, administrators, workers, or consumers, most people need to know some economics. In order to answer the question of whether physical methods and ideas may be helpful to solve economic problems, we must specify the scope of economic science.

Economics is concerned with the way in which limited resources are allocated among alternative uses to satisfy human wants. Usually, economics is divided into two parts: microeconomics and macroeconomics. Microeconomics deals with the economic behavior of individual units such as consumers, firms, and resource owners. Macroeconomics deals with the behavior of large economic aggregates in order to describe such quantities as the gross domestic product and the level of employment.

The basic procedure to answer economic questions quantitatively is the formation of models. An economic model or theory is composed of a number of more or less empirical assumptions from which conclusions and predictions are deduced. Such theories in general strongly simplify the real situation. Although the necessary assumptions and constraints that are made obviously must bear some relationship to the type of situation to which the model is applicable, it is very important to understand that economic models need not be very exact replicas of reality.

Surprisingly, many predictions obtained from economic theories are close to reality. Obviously, the rules controlling economic situations are relatively robust against the individual character of the units forming the underlying economic system. Hence, the majority of economic models offer a deterministic structure. Random processes play no role or only a subordinate one in traditional economics and also in modern economic science. That seems a surprising result at first since the key objects of economics are the markets. As we have seen in the previous chapter, fluctuations dominate the markets, in particular financial markets. In order to understand this apparent contradiction, we should analyze this problem from a physical point of view.

As we have remarked in the first chapter and the Preface to this book, economic systems show a very high degree of complexity. However, from the many physical degrees of freedom, only a small part are considered economically relevant quantities. All other degrees of freedom, including those forming the individual character of economic objects and humans, may be interpreted as irrelevant quantities contributing to apparently stochastic processes affecting the economic system.

We have studied the stochastic character of the relevant variables due to randomness caused by the irrelevant quantities in the framework of financial markets. In fact, besides the fluctuations of an arbitrary price occurs the central tendency of this price. This trend represents mainly the evolution of the underlying economic system. Assuming the simple Gaussian model and a stationary market, the price fluctuations over a certain time period Δt grow as $\Delta t^{1/2}$, while the trend leads to a change of the average price of an order of magnitude Δt. Hence, the asset price fluctuations lose their importance against the trend for sufficiently long timescales.

The scale on which the trend of a price shows significant changes is the characteristic timescale of micro- and macroeconomic processes. Usually, economic observations are carried out for time horizons of months and years. For example, the unemployment rates are published at least monthly and economic data for leading companies are available every three months.

Economic quantities, such as the trend of asset prices, imply of course a random character also at very large timescales. But these effects are apparently not very strongly induced by hidden microscopic degrees of freedom and are probably a consequence of the nonlinear couplings between the relevant economic and social degrees of freedom. In other words, modern economists believe that deterministic economic laws determine mainly the relations between markets, companies, transportation units, employees, and employers. In fact, individual conditioned fluctuations are mostly restricted on short timescales and small geographic regions and particular lines of business. These effects compensate mutually for typical economic scales. Economic indicators, such as unemployment or the inflation rate, show only a slow variability, which is assumed to be the result of a deterministic evolution on the level of the relevant economic degrees of freedom.

As discussed in Chapter 2, complex systems may show pronounced deterministic regimes for a certain level of the relevant degrees of freedom used. The corresponding equations of motion or, more precisely, the kinetic equations, are of the type (2.92). Formally, these equations are the systematic contributions to the Ito stochastic differential equations (2.139), providing also a natural way for the introduction of random processes in economic problems. A central aim of economics is the adaptation of these formal equations to reality by empirical observations and suitable assumptions.

This leads to the question of how the complicated dynamics of economic systems may be described in an appropriate manner. The economically relevant variables form a very complicated system, responsive to various climate, social, or policy influences, so that we should answer the question of why such a system usually shows a relatively stable structure and not a completely chaotic character. Probably, just a few collective modes become dominant and effectively serve the time evolution of the economic system. At the same time, these global variables govern the behavior of all other relevant variables. This synergetic principle [166] may be characteristic of the evolution of economic markets.

To be more precise, we should start with an analysis of the structure of an economic system. For a better understanding, we will not continuously use the traditional economic language; instead we introduce some physically motivated terms.

Each economic system consists of different elements. In principle, we have to distinguish between active and passive elements. Active elements, also denoted as individual units, such as firms or human consumers, use passive elements such as goods, money, labor, or resources and transform these into new passive elements. Each active element has an input and an output. The input of a firm encloses the currents of consumed passive elements, such as intermediate products, labor, energy, and the proceeds for the goods produced, while the output contains goods, wages, rents, taxes, and the cost of resources and intermediate products (Figure 4.1). The inputs of a human consumer are the wages, food, and other goods, while the outputs are the costs for food and consumer articles and the hours worked (Figure 4.2). The active elements can also have positive or negative storage functions, which represent money or product reserves or loans for further investments.

Inputs and outputs are often discrete quantities. However, arguments similar to those we have used in the context of financial time series allow us to introduce continuous values for these economic quantities also. Here, we will use the latter concept. We remark that a discussion of the pros and cons of such a continuation is useless with respect to the large characteristic

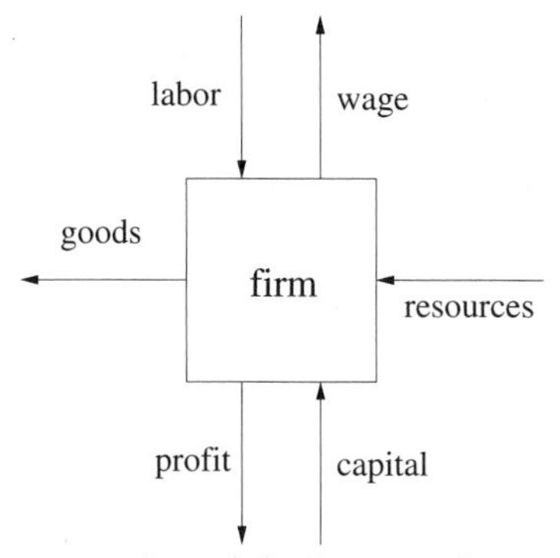

Fig. 4.1. Schematic representation of the input and output of a firm.

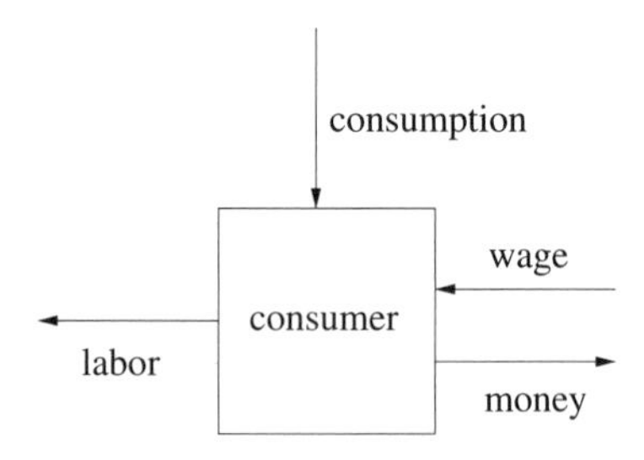

Fig. 4.2. Schematic representation of the economic input and output of a consumer.

timescales of economic observations. Suppose that the active elements are labeled by $I = 1, ..., N$. Then, let us write the output rate of the Ith active element as $\dot{\omega}_{\alpha,I} = \partial\omega_{\alpha,I}/\partial t$, where the type of the passive element is labeled by $\alpha = 1, ...M$ and the input rates are given by $\dot{\nu}_{\alpha,I} = \partial\nu_{\alpha,I}/\partial t$. The central idea of a microeconomic ansatz is that there exists a function

$$\dot{\omega}_{\alpha,I} = f_{\alpha,I}(\{\dot{\nu}_{\beta,I}\}, \{g_{b,I}\}) \tag{4.1}$$

connecting the output rate $\dot{\omega}_{\alpha,I}$ with the inputs [349] and a set $\{g_{b,I}\}$ of control parameters or production coefficients [69, 132]. The activity or production function $f_{\alpha,I}$ is a specific function that is mainly determined by the structure of the firm and the technology applied. The application of production functions relating output to its underlying factor inputs has a long history [126, 142, 158, 199, 273, 350, 382, 406, 415, 425, 426]. The input rates $\dot{\nu}_{\beta,I}$ determine the production scale, while the control parameters $g_{b,I}$ define the actual constitution of the active element. The choice of the control or production parameters $g_{b,I}$ allows firms to change their range of products by internal rearrangements of the operational procedure. In the case of individual humans, the control parameters determine their consumption behavior.

We remark that the $g_{b,I}$ may have both discrete values and continuous values. A discrete value corresponds to the case where a firm has a limited number of production processes. For instance, a small winery produces only one type of wine over a certain time period, another type over the next time period, and so on. The discrete value of the control parameter then defines the type of wine currently produced. A refinery can change the output fraction of gasoline, light oil, and heavy oil continuously in a certain range. Here, we have to deal with continuous control parameters.

Economists believe in the existence of the activity functions $f_{\alpha,I}$, although it is very hard to construct these functions from empirical observations since each measurement means a change of the production regime. Only some universal properties are known. For example, the activity function is a non-negative quantity, $f_{\alpha,I} \geq 0$. If certain inputs vanish, the function $f_{\alpha,I}$ becomes zero. Furthermore, a sufficiently large input leads to saturation effects due to the limited production capacity. As a general problem in real economics,

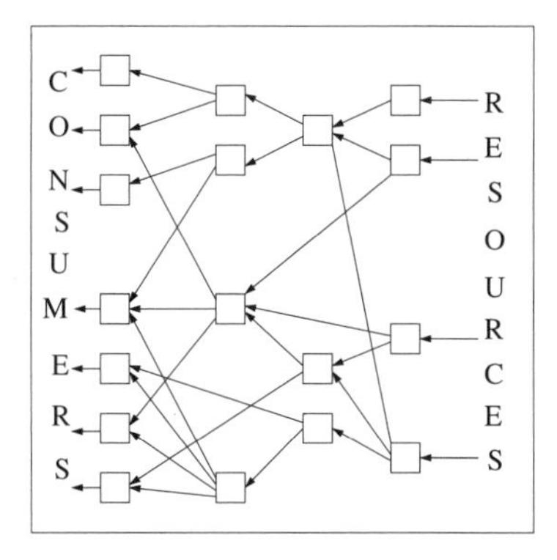

Fig. 4.3. Schematic representation of the flows of goods in a planned economy.

it remains a fact, however, that the functions $f_{\alpha,I}$ are completely unknown apart from some points.

Let us assume for the moment that we know all functions $f_{\alpha,I}$. Then, the question remains how we can determine the control parameters $g_{b,I}$ optimizing the economy. To do this, we have to take into account that economics include the exchange of products. This exchange connects all active elements of the economic system to a closed network. Modern economics know two extreme ways of realizing such an exchange. Note that other possibilities have also taken place over the course of history.

All active elements are connected directly in a planned economy. Thus, the output of one active element is the input of a following active element. Such a system has fixed flows of the passive elements (Figure 4.3), at least during a given time period.

The necessary constraints of such a system are

$$\sum_{I=1}^{N} (\dot{\omega}_{\alpha,I} - \dot{\nu}_{\alpha,I}) = 0; \tag{4.2}$$

that is, the total output with respect to a given class of passive elements is equivalent to the total input of the same passive element. If human wants or the technology are changed, the parameters must be corrected by a central planning commission. In this case, specific goals that are considered necessary for the political stability of the respective state must be reached. Such aims are the allocation of basic foods and special consumer goods, the completion of armaments, the support of the state machinery, housebuilding, or new investments. Formally, these aims can be expressed by inequalities of the type

$$Z_{\alpha}^{\max} \geq \sum_{I=1}^{N} \dot{\omega}_{\alpha,I} \geq Z_{\alpha}^{\min}. \tag{4.3}$$

The task of the commission is now to optimize the economy by defining the flows of the inputs and outputs of all firms and the set of the control parameters considering (4.1) and the restrictions (4.2) and (4.3).

Obviously, such an economic system is in a stationary state (i.e., the flow of goods, money, resources, and labor between the active elements is constant during the whole planning period). But even if the constraints (4.3) are realistically formulated, which is usually not the case in state socialism, planned economics contains considerable dangers. Every damage and every tempest, significant price changes in global markets, or change of human wants can lead to a considerable perturbation of the steady state during a planning period. But the main problem is that any optimization, especially of the whole national economy, degenerates to a rough estimation since the activity functions $f_{\alpha,I}$ are largely unknown.

The periodically realized corrections of the plan usually do not lead to a stabilization of the economic system. Instead, indications of chaotic movement may be observed. Below, we demonstrate such behavior for the case of a monopolist, which has similar properties in principle. It should be remarked that such a planned economy is not only a phenomenon of the last century. Similar control and optimization mechanisms could be observed in the later Roman empire [274], such as the "edictum Diocletiani" in 301 A.D., and during the middle ages [313].

In a market economy, the active elements are connected via the markets (Figure 4.4).

A market controls the exchange of passive elements via the competition between supply and demand. Basically, every kind of passive element can be traded in the markets. There exist financial markets, labor markets, coffee markets, steel markets, and so on. An important role in controlling the market dynamics is played by the market price. This quantity is given by a so-called price function P_α, which defines the average price of a certain amount of a given passive element α. The price function depends mainly on demands and supplies. The supply side is mainly determined by the actual total reserves

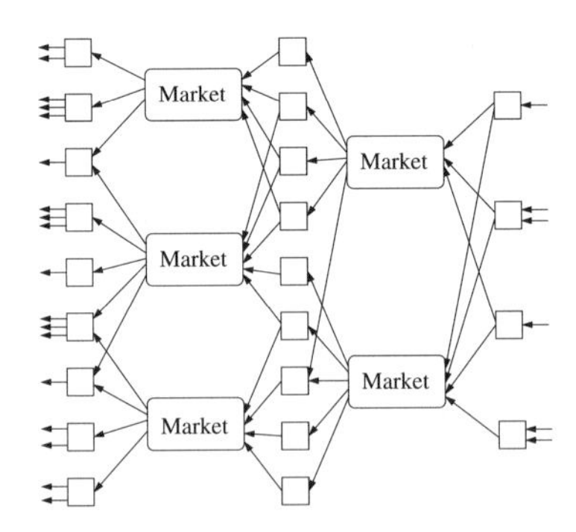

Fig. 4.4. Schematic representation of the flow of goods in a market economy.

$$R_\alpha = \int_{-\infty}^{t} dt' \sum_{I=1}^{N} \left[\dot{\omega}_{\alpha,I}(t') - \dot{\nu}_{\alpha,I}(t')\right] = \sum_{I=1}^{N} \left(\omega_{\alpha,I} - \nu_{\alpha,I}\right) \tag{4.4}$$

of the passive elements. The demand side can be represented by

$$S_\alpha = \Phi_\alpha \left(\{g_{a,I}\}, \{\dot{\nu}_{\alpha,I}\}\right), \tag{4.5}$$

where the demand function Φ_α represents the total demand for passive elements of type α. The demand function depends on the consumption rates (i.e., the inputs) and on the control parameters of the activity functions defining the actual constitution and therefore the consumption behavior of the active elements. Thus, the market price may be obtained from the general relation

$$p_\alpha = P_\alpha (R, S) \tag{4.6}$$

with $R = \{R_1, R_{2,\ldots}\}$ and $S = \{S_1, S_{2,\ldots}\}$. The occurrence of all of the supplies and demands in the price function of a certain type α of passive elements is not unexpected.

Such a dependence considers, for example, the fact that each consumer has the chance to buy alternative goods that fulfill his wants in the same way. Equation (4.6) suggests that the market price is a collective property that is determined by the individual properties of all consumers and sellers.

However, a warning should be made: the market price is not necessarily equal to the actual trading price. The latter is a result of a contract between the consumer and the seller. The market price is here only a guideline for possible negotiations between both parties. Economists explain the influence of the market price on the trading price by the application of different models, such as the monopolist, oligopolist, and polypolist models [137, 289, 373].

Similarly as the activity functions $f_{\alpha,I}$, the price functions P_α are largely unknown. Formally, they depend on the type of objects traded. The market price p_α is usually high for small reserves and increases with decreasing reserves. The importance of the market price consists in the feedback to the active elements of the economic system. High market prices are stimulations for a possible change of the production regime or a change of human wants. Thus, the control parameters $g_{a,I}$ of the activity functions and the input currents depend mainly on the actual market prices

$$g_{a,I} = G_{a,I}(p) \quad \text{and} \quad \dot{\nu}_{\alpha,I} = H_{\alpha,I}(p) \tag{4.7}$$

with $p = \{p_1, p_2, \ldots\}$. It can be expected that, for a given set of market prices p, the response functions $G_{a,I}$ and $H_{\alpha,I}$ define such control parameters $g_{a,I}$ and inputs $\dot{\nu}_{\alpha,I}$ favoring a special production regime of a firm or a special lifestyle of a human. This state may be characterized by the maximum profit in the case of a company or by the maximum private freedom in the case of a human. To be more specific, the maximum principle requires that the function

$$\mathcal{F} = \sum_{\alpha} p_\alpha \left[\dot{\omega}_{\alpha,I} - \dot{\nu}_{\alpha,I}\right] = \sum_{\alpha} p_\alpha \left[f_{I,\alpha}(\{\dot{\nu}_{\beta,I}\}, \{g_{b,I}\}) - \dot{\nu}_{\alpha,I}\right] \tag{4.8}$$

be maximized for each firm. Therefore, we obtain the conditions

$$\frac{\partial}{\partial \dot{\nu}_{\gamma,I}} \sum_{\alpha} p_{\alpha} f_{I,\alpha}(\{\dot{\nu}_{\beta,I}\}, \{g_{b,I}\}) = p_{\gamma} \tag{4.9}$$

and

$$\frac{\partial}{\partial g_{c,I}} \sum_{\alpha} p_{\alpha} f_{I,\alpha}(\{\dot{\nu}_{\beta,I}\}, \{g_{b,I}\}) = 0. \tag{4.10}$$

The solutions of (4.9) and (4.10) with respect to the production coefficients $g_{a,I}$ and the inputs $\dot{\nu}_{\gamma,I}$ define the functions $G_{I,a}$ and $H_{I,a}$. Hence, the response functions are directly obtainable from the activity functions $f_{I,\alpha}$.

The central points of a market economy are the local or, in the language of an economist, decentralized feedback mechanisms (4.7) considering a maximum approach of an active element to its internal aims and the collective interaction of all active elements via the markets.

In contradiction to the planned economy, the market economy is in a dynamical state. The equations of motion (4.1) and the constraints (4.4), (4.5), (4.6) and (4.7) form a complicated nonlinear system of differential equations. There is strong empirical evidence that the more or less open activity, price, supply and response functions are practically very robust relations so that the market economy is able to react to random disturbances and to follow long-term social and technological changes.

We remark that a real economic system is a mixed capitalism. Although decentralized decision-making based on the price system is used to organize production and consumption in most areas of a real economy, there are notable exceptions. For example, in the acquisition of new weapons by the department of defense, the market economy has not been applied. Instead, the government exercises control over sellers through the auditing of costs and through the intimate involvement of its agents in the managerial and operating structure of the sellers. Furthermore, the government decides what weapons are to be created and often decides how they are to be created and produced.

Up to now, we have roughly discussed the general concept of economy. In fact, the mathematical formulation of an economic system is much more complicated than we have shown here. An essential task of economics is the empirical estimation or theoretical determination of the open functions $f_{I,\alpha}$, Φ_{α}, and P_{α}. This problem is the content of microeconomics, which describes the individual behavior of the active elements of economic systems in a quantitative manner. One can find detailed lectures about suitable techniques and possible models in several economics textbooks [214, 258, 331, 344].

The general starting point is similar to a molecular field approach used for the description of physical many-body systems. Instead of the investigation of the whole system, one analyzes only a particular active element in its environment consisting of various average markets. The contribution of other competing elements to these markets is equivalent to the contribution of the element considered.

The analysis of the evolution of the whole economic system is the task of macroeconomics [18, 64, 121, 372]. The behavior of single elements here plays only a secondary role. In principle, the concept of macroeconomics implies a further reduction of the relevant degrees of freedom to a set of a few quantities. It may be assumed that these few, probably collective quantities dominate the time evolution of the whole economic system. Typically, macroeconomic theories show some similarities to physical field theories.

Below, we represent some ideas and concepts that illustrate the possible contribution of modern physical concepts and findings to economic problems.

4.2 Microeconomics

4.2.1 The Polypolist and Stability Analysis

A very popular case for microeconomic investigation is a set of small firms, each of them with only a small share of the market. Such a firm reacts relatively quickly to a change in the market. Because the firm shares the market with many other similar firms, it is called a polypolist. Such firms are typical candidates for a self-consistent description of market dynamics.

In order to describe the behavior of such a polypolist, let us collect the outputs of the particular firm into a vector $\omega = (\omega_1, ...\omega_N)$. Furthermore, we assume stable resources markets and a stable labor market so that every possible production regime may be adjustable. That is particularly the case if the set of small firms investigated needs only a small part of the resources supplied.

The product currents are given by $\dot{\omega}_\alpha = f_\alpha(\{\dot{\nu}_\beta\}, \{g_b\})$. Note that we have neglected the index I because we are interested in the behavior of one firm. The market price of the products follows from the current demand and supply. The supply side corresponds to the market reserve $R_\alpha = N_0\omega_\alpha - \nu_\alpha$, which considers the production of all contributing firms. The simple factor N_0 reflects the "molecular field" argument that all firms of the polypolist model show an equal behavior. Hence, the total output of the product α is $N_0\omega_\alpha$ instead of the cumulative output ω_α of one firm.

The present consumtion rate may be a constant quantity so that we can write $\nu_\alpha \approx N_0 u_\alpha t$ and consequently $R_\alpha = N_0 (\omega_\alpha - u_\alpha t)$. The market price is a function of the demand side given by the set of demand parameters S and of the supply side defined by the present reserve

$$p_\alpha = P_\alpha (R, S) = P_\alpha (\omega - ut, S) \tag{4.11}$$

with $u = (u_1, ...u_N)$. The market price determines the internal control parameters g_b of the firm and the inputs $\dot{\nu}_\beta$; see (4.7). Thus, we obtain the relations

$$g_b = G_b(p) = \widetilde{G}_b(\omega - ut, S) \tag{4.12}$$

and

$$\dot{\nu}_\beta = H_\beta(p) = \widetilde{H}_\beta(\omega - ut, S). \tag{4.13}$$

Hence, the output is described by the evolution equation

$$\frac{\partial \omega}{\partial t} = f\left(\left\{\widetilde{G}(\omega - ut, S)\right\}, \left\{\widetilde{H}(\omega - ut, S)\right\}\right) = F(\omega - ut, S), \tag{4.14}$$

where f, F, $\widetilde{G}$, and $\widetilde{H}$ are vector functions of the components f_α, F_α, $\widetilde{G}_a$, and $\widetilde{H}_\alpha$. Introducing the new vector $x = \omega - ut$, we get

$$\frac{\partial x}{\partial t} = -u + F(x, S). \tag{4.15}$$

Usually, this equation approaches a stationary state so that the output of the firm is given by the solution of the steady-state equation

$$F(x_0, S) = u. \tag{4.16}$$

This equation defines the equilibrium structure of the market. The dynamic equation (4.15) becomes important for the relaxation of the market into its stationary state after an external perturbation or after a change of the demand parameters S.

The treatment of such an ordinary differential equation is always possible by standard numerical methods. We will ask whether we can derive some general properties. Because the perturbations are typically small, the analysis is restricted to the nearest environment of a stationary state x_0. We assume that the Taylor expansion of $F(x_0, S)$ with respect to the variable $y = x - x_0$ exists. Then, we can write (4.15) in the form

$$\frac{\partial y}{\partial t} = Ay + \psi^{(r)}(y). \tag{4.17}$$

The rest function is given by $\psi^{(r)}(y) = F(x_0 + y, S) - u - Ay$, where the matrix A has the components

$$A_{\alpha\beta} = \left.\frac{\partial F_\alpha(x, S)}{\partial x_\beta}\right|_{x=x_0}. \tag{4.18}$$

The leading term of the function $\psi^{(r)}(y)$ is of an order of magnitude $|y|^r$ with $r \geq 2$. Let us now introduce a transformation $z = y + h(y)$, where h is a vector polynomial with the leading order 2 so that $h(0) = \partial h/\partial y|_{y=0} = 0$. Thus, we obtain

$$\begin{aligned}
\frac{\partial z}{\partial t} &= \frac{\partial z}{\partial y}\frac{\partial y}{\partial t} = \left(1 + \frac{\partial h}{\partial y}\right)\frac{\partial y}{\partial t} = \left(1 + \frac{\partial h}{\partial y}\right)[Ay + \psi(y)] \\
&= Ay + \frac{\partial h}{\partial y}Ay + \psi(y) + \frac{\partial h}{\partial y}\psi(y) \\
&= Az - \left[Ah(y) - \frac{\partial h(y)}{\partial y}Ay - \psi(y)\right] + \frac{\partial h}{\partial y}\psi(y).
\end{aligned} \tag{4.19}$$

We determine the open function h by setting

$$\hat{L}_A h = Ah - \frac{\partial h}{\partial y} Ay = \psi^{(r)}. \tag{4.20}$$

This equation has a unique solution if the eigenvalues of the introduced operator $\hat{L}_A$ are nonresonant. To understand this statement, we consider that the matrix A has the set of eigenvalues $\lambda = \{\lambda_1, ..., \lambda_N\}$ and the normalized eigenvectors $\{e_1, ..., e_N\}$. Then, the vector y can be expressed in terms of $y = \eta_1 e_1 + ... + \eta_N e_N$. The eigenvectors of $\hat{L}_A$ are the vector monomials

$$\varphi_{m,\gamma} = \eta_1^{m_1} ... \eta_N^{m_N} e_\gamma \tag{4.21}$$

with $m = \{m_1, ..., m_N\}$. The quantities m_α are nonnegative integers satisfying $m_1 + ... + m_N \geq 2$. Note that $\hat{L}_A$ acts in the space of functions that have an asymptotic behavior $h \sim |y|^r$ with $r \geq 2$ for $|y| \to 0$. We remark that $A\varphi_{m,\gamma} = \lambda_\gamma \varphi_{m,\gamma}$ and

$$\sum_{\alpha,\beta} \frac{\partial \varphi_{m,\gamma}}{\partial \eta_\alpha} A_{\alpha\beta} \eta_\beta = \sum_{\beta} \frac{\partial \varphi_{m,\gamma}}{\partial \eta_\beta} \lambda_\beta \eta_\beta = (m, \lambda)\, \varphi_{m,\gamma}, \tag{4.22}$$

where (m, λ) is the Euclidean scalar product between the vectors m and λ. Thus, we find

$$\hat{L}_A \varphi_{m,\gamma} = -\left[(m, \lambda) - \lambda_\gamma\right] \varphi_{m,\gamma}; \tag{4.23}$$

that is the operator $\hat{L}_A$ has the eigenvalues $(m, \lambda) - \lambda_\gamma$. If all eigenvalues of $\hat{L}_A$ have nonzero values, the equation (4.20) has a unique solution that requires $(m, \lambda) \neq \lambda_\gamma$. Otherwise, we have a so-called resonance $\lambda_\gamma = (m, \lambda)$, and $\hat{L}_A$ is not reversible.

Suppose that no resonances exist. Then, the solution of (4.20) defines the transformation function $h(y)$ so that

$$\frac{\partial z}{\partial t} = Az + \frac{\partial h}{\partial y} \psi^{(r)}(y). \tag{4.24}$$

Comparing the order of the leading terms of h and $\psi^{(r)}$, we find that the product $\psi^{(r)}(y)\, \partial h / \partial y$ is at least of an order $r + 1$ in $|y|$. Considering the transformation between z and y, we arrive at

$$\frac{\partial z}{\partial t} = Az + \psi^{(r+1)}(z), \tag{4.25}$$

where $\psi^{(r+1)}(z)$ is a nonlinear contribution with a leading term proportional to $|z|^{r+1}$. The repeated application of this formalism generates an increasing order of the leading term.

In other words, the nonlinear differential equation approaches step-by-step a linear differential equation. This is the content of the famous theorem of Poincaré [15]. In the case of resonant eigenvalues, the Poincaré theorem must be extended to the theorem of Poincaré and Dulaque [15]. Here, we get instead of (4.25) the differential equation

$$\frac{\partial z}{\partial t} = Az + w(z) + \psi^{(r+1)}(z), \tag{4.26}$$

where $w(z)$ contains the resonant monomials. The convergence of this procedure depends on the structure of the eigenvalue spectra of the matrix A. If the convex cover of all eigenvalues $\lambda_1, ..., \lambda_N$ in the complex plane does not contain the origin, the vector $\lambda = \{\lambda_1, ..., \lambda_N\}$ is an element of the so-called Poincaré region of the corresponding $2N$-dimensional complex space. Otherwise, the vector is an element of the Siegel region [374].

If λ is an element of the Poincaré region, the procedure discussed above is convergent and the differential equation (4.17) or (4.15) can be mapped formally onto a linear differential equation for nonresonant eigenvalues or onto the canonical form (4.26). If λ is an element of the Siegel region, the convergence cannot be guaranteed.

The Poincaré theorem allows a powerful analysis of the stability of systems of differential equations that goes beyond the standard method of linear approximation. In particular, this theorem can be a helpful tool in classifying the microeconomic task discussed above and many other related problems.

We have mentioned that the economic functions, such as the activity function or the response function, are largely unknown. However, some properties are obtainable from empirical investigations. For instance, the coefficients of the matrix (4.18) are connected with the so-called economic elasticity coefficients [39]. Therefore, it seems reasonable to extract as much information as possible about the stability of the underlying firm's strategy from this matrix and a few other estimations.

In the case of a single product, the variable x is a one-dimensional quantity, and only one eigenvalue, $\lambda = A$, exists. Then, the stationary production x_0 (see (4.16)), corresponds to a stable state for $\lambda < 0$ and to an unstable state for $\lambda > 0$. Special investigations considering the leading term of the nonlinear part of (4.17) are necessary for $\lambda = 0$.

Another situation occurs for a firm with two product types. Here, we have two eigenvalues, λ_1 and λ_2. If resonances are excluded, the largest real part of the eigenvalues determines the stability or instability of the state closed to the reduced output x_0. A resonance exists if $\lambda_1 = m_1\lambda_1 + m_2\lambda_2$ or $\lambda_2 = m_1\lambda_1 + m_2\lambda_2$, where m_1 and m_2 are nonnegative integers. In this case, we expect a nonlinear normal form (4.26) containing the resonant monomials.

Let us illustrate the formalism by using a very simple example. The eigenvalues $\lambda_1 = -\lambda_2 = i\Omega$, obtained from the linear stability analysis, are usually identified with a periodic motion of the frequency Ω. But this case contains two resonances, namely $\lambda_1 = 2\lambda_1 + \lambda_2$ and $\lambda_2 = \lambda_1 + 2\lambda_2$. Thus, the stationarity of the evolution of the corresponding nonlinear system of differential equations (4.17) is no longer determined by the simple linear system[1] $\dot{\eta}_1 = i\Omega\eta_1$ and $\dot{\eta}_2 = -i\Omega\eta_2$ but by the normal system

$$\dot{\eta}_1 = i\Omega\eta_1 + c_1\eta_1^2\eta_2 \quad \text{and} \quad \dot{\eta}_2 = -i\Omega\eta_2 - c_2\eta_1\eta_2^2 \,. \tag{4.27}$$

The substitutions $x_1 = \eta_1 + i\eta_2$ and $x_2 = i(\eta_1 - i\eta_2)$ and the agreement $x^2 = x_1^2 + x_2^2$ lead to the real normal form

$$\dot{x}_1 = \Omega x_2 + \frac{x^2}{4}\left[x_1 \text{Im } c - x_2 \text{Re } c\right] \tag{4.28}$$

and

$$\dot{x}_2 = -\Omega x_1 + \frac{x^2}{4}\left[x_1 \text{Re } c + x_2 \text{Im } c\right], \tag{4.29}$$

where the real structure of the differential equations requires $c_1 = c$ and $c_2 = \bar{c}$.

Such a structure is already expected after the first step of the Poincaré algorithm applied onto (4.17). Only the two parameters Re c and Im c are still open. All other nonlinear terms disappear step-by-step during the repeated application of the reduction formalism. However, it is not necessary to execute these steps because the resonance terms remain unchanged after their appearance. The stability behavior follows directly from the dynamics of x^2. We obtain from (4.29)

$$\frac{\partial x^2}{\partial t} = \frac{\text{Im } c}{2} x^4. \tag{4.30}$$

Thus, the system is stable for Im $c < 0$ and unstable for Im $c > 0$. Obviously, we need only an estimation about the sign of the quantity Im c. This may be obtained from several economic investigations or empirical observations of appropriate companies. However, the determination of such quantities is a natural task for economists.

In many cases, microeconomic problems lead to a stability analysis or a related technique to find the steady state. Such a usually temporary stationarity is practically obtained by the firm under consideration by the use of empirical observations of the supply and the demand and by suitable response steps in order to stabilize production at a sufficiently high level.

Obviously, the attempts to undersell the actual market prices, social influences due to labor disputes, and the fluctuating costs of resources generate permanent fluctuations around such a stable point. Finally, the change of technology, long-term political decisions, or social evolution are reasons that a microeconomic state stabilizes only for a finite time. After this time, the economic system converges to a completely new steady state.

[1] The linear system is written in the standard form considering the representation in terms of the eigenvectors of the matrix A.

4.2.2 The Monopolist and Chaotic Behavior

The polypolist model characterizes one possible market type. Another one is the monopolist model. In the polypolist model, we have perfect competition of individual small firms so that a certain steady state occurs. Each of the small firms cannot noticeably influence the market price on its own. They just note the current prices and demands and react accordingly with respect to their supply. Only the supply of all of the numerous firms of the polypolist model together becomes a force on the market strong enough to determine the price in a balance with the demand side.

The single monopolist, on the other hand, is assumed to deliberately choose to limit the quantity supplied so as to keep the price sufficiently high to yield monopoly profit. Therefore, the monopolist should know the complete relation between the prices and the demands in order to determine the adequate price relative to the demand. However, it is more likely that a monopolist just knows a few points on the economic functions recently visited in its more or less random search for maximum profit.

In the case of a monopoly model, the price function (4.6) reduces to $p_\alpha = P_\alpha(S)$ since the monopolist completely controls the market reserves in such a way that $R = \text{const}$. In the economics literature, our reduced price function is also called the demand function.

Let us now analyze the case of one product and one kind of demand. Thus, we have to deal with the scalar function $p = P(S)$. An essential difference between the monopolist and the polypolist models is the interpretation of this formula. In the polypolist model, the price p is a real market price that is controlled by the demand. In the monopolist model, the price is not a real market price and the monopolist controls the demand via the market price.

The general form of the demand function is assumed to be downward-sloping (i.e., the demand decreases as the price increases). In textbooks, the demand curve is just a straight line, $p = p_0(1 - S/S_0)$, but this is usually too simplistic [332].

The total revenue $\zeta(S) = pS$ defines the total market value of the products that are necessary for the fulfillment of consumer wants. The derivation with respect to the demand is the marginal revenue $\mu(S) = d\zeta(S)/dS = \zeta'(S)$.

Another important quantity characterizing the effectiveness of a firm is given by the costs of production. The total costs of production depend also on the demand, $\vartheta(S)$. The corresponding marginal costs are then given by $c(S) = \vartheta'(S)$. This function typically has a minimum for a certain demand S.

The total profit is the difference between the total revenue and the total costs $\pi(S) = \zeta(S) - \vartheta(S)$. The maximum profit corresponds to $\pi'(S) = 0$; that is, it requires the balance of marginal costs and marginal revenue $c(S) = \mu(S)$.

This equation may have various solutions. In theory, the monopolist would calculate these by solving the balance equation and then evaluating the second-order derivatives $\pi''(S)$ to check which solutions correspond to a local maximum point. Finally, the monopolist would choose between these local maxima to identify the global profit maximum.

However, in reality, the monopolist does not know more than a few points on the demand and cost functions. Thus, it would be the task of the monopolist to design a search algorithm for the maximum of the unknown profit function. Let us assume that the monopolist knows at a given time t_n the value of the present marginal revenue μ_n, the value of the marginal costs c_n, and the value of the demand S_n. Furthermore, it is clear that $\mu_n - c_n < 0$ (i.e., $\pi'(S_n) < 0$) indicates that a reduction of S_n leads to an approach to the next maximum. Consequently, the monopoly changes its price policy in order to obtain the new demand $S_{n+1} < S_n$. On the other hand, for $\mu_n - c_n > 0$, the monopolist sets a price level corresponding to the new demand $S_{n+1} > S_n$. The difference between S_{n+1} and S_n is simply chosen to be

$$S_{n+1} = S_n + \delta\left(\mu_n - c_n\right), \tag{4.31}$$

where δ is an empirical constant [332]. If the marginal revenue and costs are unknown, the monopolist must compare the measurable total profit $\pi = \zeta - \vartheta$ at the last and the next to last points in time. Then, due to $\pi' = \mu - c$, we get instead of (4.31) the monopoly equation [319]

$$S_{n+1} = S_n + \delta \frac{\pi_n - \pi_{n-1}}{S_n - S_{n-1}}. \tag{4.32}$$

Let us now discuss the consequences of the monopoly strategy. We assume that the monopolist uses the algorithm (4.31). Then, the demand is given by the recurrence equation

$$S_{n+1} = S_n + \delta\left[\mu(S_n) - c(S_n)\right]. \tag{4.33}$$

For the sake of simplicity, we assume further that both the marginal revenues $\mu(S)$ and the marginal costs $c(S)$ are truncated Taylor series up to the second order. Thus, we can write

$$S_{n+1} = \alpha_0 + \alpha_1 S_n + \alpha_2 S_n^2 \tag{4.34}$$

with the specific coefficients α_n $(n = 0, ..., 2)$. The transformation $S_n = r(x_n - b)/\alpha_2$ with $r = 2b - \alpha_1$, where the parameter b is a solution of $b^2 + (1 - \alpha_1)b + \alpha_0\alpha_2 = 0$, leads to the representation

$$x_{n+1} = r x_n \left(1 - x_n\right) = \phi_{\log}(x_n), \tag{4.35}$$

which is well-known as a logistic map. This model was introduced by Verhulst in 1845 to simulate the growth of a population in a closed area. Another applications related to economic problems is used to explain the growth of a deposit under progressive rates of interest [304].

As found by several authors [79, 118, 160, 260], the iterates x_n $(n = 1, 2, ...)$ display, as a function of the parameter r, a rather complicated behavior that becomes chaotic at large r.

The chaotic behavior is not tied to the special form of the logistic map (4.35). Thus, the following results are also valid for functions of marginal revenues $\mu(S)$ and marginal costs $c(S)$ other than the truncated Taylor expansions and for the second monopolist strategy (4.32); see [319]. In particular, the transition from a regular (but not necessarily simple) behavior to the chaotic regime during the change of an appropriate control parameter is a universal behavior for all first-order difference equations, $x_{n+1} = f(x_n)$, in which the function f has only a single maximum in the properly rescaled unit interval $0 \leq x_n \leq 1$.

It should be remarked that other difference equations with chaotic properties, including several types of second-order difference equations $x_{n+1} = f(x_n, x_{n-1})$ such as (4.32), may belong to other universality classes. However, most of the properties that are valid for the logistic map (4.35) and the corresponding universality class of first-order difference equations hold at least qualitatively for other difference equations also.

Let us briefly discuss the main properties of the logistic map. The logistic map $x_n \to x_{n+1}$ has two fixed points, $x^\star = 0$ and $x^\star = 1 - r^{-1}$, satisfying $x^\star = \phi_{\log}(x^\star)$. The stability analysis considers weak perturbations of the fixed points, $x_n = x^\star + \varepsilon_n$. Thus, we obtain $\varepsilon_{n+1} = \Lambda\varepsilon_n + o(\varepsilon_n)$, where the so-called multiplier Λ is given by $\Lambda = \phi'_{\log}(x^\star)$. Obviously, for $|\Lambda| < 1$, the fixed point is linearly stable. Conversely, if $|\Lambda| > 1$, the fixed point is unstable. The stability of the marginal case $|\Lambda| = 1$ cannot be decided in the framework of the linear stability analysis. For small control parameters, $r < 1$, the quantity x_n develops toward the stable fixed point $x^\star = 0$ because $\phi'_{\log}(0) = r < 1$. For $1 < r < 3$, we get the multiplier $\phi'_{\log}(1 - r^{-1}) = 2 - r$ so that here the fixed point $x^\star = 1 - r^{-1}$ becomes stable.

Both fixed points are unstable for $r > 3$. Now, we observe a stable oscillation of period 2 (see Figure 4.5) with the two alternating values $\overline{x}_1^\star$ and $\overline{x}_2^\star$, which together are related via the equations $\overline{x}_1^\star = \phi_{\log}(\overline{x}_2^\star)$ and $\overline{x}_2^\star = \phi_{\log}(\overline{x}_1^\star)$. Of course, both values $\overline{x}_1^\star$ and $\overline{x}_2^\star$ are stable fixed points of the second-iterate map $\phi_{\log}^{(2)}(x) = \phi_{\log}(\phi_{\log}(x))$. In fact, we obtain $\overline{x}_{1/2}^\star = \left[r + 1 \pm \sqrt{(r-3)(r+1)}\right]/2r$, and the corresponding multiplier of the 2-cycle, $\Lambda = 4 + 2r - r^2$, satisfies $|\Lambda| < 1$ for $3 < r < 1 + \sqrt{6}$.

Thus, the unique asymptotic solution $x_n = 1 - r^{-1}$ for $n \to \infty$ splits into two alternating solutions, $\overline{x}_1^\star$ and $\overline{x}_2^\star$. At $r = 3$, the values of $\overline{x}_1^\star$ and $\overline{x}_2^\star$ coincide and equal $x^\star = 1 - r^{-1} = 2/3$, which shows that the 2-cycle bifurcates continuously from $x^\star$. This bifurcation is sometimes called a pitchfork bifurcation.

Above $r = 1 + \sqrt{6}$, the 2-cycle splits into a 4-cycle. Further period doublings to cycles of period 8, 16, ..., 2^m,... occur as r increases. The values r_m, where the number of fixed points changes from 2^{m-1} to 2^m, scale as $r_m = r_\infty - C\delta^{-m}$. Here, C and r_∞ are specific parameters ($r_\infty = 3.56994...$ for the logistic map), while the Feigenbaum constant $\delta = 4.6692...$ is a universal quantity [118, 367, 391]. All of the cycles can be interpreted as attractors of a finite set of points.

For $r > r_\infty$, the asymptotic behavior of the series $x_1, ..., x_n, ...$ becomes unpredictable. More precisely, we must say that, for many values of $r > r_\infty$, the sequence never settles down to a fixed point or a periodic cycle. Instead, the asymptotic behavior is aperiodic and therefore chaotic. The corresponding attractor changes from a finite to an infinite set of points.

However, the region for $r > r_\infty$ shows a surprising mixture of periodic p-cycles ($p = 3, 5, 6, ...$) and the true chaos. The periodic cycles occur in small r-windows among other windows with chaotic behavior and also show successive bifurcations p, $2p$, $...2^n p$,... . The corresponding r-values scale as the above-mentioned Feigenbaum law except that the nonuniversal constants C and r_∞ are different.

Furthermore, periodic triplings $3^n p$, quadruplings $4^n p$, and higher bifurcations occur at $r'_n = r'_\infty - C'\overline{\delta}^{-n}$ with different nonuniversal constants r'_∞ and C' and different Feigenbaum constants, which are again universal for the type of bifurcation (e.g., $\overline{\delta} = 55.247...$ for the tripling).

After this mathematical excursion, let us come back to the monopolist problem. As demonstrated for the logistic map, a monopolist can behave in a regular, periodic, or chaotic way (see Figure 4.5). The regular way corresponds to traditional monopoly theory, which we can find in several economics textbooks.

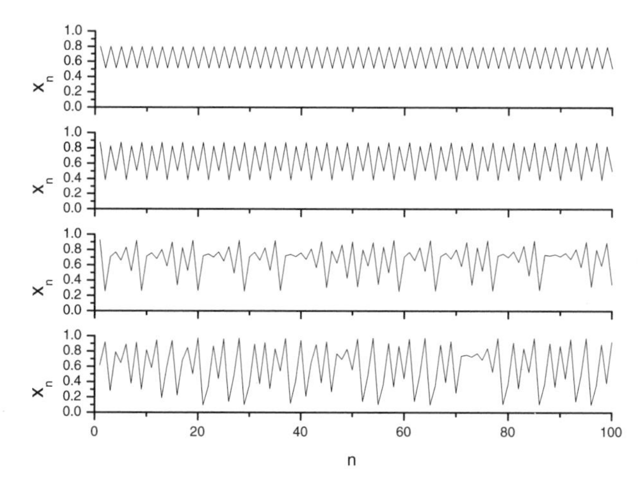

Fig. 4.5. Various dynamic regimes of the discrete logistic map recursion law: periodic oscillations with period length 2, $r = 3.2$ (top), periodic oscillations with period length 4, $r = 3.5$ (top center), chaotic behavior with $r = 3.7$ (bottom center), and chaotic behavior with $r = 3.9$ (bottom).

But the constitution of the monopoly and its interaction with the market also allow also marginal costs and marginal revenue functions corresponding to the periodic or chaotic regimes. Then the monopolist searches for the maximum profit periodically or very likely randomly.

It may be a surprise for traditional economists that the monopolist does not know all about the market and may even behave in a way that seems unpredictable. The price policy of phone monopolies or mail monopolies testifies to this theoretical result, although these monopolies could always attribute any irrational behavior to a pursuit of public benefit.

4.3 Thermodynamics and Economy

4.3.1 Thermodynamic Concepts

There is no theoretical and empirical evidence that the active elements of an economic system develop in a completely deterministic manner. Besides the self-organized chaotic regimes discussed above, there also exist a lot of perturbations due to the hidden irrelevant degrees of freedom. It is practically impossible to consider all degrees of freedom leading to an economic decision of a particular management, even if this decision is rational from the point of view of the economists involved.

This individual random character can be observed, for example, in the fluctuations of the trading prices with respect to the market price. Other sources of random processes are climate fluctuations, natural disasters, changes of public opinion, or unexpected social conflicts. Finally, the coupling between the active elements and the markets and several feedback effects lead to a self-organization of chaotic motions, as shown below.

Such a randomness at the microeconomic level suggests that the whole economic network and its dynamic evolution may be discussed in terms of classical thermodynamics. Such ideas require some essential prerequisites. First, the economic system must be in a stationary state. This condition can neither be demonstrated nor rejected for an economic system because we are not able to repeat the evolution of the same economic system under the same boundary conditions. For example, we cannot decide whether an economic trend is a long-range fluctuation or a deterministic effect, so we may or may not believe whether an economic system is in a steady state.

Furthermore, classical thermodynamics requires the existence of independent subsystems interacting with a bath and the existence of some additive quantities. These quantities may be collected in a vector $\Omega = (\omega_1, \omega_2, ...)$. In the framework of classical thermodynamics, the components of Ω represent the energy, the number of particles, or the momentum. All of these macroscopic quantities satisfy certain conservation laws if the corresponding system is closed.

All other internal degrees of freedom are interpreted as microscopic quantities without any information about their concrete configuration. The probability that a certain system is in the macroscopic state Ω may be $P(\Omega)$. Then, the combination of two subsystems requires $\Omega = \Omega_1 + \Omega_2$ due to the assumed additivity and $P(\Omega) = P(\Omega_1) P(\Omega_2)$ due to independency. These two conditions lead to the general Boltzmann representation

$$P(\Omega) \sim \exp\{-\beta\Omega\}, \tag{4.36}$$

where β is a constant vector of the same dimensionality as Ω. Indeed, the independence of the subsystems requires that boundary effects be neglected. Otherwise, we must write $P(\Omega) = P(\Omega_1) P(\Omega_2 \mid \Omega_1)$, and the underlying dynamics of one subsystem are strongly affected by the dynamics of the other one. The Boltzmann law (4.36) defines the contact between the statistical interpretation of a physical many-body system in equilibrium and classical thermodynamics.

In order to qualify an analogy between economic systems and physical equilibrium systems, one should be very careful. Our understanding of complex systems of many strongly coupled degrees of freedom is far from approaching that of systems at thermodynamic equilibrium. This holds true also for the coupled system of relevant economic degrees of freedom as discussed above. Although the separated irrelevant degrees occur at best as external random noise, the dynamics of the remaining economic network consisting of active elements and flows of passive elements are not comparable with the dynamics of a mechanical system.

To find a successful way of introducing general thermodynamics of economic systems, we have to define especially economic quantities that have the meaning of a global energy. These quantities must be functions of all microeconomic variables, and all economic configurations with the same "energy" should be equivalent from a macroeconomic point of view. Unfortunately, besides some not very serious speculations, there is no general indication that such quantities exist.

Even if we could find such a quantity, we would not expect the Boltzmann statistics to hold for the whole system of the economic network. The problem is the strong interaction of all of the active elements via the markets and the product flows. This interaction mimics an effectively long-range coupling of all participants in the economic system.

As a consequence, an economic system cannot be partitioned without a sensitive change of its dynamics and organization. A more favorable situation for the introduction of a thermodynamic concept probably existed in the middle ages or in classical antiquity, when small economic subsystems such as villages or towns were largely self-sufficient.

The temporary isolation of such a subsystem would only marginally change the economic dynamics of both the subsystem and the remaining total system. But this possible "thermodynamic" era is not comparable with modern global economics.

A partition of our economic system seems to be impossible without serious changes of the dynamics of the amputated subsystem. We believe also that the generalization of such a concept by the so-called nonextensive statistics [1, 86, 402, 403, 404] does not provide a better description of economic systems. Finally, we remark also that the introduction of thermodynamic concepts for comparatively simple nonequilibrium systems that show various kinds of self-organization (sandpile models [342, 343], turbulence [70], granular media [270, 348]), the introduction of temperature, and related thermodynamic equilibrium quantities is connected with considerable problems.

4.3.2 Consumer Behavior

The general introduction of a thermodynamic concept describing stationary economic systems seems to be excluded by the physical arguments discussed above. Nevertheless, we can apply thermodynamic ideas and relations to simulate certain aspects of the dynamics of economic systems in an artificial way. The apparently thermodynamic quantities are in reality certain control parameters that pretend to have a thermodynamic meaning. However, these quantities are not obtainable by a closed thermodynamic theory.

As a simple example, let us study the behavior of consumers who may choose between two similar kinds of products. The products are identical with respect to their quality and presentation so that both products are demanded with the same frequency. Intuitively, it is obvious that consumer behavior is of a collective nature.

To be more precise, we assume that each of the human consumers informs other consumers about the merits and the problems of the product that was purchased. An obvious relevant quantity characterizing the actual state of the consumers is the number of individuals $N_1 = \eta$ favoring the product $\alpha = 1$. The number of consumers favoring the product $\alpha = 2$ is simply $N_2 = N - \eta$ because the total number N of consumers is a constant.

As a result of the communication, a consumer can change his or her behavior and switch to the opposite product. The evolution of consumer demands is a random process that may be determined by the conditional probability $p(N_1, N_2, t \mid N_1', N_2', 0)$, which we can write also as $p(\eta, t \mid \eta', 0)$. The time transition rates $W_{\eta\eta'}$ are considered to represent only one-step processes. Then, the transition rates read

$$W_{\eta\eta'} = w^+(\eta')\delta_{\eta\eta'+1} + w^-(\eta')\delta_{\eta\eta'-1} \tag{4.37}$$

so that $W_{\eta\eta'}dt$ is the probability of a change $\eta \to \eta \pm 1$ during the infinitesimally short time period dt. The corresponding master equation is

$$\begin{aligned}\frac{\partial p(\eta, t; \eta_0, 0)}{\partial t} &= w^-(\eta+1)p(\eta+1, t \mid \eta_0, 0) \\ &+ w^+(\eta-1)p(\eta-1, t \mid \eta_0, 0) \\ &- \left[w^+(\eta) + w^-(\eta)\right] p(\eta, t \mid \eta_0, 0) .\end{aligned} \tag{4.38}$$

The first problem that we have to solve is the determination of the open functions $w^+(\eta)$ and $w^-(\eta)$. The rate $w^+(\eta)$ defines a jump $\eta \to \eta+1$ (i.e., a consumer changes from product 2 to product 1). The number N_2 before the jump was $N-\eta$ (i.e., there were $N-\eta$ possibilities for a jump). Hence, we find that $w^+(\eta)$ has a combinatorial prefactor $N-\eta$.

On the other hand, the rate $w^-(\eta)$ describes a jump $\eta \to \eta - 1$ so that $w^-(\eta)$ has the prefactor η. In this sense, $w^+(\eta)/(N-\eta)$ and $w^-(\eta)/\eta$ may be interpreted as the rate of change of the behavior of an individual. These rates are enhanced by the group of humans with opposite demands and diminished by humans of one's own group.

Let us further assume that the willingness for a change from the consumption of one product to the other product is controlled by a consumption temperature T that facilitates the changes. A high temperature suppresses the influence of other consumers on a possible demand change of a certain human, while a low temperature stresses the collective behavior (i.e., a consumer favors the majority).

Furthermore, we allow external influences, such as good publicity for one of the two alternative products. Then, we may write

$$w^+(\eta) = w_0 (N-\eta) \exp\left\{-\frac{N-2\eta}{2TN} - \frac{h}{T}\right\} \tag{4.39}$$

and

$$w^-(\eta) = w_0 \eta \exp\left\{-\frac{2\eta - N}{2TN} + \frac{h}{T}\right\}, \tag{4.40}$$

where h/T represents the preference of a product by external effects.

The functional structure of the transition rates may be much more complicated in reality. Our simple rates are mainly influenced by classical approaches to the lattice Ising model [228], where the transition rates of a given spin are energetically influenced by its neighborhood and possible external fields. In our special case, each consumer is in contact with all other consumers. In the real world, the number of contacts is limited. But we get the same transition rates also for this case in the framework of a mean-field theory, which works very well for a large number of neighbor contacts [228].

Furthermore, we remark that a similar model is used to explain the change of opinions in a human society [419, 420]. The master equation (4.38) is an ergodic one so that the probability distribution reaches a stationary state after a sufficiently long relaxation time. We can write the equation for the stationary solution

$$p_{\text{stat}}(\eta) = \lim_{t\to\infty} p(\eta, t; \eta_0, 0) \tag{4.41}$$

as

$$0 = K(\eta+1) - K(\eta) \tag{4.42}$$

with

$$K(\eta) = w^-(\eta)p_{\text{stat}}(\eta) - w^+(\eta - 1)p_{\text{stat}}(\eta - 1). \tag{4.43}$$

We now consider that η is a nonnegative integer. In fact, the master equation (4.38) and the transition rates (4.39) and (4.40) are organized in such a manner that a jump from $\eta = 0$ to $\eta < 0$ does not take place. Hence, we have $p_{\text{stat}}(-1) = 0$ and therefore $K(0) = 0$. We sum (4.42) and obtain

$$0 = \sum_{m=0}^{\eta-1} [K(m+1) - K(m)] = K(\eta) - K(0) = K(\eta). \tag{4.44}$$

Using (4.43), we arrive at the recurrence equation

$$p_{\text{stat}}(\eta) = \frac{w^+(\eta - 1)}{w^-(\eta)} p_{\text{stat}}(\eta - 1). \tag{4.45}$$

Finally, the repeated application of this relation leads to

$$p_{\text{stat}}(\eta) = \prod_{m=1}^{\eta} \frac{w^+(m-1)}{w^-(m)} p_{\text{stat}}(0). \tag{4.46}$$

The substitution of the transition rates by the use of (4.39) and (4.40) leads to the explicit form of the stationary probability distribution

$$p_{\text{stat}}(\eta) = \binom{N}{\eta} \exp\left\{-\eta\left(\frac{N - \eta - 2}{TN} + \frac{2h}{T}\right)\right\} p_{\text{stat}}(0), \tag{4.47}$$

where $p_{\text{stat}}(0)$ is obtainable from the normalization condition.

Figure 4.6 shows the probability distribution function without external influence. As we expect from our knowledge of the Ising model, there are two

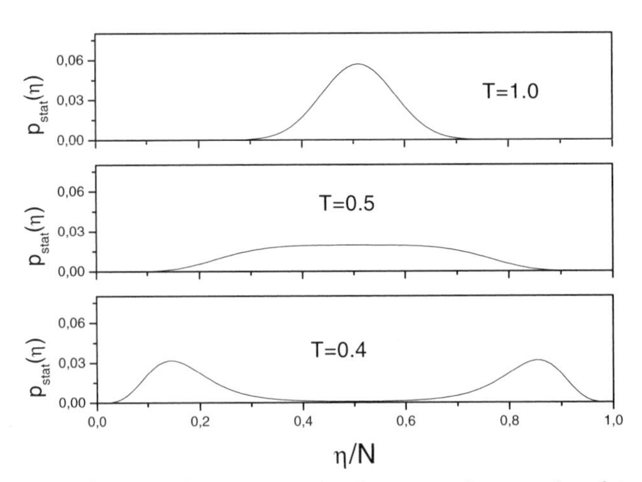

Fig. 4.6. Distribution of consumers in the case of a supply of two indifferent products without external influence $h = 0$ for different consumption temperatures.

regimes. The high-temperature regime is characterized by frequent changes of the individual behavior around the average $\overline{\eta} = N/2$. In other words, the groups favoring the first or the second product have nearly the same size.

If the consumption temperature is lowered, two pronounced configurations are possible. Most consumers favor either product 1 or product 2. Between the high and low temperature regimes, a phase transition occurs. The critical temperature $T_c = 1/2$ corresponds to the first appearance of an inflection point in the probability distribution function $p_{\text{stat}}(\eta)$. The phase transition describes, at least in a qualitative manner, the polarization of the set of consumers due to a collective communication.

As is known from the mean-field solution of the Ising model or the corresponding Ginzburg–Landau theory [223], consumer behavior becomes unstable close to the critical value. Not quite above the critical temperature, we observe strong fluctuations (i.e., large groups of consumers are formed and dissolved only slowly). Below the critical consumption temperature, one group wins, and the difference between the average group occupation number is given by $\left|\overline{N}_2 - \overline{N}_1\right| \sim (T - T_c)^{1/2}$. This is a kind of spontaneous symmetry breaking favoring one product after a sufficiently long time.

The presence of an external factor shifts the center of the probability distribution function (Figure 4.7). The importance of a finite value of $|h|$ is visible close to the critical temperature. Especially for $T = T_c$, we expect a behavior $\left|\overline{N}_2 - \overline{N}_1\right| \sim |h|^{1/3}$. In other words, a small external influence may essentially change consumer behavior. This is also the reason why the appearance of new products on the market is supported by special introductory offers.

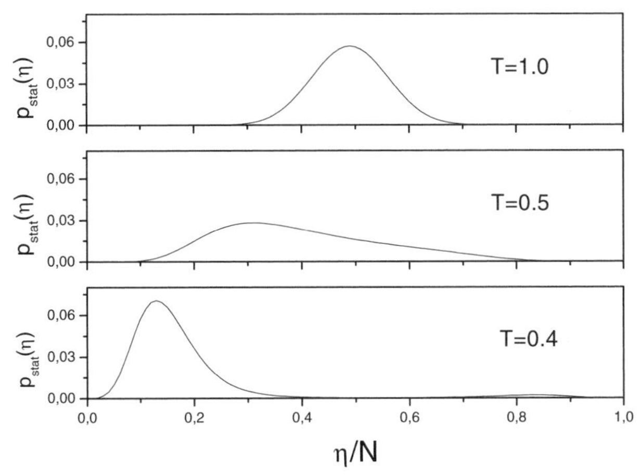

Fig. 4.7. Distribution of consumers in the case of a supply of two indifferent products under a weak external influence $h = 0.02$.

4.3.3 Thermodynamics and Environmental Economics

As discussed above, it seems not very fruitful to use thermodynamical concepts for an explanation of the dynamics of the organization of economic systems. However, the application of physical thermodynamics is always possible in order to solve special economic problems. This is especially the case for the so-called environmental economics. This discipline focuses on the economics of pollution and environmental quality [369]. In a formal way, this discipline can be described as the scientific study of economic systems in relation to their natural, physical, or residential surroundings.

In consequence, environmental economics is frequently involved in the unintended effects of human decisions on the environment [33, 87, 98, 106]. Obviously, environmental economics is an important link between economic decision-making and the solution of the environmental dilemma in order to find a continuous balance between economic development and environmental quality.

A way to obtain insight into the relationship between economic and technological activities and ecological effects is the use of physically and technologically motivated balance and transport equations [21, 89, 130, 293, 411] and the application of thermodynamic relations. Obviously, energy and materials are absorbed, transformed, and thrown out to the next step by each active element of the economic system apart from possible heat sinks and unusable waste.

The characteristic feature of all of these activities is the physical balance between inputs and outputs in the transformation processes (Figure 4.8). From an economic point of view, each transformation process leads to a qualitative or economic difference of the material in question. On the other hand, besides the economic and monetary differences there is a further essential physical difference: matter and energy enter the transformation process in a state of low entropy and come out of it in a state of high entropy. This is the second law of thermodynamics now applied to economics: the irreversibility of real transformation processes requires the application of thermodynamic process inequalities.

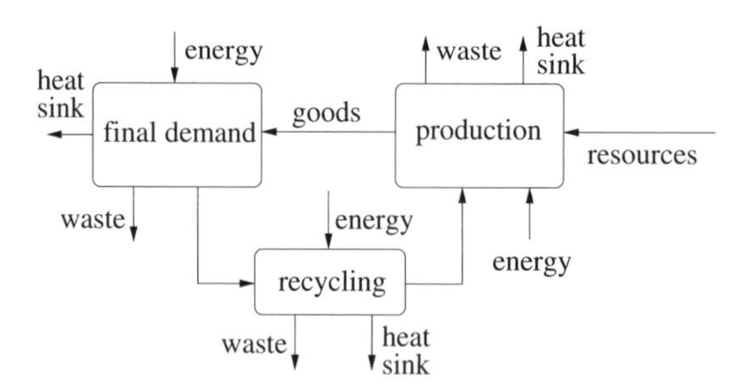

Fig. 4.8. A simple material, entropy, and energy-balance model

In principle, such an economic system is comparable with the set of processes in complex chemical companies. Similar to that in process engineering, the consideration of balance and transport equations and the application of thermodynamic inequalities allow us to estimate the production level and the emission of heat and pollutants into the surroundings in the framework of environmental economics.

There is, in particular, an important hierarchy: simple machines may be combined step-by-step in higher-production plants, companies, and aggregates. Because of their extensive character, inputs, outputs, and entropy production are additive quantities so that the thermodynamic laws and the material balance hold at each level of this hierarchy. This additivity allows a successful investigation of complex economic systems on the basis of relatively simple models. In principle, the combination of environmental economics with climatology, solar science, and ecology should give serious quantitative results about the evolution of our natural environment.

The application of classical thermodynamic relations to environmental economic problems is known as the entropy concept in the literature [90, 149, 150, 288]. We remark again that this concept is the application of thermodynamic relations in order to quantify the flows of products, energy, and entropy for a given economic system on the level of process engineering. But it is not an attempt to explain the structure and the development of economic systems in terms of thermodynamics.

4.4 Macroeconomics

4.4.1 Models and Measurements

Macroeconomics describes the general behavior of large economic systems. This economic discipline overlaps with the content of other terms such as national economics or political economics. These branches of science have a very long tradition [32, 51, 94, 172, 328, 427]. Macroeconomically relevant quantities are often denoted as economic or socioeconomic indicators.

Macroeconomic theory knows an enormous set of various models explaining the relations between these indicators, such as the relation between inflation and the unemployment rate or between the degree of technology and the production rate. It is not the aim of this book to discuss these partially contradictory models represented by different economic schools and traditions for different economic and political periods. These problems are studied in the comprehensive literature [18, 64, 121, 315, 372, 388, 410].

Here, we will discuss some general questions related to the formation of reasonable models. Furthermore, we will give a suggestion of how physical concepts may contribute to the creation of macroeconomic models. In principle, similar problems occur also in social science that are strongly connected with macroeconomic theory.

As a minimal requirement for an appropriate model, one may consider its consistency on logical grounds and on empirical experience. The main problem is that a model should be compatible with the conditions accompanying the kind of measurement. This constraint is particularly important since the attribution of numerical values to economic or social quantities is often not an obvious task.

In order to make the situation more clear, let us analyze the following problem. Suppose that we have a model that allows the derivation of the quantity z from primary quantities x and y via $z = f(x, y)$. This relation may be the mathematical kernel of a model following from a series of economic or social investigations.

If the model does not fix the origin and the units of x and y, we could also use $x' = \Omega x + \omega$ and $y' = \Theta y + \vartheta$ with $\Omega > 0$ and $\Theta > 0$ instead of x and y. This apparently arbitrary change is no rarity in economics. Every country uses different statistical methods to define the unemployment, inflation, and gross national product. Other quantities, such as economic development or the level of technology, have an even higher degree of inaccuracy.

The underlying model provides only a procedure for deriving other economic or social quantities from the original data (x, y) and (x', y'). In our case, the same model yields $z = f(x, y)$ and $z' = f(x', y')$ dependent on the definition of the scale. The model has a general meaning if the quantities derived are comparable independent of the scales used [212, 279]. In other words, a function f defines an invariant relationship if the inequality $z_1 < z_2$ implies the inequality $z_1' < z_2'$.

Unfortunately, such an invariance condition is generally impossible. In order to prove this hypothesis, we consider two points, (x_1, y_1) and (x_2, y_2), with $z_1 = z_2$. Then, the difference

$$F = F(x_1, y_1, x_2, y_2) = f(x_1, y_1) - f(x_2, y_2) \tag{4.48}$$

vanishes. After an infinitesimal change of the scales, $x' = (1 + \delta\Omega)\, x + \delta\omega$ and $y' = (1 + \delta\Theta) y + \delta\vartheta$, we arrive at

$$\begin{aligned} F' = &\frac{\partial f(x_1, y_1)}{\partial x_1}[x_1 \delta\Omega + \delta\omega] + \frac{\partial f(x_1, y_1)}{\partial y_1}[y_1 \delta\Theta + \delta\vartheta] \\ &- \frac{\partial f(x_2, y_2)}{\partial x_2}[x_2 \delta\Omega + \delta\omega] - \frac{\partial f(x_2, y_2)}{\partial y_2}[y_2 \delta\Theta + \delta\vartheta]. \end{aligned} \tag{4.49}$$

The function $F' = F(x_1', y_1', x_2', y_2')$ may now be positive, negative, or zero. If F' has a positive (negative) value, we get a negative (positive) value after a change of the sign of the infinitesimal transformation parameters $\delta\Omega \to -\delta\Omega$, $\delta\omega \to -\delta\omega$, $\delta\Theta \to -\delta\Theta$, and $\delta\vartheta \to -\delta\vartheta$. In other words, if $F' \neq 0$, the invariance condition fails.

It remains the case that $F(x_1', y_1', x_2', y_2') = 0$. Because the infinitesimal transformation parameters have arbitrary values, the four equations

$$x_1 \frac{\partial f_1}{\partial x_1} = x_2 \frac{\partial f_2}{\partial x_2}, \qquad \frac{\partial f_1}{\partial x_1} = \frac{\partial f_2}{\partial x_2}, \tag{4.50}$$

and

$$y_1 \frac{\partial f_1}{\partial y_1} = y_2 \frac{\partial f_2}{\partial y_2}, \qquad \frac{\partial f_1}{\partial y_1} = \frac{\partial f_2}{\partial y_2} \tag{4.51}$$

with $f_k = f(x_k, y_k)$ must be fulfilled for all pairs (x_1, y_1) and (x_2, y_2) that satisfy the equation $f_1 = f_2$.

These equations are simply fulfilled for $f =$ const. or if f is a strictly monotonous function of only one independent variable, such as $f = f(x)$. The latter case requires only the solution $x_1 = x_2$, and the equations (4.50) are identically fulfilled, whereas (4.51) is a trivial relation.

But each function f with $\partial f / \partial x \neq 0$ and $\partial f / \partial y \neq 0$, with the exception of some isolated points, does not represent an invariant relationship. A solution exists only for special transformations. For example, $\omega = \vartheta = 0$ leads to $f(x, y) = x^{2n+1} / y^{2m+1}$ with arbitrary integers m and n, while $\Omega = \Theta = 1$ requires $f = \exp\{cx - dy\}$ with arbitrary constants c and d.

The failure of the general invariance condition is also denoted as the impossibility theorem. It was a central point of a long debate about several socioeconomic relations established by Huntington [181]. These relationships, for example the ratio

$$\text{social frustration} = \frac{\text{social mobilization}}{\text{economic development}}, \tag{4.52}$$

were criticized by Lang [222] and Koblitz [204, 205, 206] and defended by Simon [375, 376, 377]. The problem is that any of the terms involved in (4.52) could be measured by numerical values, provided that the origin and the units of the measurement are defined.

Without a definition of the scales, the algebraic ratio is completely meaningless. There exist various interpretations of Huntington's formulas. For instance, we have to interpret the ratio in the sense that the derivatives are positive for variables in the numerator and negative for variables in the denominator or that the ordering structure behind this relation has an essentially lexicographic structure [213].

The usual way to find an economic law is a combination of statistical analysis and a model hypothesis. Although the dynamics of the economic system described by a suitable set of relevant economic variables are mainly determined by deterministic equations, the behavior of economic quantities may show a random behavior; see subsection 4.2.2.

This deterministic chaos can be observed also at the macroeconomic level. Nevertheless, characteristic correlations are detected between several economic indicators. Suppose that the scale of the quantities under consideration is well-defined. Then, the correlations can be expressed by quantitative relationships.

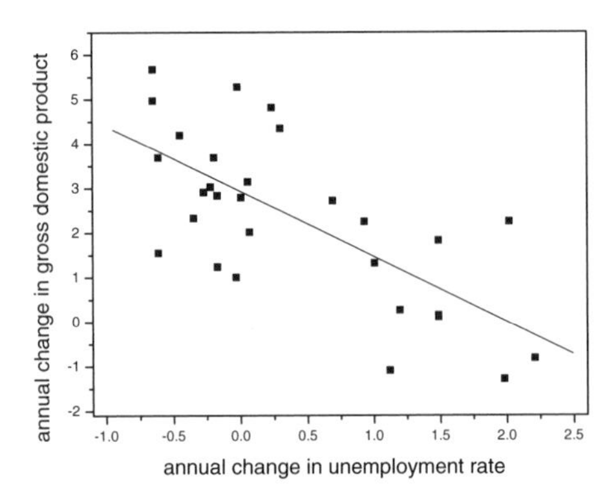

Fig. 4.9. Okun's law for the German national economy, data from http://www.sachverstaendigenrat-wirtschaft.de.

The typical macroeconomic method may be explained by an analysis of Okun's law. The American economist Arthur Okun [290] detected this economic principle in 1962. Okun's law states that to avoid the waste of unemployment, an economy must continually expand and that economic growth and unemployment are related such that decreases in unemployment increase productivity, while increasing unemployment leads to declining productivity [295].

The widespread acceptance of Okun's law is so pervasive that public officials use it as a basic rule for the determination of socioeconomic policies. Figure 4.9 shows a scatterplot of the annual change in the German gross domestic product versus the annual change in the German unemployment rate and a linear regression verifying Okun's law.

4.4.2 Scaling Laws and Scale Invariance

The transition from the microeconomic level to the macroeconomic level takes place in a similar way as in classical thermodynamics, where microscopic degrees of freedom are projected onto few macroscopic degrees of freedom. As we have pointed out, the similarities are not too strong. However, there is empirical evidence that at least some economic processes in large economic systems are comparable with phenomena close to the phase transition in physical systems. We remark especially that the occurrence of strong fluctuations at the macroeconomic level is an indication of such a comparison.

Because of the enormous complexity of an economic network, it is very complicated to analyze the complete dynamics of a large economic system, with the exception of strongly idealized models. Nevertheless, it can be assumed, but not proved directly, that the dynamics of an economic network are controlled by a kind of self-organized criticality [25, 26, 27] showing various types of self-similar structures [30, 115, 247, 413].

Self-organized criticality means that the respective system drives itself to a stable or metastable critical regime that is normally characterized by long-range correlations and scale-free power laws. Possible self-similar properties of large economic systems may occur due to political, geographical, or social circumstances.

For example, the worldwide economic system decays into national economic systems, and each of those consists of regional economic clusters. Such structures suggest a kind of self-similarity, leading to a possible scale invariance of macroeconomic properties. Roughly speaking, scale invariance means reproducing itself on different scales of observation [100, 247]. The latter imply spatial and temporal scales but also production scales, consumption scales, or cost scales in the case of economic systems.

The observation of scale-invariant economic laws depends on the problem and the available set of data. Obviously, once the quantity that we wish to study has been clearly defined, we have to answer the question of how this quantity changes with the scale of observation. A quantity U that depends on the variables $\{x_1, x_2, ..., x_n\}$ is scale-invariant under a change

$$\{x_1, x_2, ..., x_n\} \to \left\{\frac{x_1}{\lambda^{\alpha_1}}, \frac{x_2}{\lambda^{\alpha_2}}, ..., \frac{x_n}{\lambda^{\alpha_n}}\right\} \tag{4.53}$$

with-well defined exponents α_k $(k = 1, ..., n)$ if

$$U(x_1, x_2, ..., x_n) = h(\lambda) U\left(\frac{x_1}{\lambda^{\alpha_1}}, \frac{x_2}{\lambda^{\alpha_2}}, ..., \frac{x_n}{\lambda^{\alpha_n}}\right). \tag{4.54}$$

The choice $\lambda^{\alpha_g} = x_g$ with $g \in (1, ..., n)$ connects λ and x_g. Thus, we get

$$U(x_1, x_2, ..., x_n) = H_g(x_g) U\left(\frac{x_1}{x_g^{\alpha_1/\alpha_g}}, ..., 1, ... \frac{x_n}{x_g^{\alpha_n/\alpha_g}}\right) \tag{4.55}$$

with $H_g(x_g) = h(x_g^{1/\alpha_g})$. For the special case $x_k = 0$ for all $k \neq g$, the generalized homogeneity relation (4.54) yields

$$U(0, ..., x_g, ..., 0) = h(\lambda) U\left(0, ..., \frac{x_g}{\lambda^{\alpha_g}}, ..., 0\right). \tag{4.56}$$

The solution of this equation is simply

$$U(0, ..., x_g, ..., 0) = C_g x_g^{\beta_g}, \tag{4.57}$$

and we obtain $h(\lambda) = \lambda^{\alpha_g \beta_g}$ so that $H_g(x_g) = x_g^{\beta_g}$. These relations are the fundamental properties that associate power laws with scale invariance, self-similarity, and self-organized criticality.

A typical economic power law is the Cobb–Douglas production function [78, 240, 292, 346, 347, 409]. This function has both a macroeconomic and a microeconomic interpretation. The microeconomic Cobb–Douglas function connects input and output rates of an active element of an economic network by simple algebraic relations. Hence, (4.1) can be written as

$$\dot{\omega}_{\alpha,I} = f_{\alpha,I}(\{\dot{\nu}_{\beta,I}\}, \{g_{b,I}\}) = C_{\alpha,I}(\{g_{b,I}\}) \prod_{\beta=1}^{M} [\dot{\nu}_{\beta,I}]^{c^I_{\alpha\beta}} \tag{4.58}$$

with fixed individual exponents $c^I_{\alpha\beta}$. This power law implies the important property

$$d \ln \dot{\omega}_{\alpha,I} = \sum_{\beta=1}^{M} c^I_{\alpha\beta} d \ln \dot{\nu}_{\beta,I}; \tag{4.59}$$

that is, the percentage change of a given input rate is proportional to the percentage change of the output rate.

The macroeconomic version of a Cobb–Douglas function defines the average technological level of an economic system. A frequently used representation[2] defines the production rate of goods [2, 140, 285, 318]

$$\dot{\omega} = CK^{\alpha} L^{\beta} M^{\gamma} E^{\delta} \tag{4.60}$$

for every economy as a function of capital services K, labor services L, land services M, and energy consumed E. However, there is an important difference between the microeconomic and the macroeconomic versions of the Cobb–Douglas law.

The latter is an observable law representing the average production and consumption processes that take place in a large economic network. On the other hand, the microeconomic Cobb–Douglas function is at most a nice model that describes an idealized firm or an idealized consumer.

Another kind of scaling behavior may be observed with respect to the Phillips curve [307]. This curve postulates an empirical relationship between inflation and unemployment. The basic concept is that as an economy approaches full employment there is upward pressure on wages that increases costs and thus prices. In addition, more people working implies more demand for goods and more upward pressure on prices.

This seems to offer policy-makers a simple choice: they have to accept either inflation or unemployment. The Phillips curve, however, began to break down in the late 1960s and early 1970s. Today, the curve is a complicated trajectory with serious differences for different countries (Figure 4.10). We may ask whether this "motion" offers universal properties that are valid for all national economies. To this end, let us assume that the inflation rate I depends statistically on the unemployment rate u via the conditional probability distribution function $p(I \mid u)$. Furthermore, we make the hypothesis that this probability is controlled by a scaling law that may be written as

$$p(I \mid u) = \theta u^{\beta} g\left(\theta I u^{\beta}\right) \tag{4.61}$$

---

[2] There exist a lot of similar formulas in the literature that are also called Cobb–Douglas functions. The common property is that output and input variables are connected by a power law.

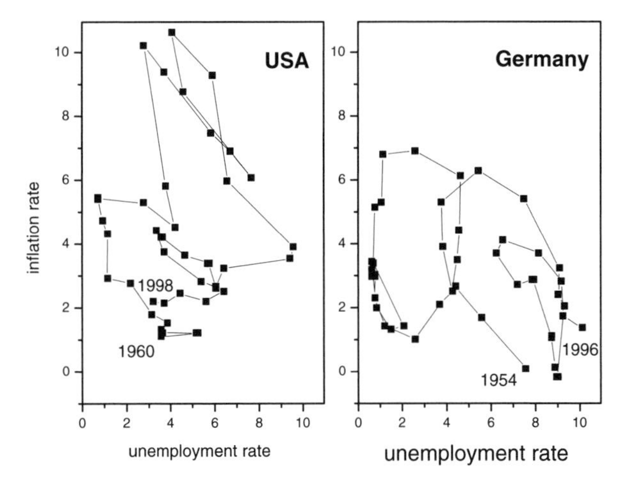

Fig. 4.10. The Phillips curve for the US national economy and for the German national economy.

with the universal function g, the universal exponent β, and a nonuniversal factor θ, which compensates for possible differences in the scales of u and I. Then, we obtain the probability

$$P\left(I < Cu^{-\beta}\right) = \int_{-\infty}^{Cu^{-\beta}} p(I \mid u)dI = \int_{-\infty}^{C\theta} g(z)\, dz = \Phi(C\theta). \tag{4.62}$$

Because of the normalization, we require $\Phi(\infty) = 1$, whereas $\Phi(-\infty) = 0$. Equation (4.62) may be written also as

$$P\left(\theta I u^{\beta} < H\right) = P_{<}(H) = \Phi(H). \tag{4.63}$$

This result and consequently the scaling hypothesis may be tested by application of the rank-ordering statistics. As we have seen in subsection 3.6.1, the likely value of the nth element of a set $S = \left\{\xi^{(1)}, \xi^{(2)}, ..., \xi^{(N)}\right\}$ of N observations ordered by increasing values is given by $P_{<}\left(\xi^{(n)}\right) = n/N$; see equation (3.188). Hence, the likely rank of the value $\xi^{(n)} = I_{(n)} u_{(n)}^{\beta}$ of a given observation $\left(I_{(n)}, u_{(n)}\right)$ is determined by $N\Phi\left(\theta I u^{\beta}\right)$. Suppose that we compare different economic systems. Each of these systems –for example, the US national economy or the German national economy– has its own series S.

If our scaling hypothesis is correct, the rank-ordered sets of all national economies should collapse to one common curve after rescaling the values $\xi^{(n)}$ of a given economic system with one (national) factor θ. In fact, using the exponent $\beta = 0.25 \pm 0.05$, a suitable collapse is observable (Figure 4.11). The differences for the region of extreme values are probably caused by the

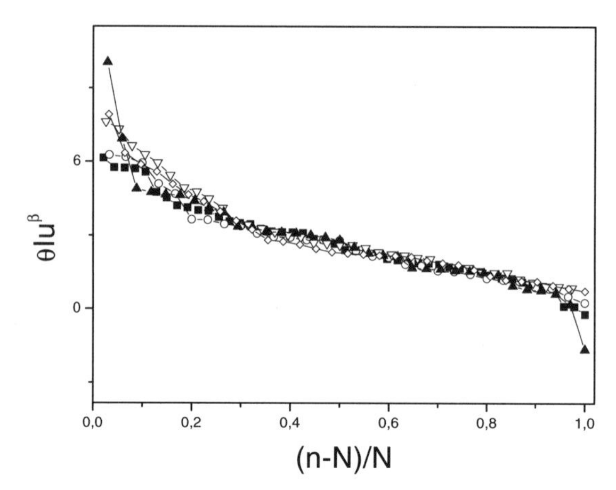

Fig. 4.11. Collapse of the rank-ordered data sets obtained from the Phillips curves of Germany (filled squares), Switzerland (open circles), US (open down triangles), Australia (filled up triangles), and United Kingdom (open diamonds).

relatively small number of observations per series, $N \leq 50$.

Obviously, the scaling law (4.61) and the validity of a universal behavior observed for various economies suggest the existence of universal macroeconomic processes behind the strange dynamics of the national Phillips curves.

4.4.3 Economic Field Theories

A natural way to handle an economic system at large scales consists in embedding economic networks into a two-dimensional geographic space and subsequently a continuous description. Such a procedure reduces the detailed microeconomic relations to several economic and social fields satisfying several local and global conservation laws.

Various publications [13, 40, 42, 316, 318, 320] deal with such macroeconomic mean-field concepts, which may called economic field theories. Other names, such as urban fields, continuous flow models, or spatial economy are also popular.

Historically, the first problems of this kind were analyzed by Thünen in 1826 [397]. He found that agricultural production was distributed among a set of concentric rings around the central town (i.e., a singular consumption region) according to the cost of transportation. Heavy or bulky goods, such as wood for energy production and the building trade, were produced closer to the city, while goods more easily transportable were produced farther away. This special theory is also known as the land use model.

Other early publications [227, 418], so-called location theories, consider the location of production plants instead of consumers. The standard theory of spatial economics was developed in the early 1950s [40, 317] using Euler–Langrange variation principles and hydrodynamic concepts. As such, the model is extremely elegant and versatile, being able to represent all previously known continuous models as special cases. Here, we will give a simple example [320] in order to demonstrate the basic ideas and the relation to physical field theories.

In a two-dimensional space, the trade flow can be represented by a vector field $\mathbf{v}(\mathbf{x}, t) = \{v_1(\mathbf{x}, t), v_2(\mathbf{x}, t)\}$ and $\mathbf{x} = \{x_1, x_2\}$. The absolute value of the trade flow, $|\mathbf{v}(\mathbf{x}, t)|$, represents the quantity of goods traded, whereas the unit direction field $\mathbf{n} = \mathbf{v}/|\mathbf{v}|$ defines the local direction of the flow. Similar to how the flow of a liquid satisfies various balance equations controlling the local conservation of mass, momentum, or energy, the flow of traded commodities is defined by a continuous equation

$$\frac{\partial c(\mathbf{x}, t)}{\partial t} + \mathrm{div}\ \mathbf{v}(\mathbf{x}, t) = q(\mathbf{x}, t). \tag{4.64}$$

The source term $q(\mathbf{x}, t)$ is related to the local excess supply of production over consumption. Positive values of $q(\mathbf{x}, t)$ correspond to local sources of commodities due to a production center, while negative values of $q(\mathbf{x}, t)$ represent a local excess of consumers. The concentration $c(\mathbf{x}, t)$ defines the local stock on hand. In general, this quantity depends on the local number of traded goods. Hence, we expect a relation of the type

$$c(\mathbf{x}, t) = c(|\mathbf{v}(\mathbf{x}, t)|). \tag{4.65}$$

For example, a possible assumption is the power law $c = c_0 |\mathbf{v}(\mathbf{x}, t)|^{\beta}$ with the reserve exponent β. In particular, if we assume an economic system without stockkeeping, we have to set $c(\mathbf{x}, t) = 0$. In this particular case, we arrive at the balance equation

$$\mathrm{div}\ \mathbf{v}(\mathbf{x}, t) = q(\mathbf{x}, t). \tag{4.66}$$

This relation may also be interpreted as a condition for quasistationary interregional trade, which has the same meaning as the quasistationary regime in electrodynamics. Quasistationarity means that the whole system follows a change of the boundary conditions and possible external disturbances without any relaxation. This is possible for slow fluctuations so that the local stocks on hand remain approximately constant.

Fast fluctuations must be compensated by a change of the capacities of the commodities in storage (i.e., we have to deal with the more general balance equation (4.64)). The integral form of (4.64) leads to the global conservation law

$$Q - \frac{\partial C}{\partial t} = \int_G \mathrm{div}\ \mathbf{v}(\mathbf{x}, t)\ dA = \oint_{\partial G} \mathbf{v}(\mathbf{x}, t)\ ds \tag{4.67}$$

with

$$Q = \int_G q\ dA \qquad \text{and} \qquad C = \int_G c\ dA. \tag{4.68}$$

Here, $d\mathbf{s}$ is the integral element of the boundary ∂G of the region G. Obviously, the curve integral on the right-hand side of (4.67) is the export from or, depending on sign, the import to the region G through the borderline ∂G. The quantity Q is the excess supply of production over consumption for the whole region G, while C is the total reserve of the traded commodities. In the case of a stationary state (4.66), we obtain instead of (4.67) the relation

$$Q = \oint_{\partial G} \mathbf{v}(\mathbf{x}, t)\, d\mathbf{s}. \tag{4.69}$$

In order to determine the trade flow, we need a further equation. This equation follows from the economic principle of minimum transportation costs. We assume a transportation cost field $\kappa(\mathbf{x})$, which is determined by the local state of the roads and the structure of the ground. Then, the total transportation costs $K(t)$ at a given time t are given by [40, 41, 42]

$$K(t) = \int_G |\mathbf{v}(\mathbf{x}, t)|\, \kappa(\mathbf{x})\, dA, \tag{4.70}$$

and the total costs over a given time period T are

$$\widetilde{K} = \int K(t) dt. \tag{4.71}$$

We carry out the minimization procedure by application of the Euler–Lagrange variation principle considering the constraint (4.67). Thus, we get the economic action

$$S = \int_G \mathcal{L} dA dt \tag{4.72}$$

with

$$\mathcal{L} = |\mathbf{v}(\mathbf{x}, t)|\, \kappa(\mathbf{x}) + \lambda(\mathbf{x}, t) \left[\frac{\partial c(|\mathbf{v}(\mathbf{x}, t)|)}{\partial t} + \operatorname{div} \mathbf{v}(\mathbf{x}, t) - q(\mathbf{x}, t) \right], \tag{4.73}$$

where $\lambda(\mathbf{x}, t)$ is a Lagrange multiplier associated with the constraint (4.67). The variation with respect to $\mathbf{v}(\mathbf{x}, t)$ yields

$$\left[\kappa(\mathbf{x}) - c'(|\mathbf{v}(\mathbf{x}, t)|) \frac{\partial \lambda(\mathbf{x}, t)}{\partial t} \right] \frac{\mathbf{v}(\mathbf{x}, t)}{|\mathbf{v}(\mathbf{x}, t)|} = \nabla \lambda(\mathbf{x}, t), \tag{4.74}$$

while the variation with respect to $\lambda(\mathbf{x}, t)$ leads to (4.64). The quantity $\lambda(\mathbf{x}, t)$ has the interpretation of the commodity price. To understand this interpretation, we focus on a steady state or on a storage-free case. Then, (4.74) becomes

$$\kappa(\mathbf{x}) \frac{\mathbf{v}(\mathbf{x})}{|\mathbf{v}(\mathbf{x})|} = \nabla \lambda(\mathbf{x}). \tag{4.75}$$

Obviously, the unit flow field $\mathbf{v}(\mathbf{x})/\,|\mathbf{v}(\mathbf{x})|$ is parallel with the gradient field $\nabla \lambda(\mathbf{x})$ for the commodity price. Hence, commodities always flow in the direction of steepest price increase. Furthermore, the norm of the price gradient equals the local transportation cost rate. In other words, the commodity price increases along the flow with accumulated transportation costs.

The equations (4.64) and (4.74) form, together with the boundary conditions at the borderline of the region A, a complete field-theoretical problem. Unfortunately, these equations are dominated by strong nonlinearities. This problem remains also for the steady state. In fact, the decomposition of the transport field $\mathbf{v}(\mathbf{x})$ into a vortex-free part $\nabla\varphi$ and a divergence-free part $\mathbf{w}(\mathbf{x})$ with div $\mathbf{w}(\mathbf{x}) = 0$ transforms (4.66) into the two-dimensional Poisson equation

$$\Delta\varphi(\mathbf{x}) = q(\mathbf{x}). \tag{4.76}$$

The solution of this equation is possible with well-known standard methods. The distribution of the local excess supply of production over consumption $q(\mathbf{x})$ is an external quantity characterizing the economic system.

In a certain sense, the field $q(\mathbf{x})$ is comparable with the charge density in electromagnetic theory. The second step is the determination of the scalar price field $\lambda(\mathbf{x})$. If we take squares of both sides of (4.75), we obtain the closed nonlinear field equation

$$(\nabla\lambda)^2 = \left(\frac{\partial\lambda}{\partial x_1}\right)^2 + \left(\frac{\partial\lambda}{\partial x_2}\right)^2 = \kappa^2(\mathbf{x}) \tag{4.77}$$

for the commodity price field. The local transportation costs $\kappa(\mathbf{x})$ are, similar to the quantity $q(\mathbf{x})$, an external field that may be empirically determined by suitable observations or estimations. Finally, inserting the solution of (4.77) into (4.75), we obtain a nonlinear algebraic equation for the field $\mathbf{w}(\mathbf{x})$ that depends on the local structure of the fields $\varphi(\mathbf{x})$ and $\lambda(\mathbf{x})$.

The remaining problem is the adjustment of the solutions $\mathbf{w}(\mathbf{x})$ and $\varphi(\mathbf{x})$ at the boundary conditions. Unfortunately, this is a very complicated task because the boundary conditions are usually defined for the common field $\mathbf{v}(\mathbf{x}) = \nabla\varphi(\mathbf{x}) + \mathbf{w}(\mathbf{x})$. Considering the nonlinear relations between the field components, we get a complicated functional from which we may determine the proper field $\mathbf{w}(\mathbf{x})$ at the boundary.

The present theory allows the construction of optimal roads. Let us assume that we know the scalar field of the transportation costs $\kappa(\mathbf{x})$. As discussed above, the vector field $\mathbf{v}(\mathbf{x})$ defines the direction of the local flow. A road may be defined by the curve $\mathbf{y}(s)$, where s is an arbitrary curve parameter. Then, the tangent of the curve

$$\mathbf{t}(s) = \frac{d\mathbf{y}(s)}{ds} \tag{4.78}$$

always shows the direction of the local flow. This means that we have the relation

$$\mathbf{t}(s) = \frac{\mathbf{v}(\mathbf{y}(s))}{|\mathbf{v}(\mathbf{y}(s))|} \tag{4.79}$$

along the road. Therefore, we expect that an optimal road fulfills the equation

$$\kappa(\mathbf{y}(s))\mathbf{t}(s) = \kappa(\mathbf{y}(s))\frac{\mathbf{v}(\mathbf{y}(s))}{|\mathbf{v}(\mathbf{y}(s))|} = \nabla\lambda(\mathbf{y}(s)). \tag{4.80}$$

Let us try to eliminate the relatively complicated price field λ from (4.80). To do this, we differentiate with respect to the curve parameter

$$\frac{d}{ds}\left[\kappa(\mathbf{y}(s))\frac{d\mathbf{y}(s)}{ds}\right] = \frac{d}{ds}\nabla\lambda(\mathbf{y}(s)). \tag{4.81}$$

The vector components on the right-hand side can be written

$$\begin{aligned}\frac{d}{ds}\frac{\partial\lambda(\mathbf{x})}{\partial x_\alpha}\bigg|_{\mathbf{x}=\mathbf{y}(s)} &= \sum_{\beta=1}^{2}\frac{\partial^2\lambda(\mathbf{x})}{\partial x_\alpha\partial x_\beta}\bigg|_{\mathbf{x}=\mathbf{y}(s)}\frac{\partial y_\beta(s)}{\partial s} \\ &= \sum_{\beta=1}^{2}\frac{\partial^2\lambda(\mathbf{x})}{\partial x_\alpha\partial x_\beta}\bigg|_{\mathbf{x}=\mathbf{y}(s)} t_\beta(s).\end{aligned} \tag{4.82}$$

We multiply this expression with $\kappa(\mathbf{y}(s))$ and apply (4.80) in order to obtain

$$\kappa(\mathbf{y}(s))\frac{d}{ds}\frac{\partial\lambda(\mathbf{x})}{\partial x_\alpha}\bigg|_{\mathbf{x}=\mathbf{y}(s)} = \sum_{\beta=1}^{2}\frac{\partial^2\lambda(\mathbf{x})}{\partial x_\alpha\partial x_\beta}\bigg|_{\mathbf{x}=\mathbf{y}(s)}\frac{\partial\lambda(\mathbf{x})}{\partial x_\beta}\bigg|_{\mathbf{x}=\mathbf{y}(s)} \tag{4.83}$$

$$= \frac{1}{2}\frac{\partial}{\partial x_\alpha}\sum_{\beta=1}^{2}\left(\frac{\partial\lambda(\mathbf{x})}{\partial x_\beta}\right)^2\Bigg|_{\mathbf{x}=\mathbf{y}(s)} \tag{4.84}$$

or with (4.77)

$$\kappa(\mathbf{y}(s))\frac{d}{ds}\nabla\lambda(\mathbf{y}(s)) = \frac{1}{2}\left.\nabla\kappa^2(\mathbf{x})\right|_{\mathbf{x}=\mathbf{y}(s)}. \tag{4.85}$$

Hence, we obtain the road equation

$$\frac{d}{ds}\left[\kappa(\mathbf{y}(s))\frac{d\mathbf{y}(s)}{ds}\right] = \nabla\kappa(\mathbf{y}(s)). \tag{4.86}$$

We remark that the last equations are close to Fermat's law in optics. In particular, roads are equivalent to light rays, the commodity price field corresponds to the eikonal function, and the transportation cost may be interpreted as the refraction index.

For example, if we have spatially separated types of transportation, such as transportation over sea and over land, the trading routes are straight lines, broken at the coastline via the well-known refraction law [294, 387]. A similar phenomenon applies to roads through high mountain regions. For instance, the highways from Rome to Milan pass the Appennines similar to light rays through a glassy plate; see Figure 4.12.

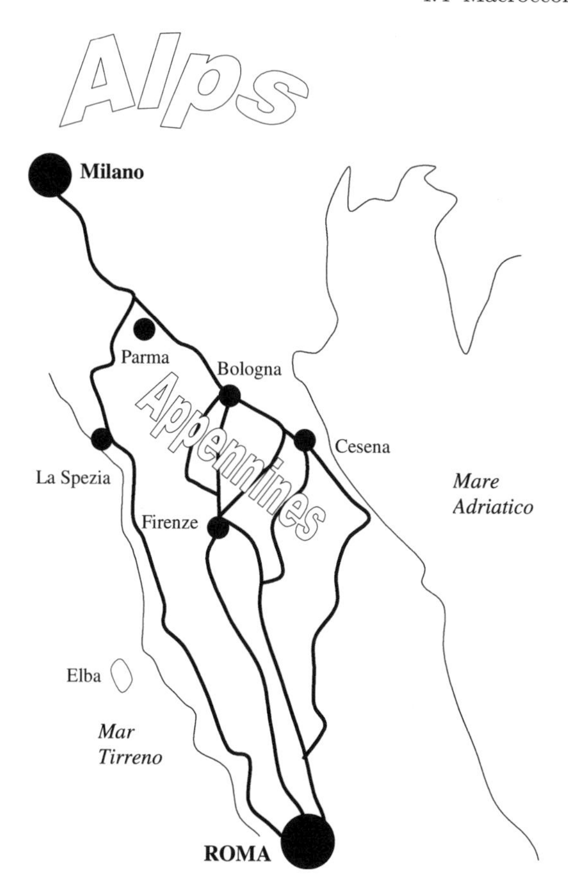

Fig. 4.12. Schematic representation of the highways from Rome to Milan.

5. Computer Simulations

5.1 Models and Simulations

Computer simulations are powerful tools supporting economic theories by bringing precision and rigor into economic and socioeconomic theories [97, 264, 370, 424]. By utilizing the speed and power of modern computers, they allow us to examine the consequences of complex rules and to study the dynamics of large systems. This makes them an important tool for dealing with the complex structure of economic systems. From a physical point of view, computer simulations allow the introduction of several natural scientific concepts and methods into the theoretical framework of economics, finance, and social science.

Of course, computer simulations can never fully describe the richness of economic processes. In fact, each element of an economic system considered in a certain computer simulation represents a small number of relevant degrees of freedom. In general, these elements, especially the humans involved, are too complex to be completely captured by a numerical procedure. However, in spite of the differentiation between relevant and irrelevant degrees of freedom, we may describe the behavior of the elements of an economic system by probabilistic rules. This means that an element has different alternatives when reacting to a change of its environment.

The observation and formulation of the rules describing social and economic action and reaction processes is the task of social science, microeconomics, and psychology. We remark that the probability of considering an incorrect rule is much higher than in the natural sciences.

Furthermore, numerical simulations depend on the level of the relevant quantities taken into account. A "human," or more generally an agent, with more than one hundred internal states is much more flexible than an agent with only two alternative freedoms of action. The term agent is very general: An agent may be a human but also may be a computer controlling the overnight financial transactions of a bank, a production firm, or elements of transportation systems and communication networks.

In physics, an enormous progress was achieved due to widespread use of computer simulations. Especially the development of the theory of nonlinear dynamical systems and of complex systems far from the thermodynamic equilibrium are strongly influenced by various kinds of simulations. The nature of those simulations is usually very different from simulations used in engineering or material science. These scientific disciplines use existing theories in order to solve a well-defined problem quantitatively.

In physics, computer simulations are often used to test theories or to refine analytical results. The typical physical strategy of a computer simulation is to isolate the most important characteristics of the underlying phenomenon and to build the simplest possible model instead of trying to model the phenomenon in its natural complexity.

For example, the Ising model [228], built to explain magnetic phenomena, considers only two orientations of local magnetic moments. In fact, this model has very little in common with a real magnetic material. However, numerical simulations on the basis of this model allow us to understand qualitatively the physics of phase transitions and the intrinsic dynamics of critical phenomena. But this simple model also has quantitative aspects. For instance, computer simulations of the Ising model allow a serious estimation of the critical exponents much better than any known analytical approach with the exception of the well-known rigorous solutions.

As discussed in the previous chapter, there is certain evidence that the general behavior of economic systems is determined by just a few dominant collective modes. In other words, the general behavior of large economic systems often should not depend on details involving the behavior of individual elements.

This empirical knowledge provides the fundamental concept of numerical simulations of economic problems. It is the general concept of creating a numerical algorithm to build simple models that capture only the essential properties of the interactions in the system that nevertheless allow for the proper description of aggregate behavior.

In other words, we should try to find a numerical procedure of a minimum length that considers the complexity of the problem just enough. We have mentioned in Chapter 1 that a general solution of this problem is avoided due to a theorem of Gödel. However, an intuitive approach to this minimum algorithm is always possible. The remaining problem is how to find such an algorithm and therefore a minimum model describing the economic problem in mind.

One possible way is to use a reduction mechanism. In other words, we start from a very comprehensive model and reduce the complexity step-by-step by eliminating those variables and interactions that are proven not to be relevant for the general behavior of the economic system. Following this way, we probably arrive at a gradual simplification of the original model.

Occasionally, we can construct a shortcut. This inductive way starts from the simplest possible model that would have the qualitative properties of the phenomena in mind. The problem is that we must intuitively capture just the main features of the underlying problem while neglecting all of the details. Of course, Gödel's theorem shows that there is no simple algorithm indicating

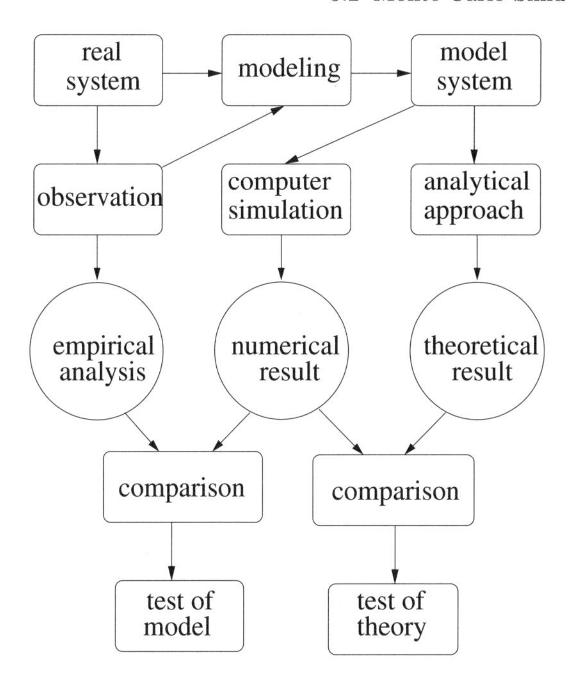

Fig. 5.1. The relations among theory, empirical observations, and computer simulations.

how to achieve it. However, if we have found a suitable basic model, we may refine the algorithm in order to approach reality. In physics and some other disciplines of natural science, computer simulation occupies an intermediate position between experiment and theory. In economics and finance, the source of knowledge is empirical observations, while experiments in the narrower sense are impracticable since a repetition is usually excluded. Here, computer simulations increasingly take over the role of the experiment (Figure 5.1).

5.2 Monte Carlo Simulations

5.2.1 Monte Carlo and Random Generators

The expression "Monte Carlo simulation" is actually very general. Monte Carlo methods [28, 60, 148, 236, 329] are stochastic techniques — meaning they are based on the use of random numbers and probability statistics to investigate problems. We can find Monte Carlo methods used in everything from economics to nuclear physics to regulating the flow of traffic.

The method is named after the city in the Monaco principality because of roulette, a simple random number generator. The name and the systematic development of Monte Carlo methods dates from about 1944. In particular, the first use of Monte Carlo methods as a research tool stems from work on the atomic bomb during World War II. This work involved a direct simulation of statistical problems connected with the random neutron diffusion in matter. The first Monte Carlo paper was published in 1949 [271].

Today, the Monte Carlo method is a very powerful tool with a large field of applications for both physics and economics. Generally, to call an algorithm Monte Carlo simulation, all we need to do is use random numbers to examine some problem.

The use of Monte Carlo methods to solve physical problems allows us to examine more complex systems than we otherwise could [46, 47, 48]. In particular, solving the equations of motion for one or two degrees of freedom is a fairly simple analytical problem. The solution of the same equations for thousands or more degrees of freedom is, with the exception of some special cases such as bead-spring models, an impossible task.

Of course, the way Monte Carlo techniques are applied varies widely from field to field. Physical research uses a lot of different Monte Carlo algorithms that can be applied on various problems such as in quantum mechanics or quantum field theory, molecular physics, statistical physics, or in astrophysics. The long list of literature regarding this topic grows rapidly.

Each Monte Carlo procedure requires a random generator. Unfortunately, computer-generated random numbers are not really stochastic since computer programs are deterministic algorithms. But, given an initial number, generally called the seed, a number of mathematical operations can be performed on the seed to generate apparently unrelated pseudorandom numbers. The output of random number generators is usually tested with various statistical methods to ensure that the generated number series are really random in relation to one another with respect to the desired accuracy. There is an important caveat: if we use a seed more than once, we will get identical random numbers every time. However, several commercial programs pull the seed from somewhere within the system, so the seed is unlikely to be the same for two different simulation runs.

A given random number algorithm generates a series of random numbers $\{\eta_1, \eta_2, ..., \eta_N\}$ with a certain probability distribution function. If we know this distribution function $p_{\text{rand}}(\eta)$, we know from the rank-ordering statistics that the likely rank of a random number η in a series of N numbers is

$$n = NP_<(\eta) = \int_{-\infty}^{\eta} dz\, p_{\text{rand}}(z). \tag{5.1}$$

In other words, if the random generator creates random series that are distributed with $p_{\text{rand}}(\eta)$, the corresponding series $\{P_<(\eta_1), P_<(\eta_2), ..., P_<(\eta_N)\}$ is uniformly distributed over the interval $[0, 1]$. On the other hand, if a series $\{\eta_1, \eta_2, ..., \eta_N\}$ is uniformly distributed, we can generate a series corresponding to another distribution function $\widetilde{p}$ via the calculation of the inverse function (i.e., $\{P_<^{-1}(\eta_1), P_<^{-1}(\eta_2), ..., P_<^{-1}(\eta_N)\}$.

5.2.2 Dynamic Monte Carlo

A typical application of Monte Carlo algorithms in economics and finance [17, 65, 77, 103, 125, 141, 185, 338, 339, 407] is the solution of stochastic differential equations of the type (3.65) or the more complicated form (3.252). The Monte Carlo method solves these equations by repeated simulations of the underlying stochastic processes and a subsequent statistical analysis of the results obtained. This allows us to avoid probably very complicated solutions of the corresponding Fokker–Planck equations or master equations. Of course, for simple financial or economic models, Monte Carlo is not the better solution because it is very time-consuming in terms of computation, and the general dependence on possible control parameters still remains open. The general interest in a Monte Carlo simulation approach is related to solving very complex and relatively realistic models. For example, Monte Carlo simulations are very helpful in the solution of American option problems [12, 20, 34, 63, 156, 157, 308, 326]; see also subsection 3.4.3.

Here, we want only to represent the most important elements of a Monte Carlo simulation without addressing the details. Special concepts and algorithms may be obtained from the comprehensive literature.

The standard financial or economic problem solved by Monte Carlo methods consists in the numerical solution of a system of stochastic difference equations. Such equations may be derived from Ito stochastic differential equations (2.153) or, in the case of a pronounced memory effect, from the more general Mori–Zwanzig equation (2.121) by an approximate integration over a short time interval δt. But it is also possible to obtain stochastic difference equations from direct empirical investigations.

Suppose that we model the time evolution of a set of economic or financial data $Y = \{Y_1, Y_2, ...Y_N\}$ and have M sources Z_k of randomness. Then, the discrete time evolution of the system may be written in the form

$$\begin{aligned} Y_\alpha(t_{n+1}) &= Y_\alpha(t_n) + A_\alpha(\{Y(t_n), Y(t_{n-1}), ...\}) \\ &\quad + \sum_{k=1}^{M} B_\alpha^k(\{Y(t_n), Y(t_{n-1}), ...\}) Z_k(t_n) \end{aligned} \tag{5.2}$$

with $\alpha = 1, ..., N$ and $t_{n+1} = t_n + \delta t$, where δt is a given time horizon. We remark that the random functions Z_k are not necessarily Gaussian-distributed.

The numerical procedure is very simple. Starting from a given state $Y(t_n)$ and probably certain information about the history given by $Y(t_{n-1})$, $Y(t_{n-2})$, ... , we are able to calculate the subsequent state $Y(t_{n+1})$ via (5.2). The required random values of the functions $Z_k(t_n)$ follow from the use of appropriate random generators. The recursion law (5.2) yields a certain path $\{Y(t_1), Y(t_2), ...\}$. Such a path may be interpreted as one possible event of the underlying model.

We can now create a sufficiently large set of such events by the use of different seeds for the random generator. This allows us to calculate the empirical frequency distribution function, which converges to the true probability distribution function for a sufficiently large number of independent runs. Usually, one refrains from a direct calculation of the probability distribution function but computes directly various moments and correlation functions.

Note that the transformation of Ito stochastic differential equations (2.153) into stochastic difference equations (5.2) requires an important remark. Except for simple Ito stochastic differential equations with constant coefficients, the substitutions $dY \to \Delta Y$, $dt \to \Delta t$, and $dW \to \Delta W = Z$ yield no automatically corresponding stochastic difference equation. For example, the simple geometric Brownian motion (3.110) for one degree of freedom may be written as

$$dY = aY dt + bY dW(t). \tag{5.3}$$

We define the new variable $x = \ln Y$ so that

$$dx = \frac{dY}{Y} - \frac{(dY)^2}{2Y^2} = \left[a - \frac{b^2}{2}\right] dt + bdW. \tag{5.4}$$

This equation can now be directly integrated. Therefore, we obtain

$$Y(t + \delta t) = Y(t) \exp\left\{\left[a - \frac{b^2}{2}\right] \delta t + b\left(W(t + \delta t) - W(t)\right)\right\}. \tag{5.5}$$

Hence, we arrive at the discrete time evolution

$$Y(t_{n+1}) = Y(t_n) \exp\left\{\left[a - \frac{b^2}{2}\right] \delta t + b\Delta W(t_n)\right\}, \tag{5.6}$$

which is much more complicated than the original Ito stochastic differential equation (5.3) and the corresponding "naive" stochastic difference equation

$$\Delta Y(t_n) = Y(t_{n+1}) - Y(t_n) = aY(t_n)\delta t + bY(t_n)\Delta W(t_n) . \tag{5.7}$$

Finally, we remark that the techniques used for the solution of stochastic differential equations are comparable with physically motivated numerical solutions of Langevin equations [355].

5.2.3 Quasi-Monte Carlo

Each Monte Carlo simulation is equivalent to a problem of integral evaluation. That is obvious if we compute moments or correlation functions using Monte Carlo techniques, but the solution of stochastic difference equations also can be transformed into an integral problem in a multidimensional space.

For example, the probability distribution function for the realization of the path $\{Y(t_1), Y(t_2), ..., Y(t_L)\}$ as a result of the stochastic difference equation (5.2) can be written as the integral

$$P\left(Y^{(L)}, t_L; ...; Y^{(1)}, t_1\right) = \int \prod_{n=0}^{L-1} \prod_{k=1}^{M} \left[p_k\left(Z_k^{(n)}\right) dZ_k^{(n)}\right]$$
$$\prod_{n=0}^{L-1} \prod_{\alpha=1}^{N} \delta\left(Y_\alpha^{(n+1)} - Y_\alpha^{(n)} - A_\alpha^{(n)} - \sum_{k=1}^{M} B_\alpha^{k,(n)} Z_k^{(n)}\right), \tag{5.8}$$

where p_k is the distribution function of the stochastic variable $Z_k^{(n)} = Z_k(t_n)$. As mentioned above, each probability distribution function is directly connected with the uniform distribution over the interval $[0, 1]$. Thus, each possible integral can be transformed into an integral over a unit hypercube $\mathcal{C}$. In the case of our example, we may use the rules

$$p_k(Z_k)\, dZ_k = dU_k \qquad \text{and} \qquad P_{k,<}(Z_k) = U_k \tag{5.9}$$

in order to transform (5.8) into the unit representation. Formally, each integral over the unit cube can be estimated by the formula

$$\int dU \Phi(U) \approx \frac{1}{Q} \sum_{m=1}^{Q} \Phi\left(U^{(m)}\right), \tag{5.10}$$

where the $U^{(m)} \in \mathcal{C}$ $(m = 1, ..., Q)$ define a representative set of more or less homogeneously distributed points inside the hypercube. One possible way to calculate the sum in (5.10) uses points $U^{(m)}$ created by a random generator. This kind of computation of multidimensional integrals by random sampling techniques is sometimes considered the main problem of a Monte Carlo simulation. Unfortunately, this concept exhibits a slow rate of convergence for the main problem of a Monte Carlo simulation. Usually, we find that the difference between the integral and the sum decreases with $Q^{-1/2}$.

An alternative way is the application of a quasi-Monte Carlo simulation [35, 66, 116, 170, 192, 233, 275, 282, 380, 396]. This is the traditional Monte Carlo simulation but using quasirandom sequences instead of pseudorandom numbers. The quasirandom sequences, sometimes also called low-discrepancy sequences, usually permit improvment of the performance of Monte Carlo simulations, offering shorter computational times and higher accuracy.

We remark that the low-discrepancy sequences are deterministic series, so the popular notation quasirandom can be misleading. The discrepancy property is a measure of uniformity for the distribution of the points. It is defined by

$$D_Q = \sup_{R \in \mathcal{C}} \left| \frac{n(R)}{Q} - v(R) \right|, \tag{5.11}$$

where R is a compact region of the unit hypercube, $v(R)$ is the volume of this region, and $n(R)$ is the number of points in this region. The discrepancy vanished for $Q \to \infty$ in the case of a homogeneous distribution of points over the whole hypercube.

Mainly for the multidimensional case, a low discrepancy corresponds to no large gaps and no clustering of points in the hypercube (Figure 5.2). Similar to a pseudorandom generator, a quasirandom generator originates from number theory. But in contrast to the pseudorandom series, quasirandom sequences offer a pronounced deterministic behavior. A quasirandom generator transforms an arbitrary positive integer I into a quasirandom number ξ_I via the

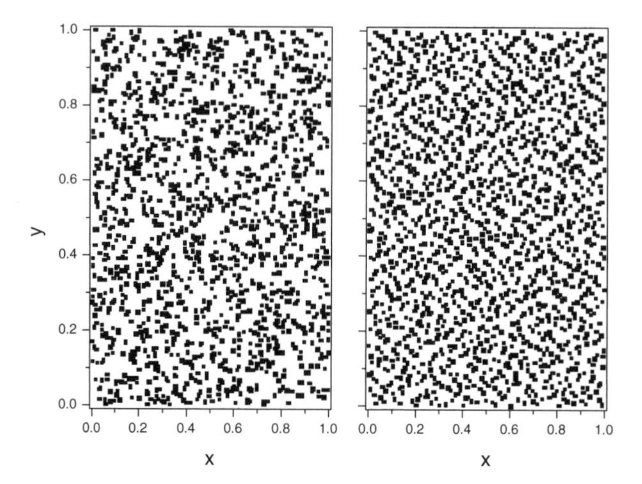

Fig. 5.2. Two-dimensional plot of pseudorandom number pairs (left) and quasirandom number pairs (right). The quasirandom number series are created with the base 2 (x-axis) and with the base 3 (y-axis).

following two steps. First, the integer I will be decomposed into the integer coefficients a_k with respect to the base b

$$I = \sum_{k=0}^{\infty} a_k b^k \tag{5.12}$$

with $0 \leq a_k \leq b - 1$. The coefficients simply form the representation of I within the base b. The second step is the computation of the quasirandom number by calculation of the sum

$$\xi_I = \sum_{k=0}^{\infty} a_k b^{-k-1}. \tag{5.13}$$

For example, the first quasirandom numbers corresponding to the base 2 are 1/2, 1/4, 3/4, 1/8, 5/8, ..., while the sequence of base 3 starts with 1/3, 2/3, 1/9, 4/9, 7/9. The main problems [170, 282] of the quasi-Monte Carlo techniques are: (i) this method may not be directly applicable to simulations of single events because of the correlations between the points of a quasirandom sequence, and (ii) the improved accuracy of quasi-Monte Carlo methods is generally lost for problems of high dimension or problems in which the integrand is not smooth.

The merit of the quasi-Monte Carlo method is the fast convergence. The theoretical upper bound rate of convergence for the estimation (5.10) is $\ln^d Q/Q$, where d is the number of dimensions of the integral problem [281].

5.2.4 Reverse Monte Carlo

Suppose that we have a model controlled by a given set of parameters $G = \{g_1, g_2, ..., g_M\}$ that produces the output data $Y = \{y_1, y_2, ..., y_N\}$. Furthermore, we have a set of empirical observations in a real economic system, $O = \{o_1, o_2, ..., o_N\}$, corresponding to the output data of the model.

Then, we can refine the model by changing the control parameter in such a way that the distance between the observations and the results of the model decreases. In other words, we modify the structure of the model in order to bring it into correspondence with reality. This does not mean that the model reflects reality. Only those observations that we have considered during the tuning of the model parameters are reproduced by the model.

A numerical standard technique consists in the formation of a positive--semidefinite functional

$$\mathcal{F}(Y - O) = \mathcal{F}(y_1 - o_1, y_2 - o_2, ...) \tag{5.14}$$

with $\mathcal{F}(0) = 0$ for $Y = O$ and $\mathcal{F} > 0$ otherwise. The functional $\mathcal{F}$ depends on the control parameters g_α via the model results $Y = Y(G)$. Therefore, we write also $\mathcal{F}[G]$ instead of (5.14). We now calculate the value of $\mathcal{F}[G_0]$ for a certain set G_0 of control parameters. Then, we change the control parameters by a small shift δG and compute the corresponding value $\mathcal{F}[G_0 + \delta G]$ of the functional $\mathcal{F}$.

Let us now consider a monotonously decreasing function $P(x)$ with $P(-\infty) = 0$ and $P(\infty) = 1$. Then, we may determine

$$P_{\text{change}} = P\left(\frac{\mathcal{F}[G_0] - \mathcal{F}[G_0 + \delta G]}{T}\right) \tag{5.15}$$

with a positive convergence parameter $T \geq 0$. We interpret the value P_{change} as a transition probability. This means that we replace the set of control parameters G_0 by the new set $G_0 + \delta G$ with a probability P_{change}.

This statistically induced replacement procedure makes the whole algorithm a Monte Carlo method. The repeated application of these steps leads to an increasing adaptation to reality. Usually, the convergence parameter T will be reduced very slowly during the Monte Carlo procedure. Thus, we arrive at a local or probably global minimum of the functional $\mathcal{F}[G]$ after a sufficiently long computation time.

The successive reduction of T is called the simulated annealing technique [203] and has a wide field of applications, especially in engineering.

In principle, the problem is comparable with a special class of Monte Carlo methods in computational physics that is also denoted as reverse Monte Carlo simulation.

In order to explain this term from a physical point of view, we have to consider that a standard Monte Carlo procedure is the importance sampling [46, 47, 48, 272]. Roughly speaking, in this case the parameter set G corresponds to the configuration C of an underlying many-body system, while the functional $\mathcal{F}$ is the Hamiltonian $\mathcal{H}(C)$ of the system. The convergence parameter T may be identified with the temperature of the system. In contradiction to the simulated annealing method, the temperature is fixed during the whole simulation. Then, all configurations C_1, C_2, ... that were occupied in the course of the Monte Carlo procedure form a weighted set that allows the determination of various thermodynamic quantities and other information about the structure of the system such as scattering curves. The values of these quantities depend on the temperature and on the microscopic interaction parameters.

We remark that the thermodynamic behavior appears due to the constraint that each statistically induced jump $C_n \to C_{n+1}$ satisfy the principle of detailed balance. This can be done by application of the Metropolis algorithm [272] with $P(x) = 1$ for $x \geq 0$ and $P(x) = \exp\{x/T\}$ for $x < 0$.

As apposed to this procedure, the reverse Monte Carlo simulation [23, 49, 198, 262, 381, 390, 416] uses no microscopically founded Hamiltonian but a functional $\mathcal{F}$ considering the experimental findings of various scattering experiments. Then, the algorithm discussed above allows the determination of configurations that generate the experimental results. The knowledge of these configurations may allow the calculation of the microscopic interaction parameters.

Hence, the standard Monte Carlo simulations in physics allow the explanation of macroscopic effects on the basis of microscopic interactions for which the reverse Monte Carlo methods give some information about the microscopic structure on the basis of experimental measurements. But there is an important warning: The reverse Monte Carlo simulation can lead to a subset of configurations without a physical meaning.

In other words, not all configurations detected with a reverse Monte Carlo simulation are consistent with physical reality. However, the more independent experimental data are considered in the functional $\mathcal{F}$, the lower is the danger of a false statement about the intrinsic structure of the underlying system.

5.3 Cellular Automata

Cellular automata provide a formal framework for investigating the behavior of complex systems [271]. Cellular automata systems are dynamical systems with discrete space and time scales. The behavior of a cellular automaton is completely specified in terms of a local relation. Each cell of a cellular automaton is connected with its nearest neighbors via input and output channels (Figure 5.3).

We assume that the cell α is in one of a finite number of K_α possible states. At a discrete time t_n, the cell receives information about the state of its neighbors via the input channels and simultaneously sends the information about its own state to its neighbors via the output channels. We remark that the flows of input and output are not necessarily symmetric. The new state of each cell at time $t_{n+1} = t_n + \delta t$ is governed by a Boolean transition function T_α of the inputs and the state of the cell in question. Generally, we distinguish between homogeneous and inhomogeneous cellular automata. A homogeneous automaton consists of identical cells with symmetric information flows, while the behavior of the cells of an inhomogeneous cellular automaton are influenced by their position in the whole network of cells. Furthermore, the local rules controlling the behavior of a cellular automaton may be time-dependent on the automaton.

Obviously, the main properties of a cellular automaton model, especially the discretization of the space and the states, the parallel update at discrete points in time, and the relatively short-range interaction with a finite number of neighbors, suggest the application of parallel computing techniques and therefore a very high performance.

Historically, the earliest automata were mechanical devices such as the town hall clock of Prague or the Parisian artificial duck of Jacques de Vaucanson dating from 1738. The first numerically working cellular automata were originally conceived by Ulam and von Neumann in the 1940s [271].

Due to the discrete structure and the elementary rules discussed above, cellular automaton algorithms offer a very high speed even for a high degree of complexity. This property makes cellular automata models important for physical and economical research.

Relatively simple interaction rules allow the description of complex phenomena [135, 193, 379, 389, 431, 432, 433] such as self-organized criticality [29, 291], evolution of chemically induced spiral waves [257], oscillations and chaotic behavior of states [257, 287, 431], forrest fires [31], earthquakes [36, 85, 384], discrete mechanics [24], statistical mechanics [414], the dynamics of granular matter [29, 305], soliton excitations [365], and fluid dynamics [73, 134]. Other applications belong to various domains in biology

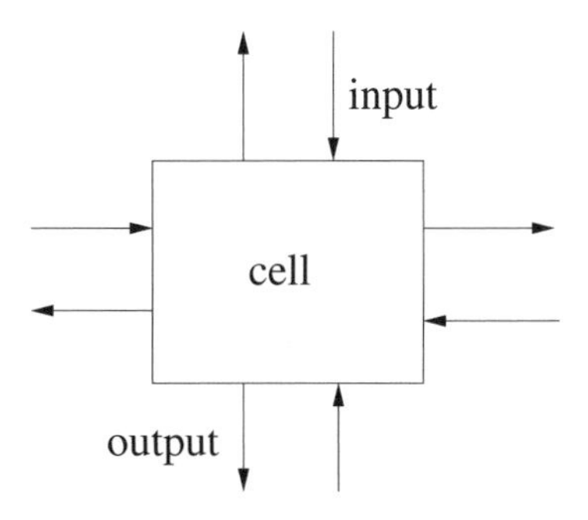

Fig. 5.3. Schematic representation of the input and the output channels of a cell of a cellular automaton

[113], including neuroscience [10, 37], and the dynamics of traffic systems [45, 75, 83, 277].

On the other hand, cellular automata models are increasingly used in social science [169, 286] and economics. In the latter, cellular automata are sometimes used as simple models of multiagent systems [162, 231, 306].

The investigation of multiagent systems focuses on systems in which many agents interact with each other. The agents are considered to be autonomous entities, such as computers and software programs, humans, or companies. In the language of cellular automata theory, each cell corresponds to an agent. Their interactions can be either cooperative or selfish. That is, the agents can share a common goal such as the member of a party or the computer of a large computation center, or they can pursue their own interests such as the companies in a free market economy.

The most important aspect of a cellular automaton is the transition rules defined by the transition function T_α. The transition function defines the state

$$S_\alpha(t_{n+1}) = T_\alpha\left(S_\alpha(t_n), \{S_\beta\left(t_n\right)\} \mid \beta \in N\left(\alpha\right)\right) \tag{5.16}$$

of cell α at time t_{n+1} as a function of the previous state $S_\alpha(t_n)$ and the states of all neighboring cells β. Even though the transition functions mainly determine the evolution, it is very hard to predict the evolution of a cellular automaton other than by explicitly simulating it. Equation (5.16) is a deterministic law. We remark that often a small change in the transition rule can have very dramatic consequences. Some rules need no detailed information about the individual states of the neighboring cells but only on the number of neighbors in a certain state S,

$$N_\alpha(S) = \sum_{\beta \in N(\alpha)} \delta_{S_\beta(t_n),S}. \tag{5.17}$$

In classical cellular automata theory, a rule is called totalistic if it only depends on the sum of the states of all cells in the neighborhood. It is called outer totalistic if it also depends on the state of the cell $S_\alpha(t_n)$ to be updated. In a formula a totalistic rule may be described as

$$S_\alpha(t_{n+1}) = T_\alpha\left(S_\alpha(t_n), \{N_\alpha(S)\}\right). \tag{5.18}$$

Another important class of transition rules are probabilistic rules. In this case, the transition rule is not a function that has exactly one result for each input configuration but a rule that provides the outcomes with associated probabilities. The normalization requires that the sum of probabilities of all outcomes must be one for each input configuration. In fact, such systems can be mapped onto a master equation and solved by several numerical [357, 359, 360] but also analytical methods [351, 363, 364].

6. Forecasting

6.1 Regression and Autoregression

Forecasting means the ability to extrapolate the future dynamics of a given system on the basis of its current state and its history. If we have a suitable model, the forecast may be extended to arbitrarily long times. Typical candidates for a nearly unlimited prognosis are the models of classical mechanics. Knowledge of the initial conditions allows the determination of any further development due to the deterministic equations of motion. Obviously, a prediction becomes better if we have a higher accuracy of the initial conditions or of the past and if we have precise equations of motion.

The situation is changed when the system is described by probability equations such as Ito stochastic differential equations or Fokker–Planck equations. Here, the dynamics of the hidden, irrelevant degrees of freedom are collected in stochastic and dissipative terms. Thus, the predictions contain an intrinsic error that increases with increasing forecasting time . The accuracy of a probability prediction depends on the macroscopic level and therefore the number of relevant degrees of freedom involved. If we knew more about the microscopic states of the system, we could predict the future development more precisely.

Physical models with stochastic character usually have a widely accepted and well-justified model. Thus, prediction of the future evolution consists in the estimation of the model parameters and in the characterization of the stochastic sources. On the other hand, we usually have insufficient models of economic systems and processes. The most unfavorable case is a black box situation: we have only observations about the input and the output, while the economic process is completely unclear. Furthermore, a real economic or financial system allows only a limited number of observations because any repetition of the same process under the same boundary conditions and initial conditions is impossible.

At the beginning of the last century, standard predictions were undertaken simply by extrapolating the time series through a global fit procedure. The principle is very simple. Suppose that we have a time series $\{y_1, y_2, ..., y_N\}$ with the corresponding points in time $\{t_1, t_2, ..., t_N\}$. Then, we can determine a regression function f in such a way that the distance between the observations y_n and the corresponding values $f(t_n)$ becomes sufficiently small.

The main problem is the definition of a suitable measure for the distance. Standard techniques such as least mean square methods minimize a certain utility function, for example

$$F = \sum_{n=1}^{N} (y_n - f(t_n))^2 , \tag{6.1}$$

by varying the parameters of the function f. For instance, the well-known linear regression requires the determination of the parameters A and B, which define the regression function f via $f(t) = A + Bt$. Obviously, the choice of the utility function is very important for the determination of the parameters of the regression function. For example, the very simple regression function $f(t) = Bt$ may be estimated by

$$F_1 = \sum_{n=1}^{N} (y_n - Bt_n)^2 \qquad \text{and} \qquad F_2 = \sum_{n=1}^{N} \left(\frac{y_n}{Bt_n} - 1 \right)^2 . \tag{6.2}$$

The first function stresses the absolute deviation between the observation and the regression function, while the second expression stresses the relative error. The first function leads to the estimation $B = \langle yt \rangle_N / \langle t^2 \rangle_N$, while the second one yields $B = \langle y^2 t^{-2} \rangle_N / \langle yt^{-1} \rangle_N$, where we have used the definition

$$\langle g \rangle_N = \frac{1}{N} \sum_{n=1}^{N} g_n . \tag{6.3}$$

It is very important to define both the regression function and the utility function very well and in agreement with the underlying problem. After determination of the regression parameters, the predictions are simply given by

$$\hat{y}_{N+k} = f(t_{N+k}). \tag{6.4}$$

The beginning of modern time series prediction was in 1927, when Yule [436] introduced the autoregressive model in order to predict the annual number of sunspots. Such models are usually linear or polynomial and are driven by white noise. In this context, predictions are carried out on the basis of parametric autoregressive (AR), moving average (MA), or autoregressive moving average (ARMA) models [62, 239, 245].

The autoregressive process AR(p) is defined by

$$y(t_n) = a_0 + \sum_{k=1}^{p} a_k y(t_{n-k}) + \eta(t_n), \tag{6.5}$$

where the a_k $(k = 0, ..., p)$ are fixed model parameters and η_n represents the current noise. We can use an appropriate method of estimation, such as ordinary least squares, to get suitable approximations $\hat{a}_k$ of the initially unknown parameters a_k. After estimating these model parameters, we get the fitted model

$$\hat{y}(t_n) = \hat{a}_0 + \sum_{k=1}^{p} \hat{a}_k y(t_{n-k}) \ . \tag{6.6}$$

Clearly, different regression methods give different estimates, but they are all estimates on the basis of the same more or less unknown but true distribution of $y(t_n)$. In this sense, $\hat{y}(t_n)$ is an estimation of the true conditional mean of $y(t_n)$, which may be generally denoted as $E\left(y(t_n) \mid \omega_{n-1}\right)$, where ω_{n-1} is the information set available at time t_{n-1}. In the case of the autoregressive process AR(p) introduced above, we have $\omega_{n-1} = \{y(t_{n-1}), ..., y(t_{n-p})\}$. This notation makes explicit how the conditional mean and therefore the prediction is constructed on the assumption that all data up to that point are known deterministic variables.

A natural way to estimate the coefficients a_k considers the Mori–Zwanzig equations (2.121). As pointed out, this equation is an exact linear relation. In a discrete version, this equation reads

$$y_\alpha(t_{n+1}) = y_\alpha(t_n) + \sum_{\beta=1}^{M} \sum_{k=0}^{n} \Xi_{\alpha\beta}(t_n - t_k) y_\beta(t_k) + \eta_\alpha(t_{n+1}), \tag{6.7}$$

where we have considered $\alpha = 1, 2, ...M$ simultaneously given time series. Note that we have replaced the notations for the relevant quantities, $G_\alpha \to y_\alpha$, and for the residual forces, $f_\alpha \to \eta_\alpha$, while the frequency matrix and the memory kernel are collected in the matrix $\Xi_{\alpha\beta}(t_n - t_k)$. Of course, the residual forces, the memory, and the frequency matrix contained in the original Mori–Zwanzig equations are implicitly dependent on the initial state at t_0. Thus, for a stationary system, the matrix $\Xi_{\alpha\beta}(t)$ is independent of the initial state, and the residual forces may be interpreted as a stationary noise. In order to determine the matrix $\Xi_{\alpha\beta}(t)$, we remember that the correlation functions of the relevant quantities are exactly defined by (2.127). This equation reads in its discrete form

$$\overline{y_\alpha(t_{n+1}) y_\gamma(t_0)} = \overline{y_\alpha(t_n) y_\gamma(t_0)} + \sum_{\beta=1}^{M} \sum_{k=0}^{n} \Xi_{\alpha\beta}(t_n - t_k) \overline{y_\beta(t_k) y_\gamma(t_0)}. \tag{6.8}$$

Besides the error due to the discretization, (6.8) is an exact relation. In the case of a stationary system, (6.8) holds for all initial times t_0 with the same matrix function $\Xi_{\alpha\beta}(t)$. Thus, we can replace the correlation functions $\overline{y_\alpha(t_n) y_\gamma(t_0)}$ by the estimations

$$C_{\alpha\gamma}(t_n - t_0) = \langle y_\alpha(t_n) y_\gamma(t_0) \rangle = \frac{1}{N} \sum_{k=0}^{N-1} y_\alpha(t_{n+k}) y_\gamma(t_k), \tag{6.9}$$

which are obtainable from empirical observations. Thus, we arrive at the matrix equation

$$C_{\alpha\gamma}([n+1]\,\delta t) = C_{\alpha\gamma}(n\delta t) + \sum_{\beta=1}^{M} \sum_{k=0}^{n} \Xi_{\alpha\beta}([n-k]\,\delta t) C_{\beta\gamma}(k\delta t), \tag{6.10}$$

where we have used $t_{n+1} = t_n + \delta t$. Equation (6.10) allows the determination of the matrix $\Xi_{\alpha\beta}(t)$ on the basis of the empirically estimated correlation functions $C_{\alpha\gamma}(t)$. After estimating the matrix functions $\Xi_{\alpha\beta}(t)$, we get the prediction formula

$$\hat{y}_\alpha(t_{n+1}) = y_\alpha(t_n) + \sum_{\beta=1}^{M} \sum_{k=0}^{n} \Xi_{\alpha\beta}(t_n - t_k) y_\beta(t_k). \tag{6.11}$$

We remark that repeated application of such prediction formulas allows also the forecasting of the behavior at later times, but of course there is usually an increasing error.

The prediction formulas of moving averages and autoregressive processes are related. A moving average is a weighted average over the finite or infinite past. In general, a moving average can be written as

$$\overline{y}(t_n) = \frac{\sum\limits_{k=0}^{N} a_k y(t_{n-k})}{\sum\limits_{k=0}^{N} a_k}, \tag{6.12}$$

where the weights usually decrease with increasing k. Typical moving averages are (3.14), (3.17), or with respect to the IGARCH process (3.332). The weight functions are chosen heuristically under consideration of possible empirical investigations. The prediction formula is simply given by

$$\hat{y}(t_{n+1}) = \overline{y}(t_n). \tag{6.13}$$

The main difference between autoregressive processes and moving averages is the interpretation of the data with respect to the prediction formula. In an autoregressive process, the input is always understood as a deterministic series in spite of the stochastic character of the underlying model. On the other hand, the moving average concept assumes that all observations are realizations of a stochastic process that is a stationary process at least over a timescale

$$\tau \sim \frac{\delta t}{a_0} \sum_{k=0}^{N} a_k. \tag{6.14}$$

Autoregressive moving averages (ARMA) are combinations of moving averages and autoregressive processes. Such processes play an important role in the analysis of modified ARCH and GARCH processes [5].

6.2 The Bayesian Concept

6.2.1 Public Surveys and Decision-Making

Instead of a long introduction, let us illustrate the Bayesian concept by a simple example. An economically reasonable decision may lead to serious consequences since each decision induces a response from consumers, competitors, or the government. In particular, this may be important for the price policy of large companies or monopolies because any increase in product prices is an unpopular step leading to a loss of confidence. Such decisions can be supported by a suitable public survey . The analysis of a survey can be carried out on the basis of Bayes' theorem. Suppose that the decisions are controlled by a set of hypotheses B_i $(i = 1, ..., N)$. The possible hypotheses are mutually exclusive (i.e., in the language of set theory, we have to write $B_i \cap B_j = \emptyset$) and exhaustive. The probability that the hypothesis B_i appears is $P(B_i)$. Furthermore, we consider an event A that may be conditioned by the hypotheses. Thus, (2.53) can be written as

$$P(A \mid B_i)P(B_i) = P(B_i \mid A)P(A) \tag{6.15}$$

for all $i = 1, ..., N$. Furthermore, (2.58) leads to

$$P(B_i \mid A) = \frac{P(A \mid B_i)P(B_i)}{\sum_{i=1}^{N} P(A \mid B_i)\, P(B_i)}. \tag{6.16}$$

This is the standard form of Bayes' theorem. In the present context, we denote $P(B_i)$ as the a priori probability, which is available before the event A appears. The likelihood $P(A \mid B_i)$ is the conditional probability that the event A occurs under the hypothesis B_i. The quantity $P(B_i \mid A)$ may be interpreted as the probability that the hypothesis B_i was true under the condition that the event A occurs. Therefore, $P(B_i \mid A)$ is also called an a posteriori probability, which may be empirically determined after the appearance of A.

To better understand the analysis of an appropriate survey, let us discuss a small example. In the year 2002, a common currency was introduced in a number of European states. The costs involved with this change were insignificant for the majority of European companies. Hence, a change in prices for industrial goods or services was not justifiable. On the other hand, many companies usually carry out necessary price adjustments at the end of the year in order to balance changes in wages and the prices of resources. How should the management decide on the amount of the adjustment?

A company Σ can generally assume that a certain number of consumers trust in the fairness of the company's product prices (hypothesis B_+). The rest of the consumers prefer products of other companies because they believe that the prices of the products of the company Σ are unfair (hypothesis B_-). The corresponding frequencies can be obtained from a public survey. For the sake of simplicity, we assume the a priori probabilities $P(B_+) = 0.5$ and $P(B_-) = 0.5$. Furthermore, the survey may give an estimation about the public opinion on whether the firm applies a fair price policy in the event that its product prices increase at the turn of the year 2001/2002. An inquiry among students yielded a probability $P(\text{stable} \mid B_+) = 0.8$ and $P(\text{stable} \mid B_-) = 0.02$ (i.e., a consumer expects with a frequency of 80% that a fair company keeps the prices stable and is convinced with a probability of 98% from the fact that unfair firms increase their prices). Now, we can estimate consumer confidence in the case of a price change. We obtain from (6.16) the following a posteriori probabilities

$$P(B_+ \mid \text{stable}) = \frac{P(\text{stable} \mid B_+)P(B_+)}{P(\text{stable} \mid B_-)P(B_-) + P(\text{stable} \mid B_-)P(B_-)} \quad (6.17)$$

and

$$P(B_- \mid \text{stable}) = \frac{P(\text{stable} \mid B_-)P(B_-)}{P(\text{stable} \mid B_-)P(B_-) + P(\text{stable} \mid B_-)P(B_-)} \quad (6.18)$$

(i.e., we get $P(B_+ \mid \text{stable}) \approx 0.97$). In other words, if the company were to keep prices stable, it would be expected that the consumers believe with a probability of 97% that the company provides a fair price policy. On the other hand, if the company increases prices, we obtain $P(B_+ \mid \text{increase}) \approx 0.17$ (i.e., only 17% of the consumers would believe that the company's price policy is fair).

This example demonstrates the general importance of the Bayesian concept . Simple relations can be used fruitfully for the selection of a hypothesis or, equivalently, for the selection of a model.

6.2.2 Bayesian Theory and Forecasting

The Bayesian theory of model or decision selection [144, 145, 442, 443] discussed above generates insights not only into the theory of decision making but also in the theory of predictions. The Bayesian solution to the model selection problem is well-known: It is optimal to choose the model with the highest a posteriori probability. On the other hand, knowledge of the a posteriori probabilities is not only important for the selection of a model but also gives essential information for a reasonable combination of forecast results. The a posteriori probabilities may be associated with various forecasting models F_i. For the sake of simplicity, we consider only two models. Then, we have the a posteriori probabilities $P(F_1 \mid \omega)$ that model 1 is true and $P(F_2 \mid \omega)$ that model 2 is true under the condition that a certain event ω occurs. The estimation of these a posteriori probabilities is obtainable from the scheme discussed in the previous chapter. Furthermore, we have the mean square deviations

$$\left.\overline{(y-\hat{y})^2}\right|_{F_1} = \int dy (y-\hat{y})^2 p\left(y \mid F_1\right) \quad (6.19)$$

and

$$\left.\overline{(y-\hat{y})^2}\right|_{F_2} = \int dy (y-\hat{y})^2 p\left(y \mid F_2\right) \quad (6.20)$$

describing the expected square difference between an arbitrary forecast $\hat{y}$ and outcome y of the model. Because

$$p(y \mid \omega) = p(y \mid F_1) P(F_1 \mid \omega) + p(y \mid F_2) P(F_2 \mid \omega), \tag{6.21}$$

we get the total mean square deviation

$$\left.\overline{(y-\hat{y})^2}\right|_{\omega} = \left.\overline{(y-\hat{y})^2}\right|_{F_1} P(F_1 \mid \omega) + \left.\overline{(y-\hat{y})^2}\right|_{F_2} P(F_2 \mid \omega) \tag{6.22}$$

that is expected under the condition that the event ω appears. The prediction $\hat{y}$ was up to now a free value. We chose this value by minimizing the total mean square deviation. We get

$$\begin{aligned} \frac{\partial}{\partial \hat{y}} \left.\overline{(y-\hat{y})^2}\right|_{\omega} &= 2\left[\overline{y}|_{F_1} - \hat{y}\right] P(F_1 \mid \omega) + 2\left[\overline{y}|_{F_2} - \hat{y}\right] P(F_2 \mid \omega) \\ &= 0 \end{aligned} \tag{6.23}$$

and therefore the optimal prediction

$$\hat{y} = \overline{y}|_{F_1} P(F_1 \mid \omega) + \overline{y}|_{F_2} P(F_2 \mid \omega). \tag{6.24}$$

This relation allows us to combine predictions of different models in order to obtain a likely forecast. For example, the averages $\overline{y}|_{F_1}$ and $\overline{y}|_{F_2}$ may be the results of two moving average procedures . At least one of these forecasting models fails. The a posteriori probabilities $P(F_i \mid \omega)$ can be interpreted as the outcome of certain tests associated with the event ω that should determine the correct moving average model. The model selection theory requires that we consider only the model that has the largest a posteriori probability (i.e., we get either $\hat{y} = \overline{y}|_{F_1}$ or $\hat{y} = \overline{y}|_{F_2}$). However, the Bayesian forecast concept also allows the consideration of unfavorable models with small but finite weights.

6.3 Neural Networks

6.3.1 Introduction

As discussed above, time series predictions have usually been performed by using of parametric regressive, autoregressive, moving average, or autoregressive moving average models. The parameters of the prediction models are obtained from least mean square algorithms or similar procedures. A serious problem is that these techniques are basically linear. On the other hand, many time series in finance and economics are probably induced by strong nonlinear processes due to the high degree of complexity of the underlying system.

In particular, this nonlinearity controls the stochastic contributions, which in a linear forecasting theory are assumed to have a Markov character. However, we know from the discussion of the memory kernel (2.122) of the Mori–Zwanzig equation (2.121) that the characteristic timescale of the apparently stochastic terms is of an order of magnitude of the relaxation time of the memory. Thus, if we have empirical evidence for an autoregressive process with a large number p of previous observations considered, we may conclude that in a real complex system[1] the stochastic terms have also a pronounced memory, which is not considered in the linear forecasting equations.

In this case, neural networks provide alternative nonlinear methods for forecasting the further development of time series. Neural networks are powerful when applied to problems whose solutions require knowledge about a system or a model that is difficult or impossible to specify but for which there is a large set of past observations available [95, 147, 398]. The neural network approach to time series prediction is parameter-free in the sense that such methods do not need any information regarding the system that generates the signal. In other words, the system can be interpreted as a black box with certain inputs and outputs. The aim of a forecast using neural networks is to determine the output with a suitable accuracy when only the input is known. This task is carried out by a process of learning from so-called training patterns presented to the network and changing the network structure and weights in response to the output error.

From a general point of view, the use of neural networks may be understood as a step back from rule-based models to data-driven methods [151].

6.3.2 Spin Glasses and Neural Networks

Let us discuss why neural networks are useful for the prediction of the evolution of economic or financial time series. Such systems can store patterns and can recall these items on the basis of an incomplete input. For example, if such a network detects similarities between a current time series and an older one related to the same economic process, it may extrapolate the possible time evolution of the current time series on the basis of the historical experience. Usually, the similarities are not very trivially recognizable. The weights of the stored properties used for the comparison of different patterns depend on the architecture of the underlying network. First, we will explain why neural networks have a so-called adaptive memory.

Neural networks have some similarities with a real nervous system consisting of interacting nerve cells [195, 217]. Therefore, let us start our investigation from a biological point of view. The human nervous system is very large. It consists of approximately 10^{11} highly interconnected nerve cells. Electric signals induce transmitter substances to be released at the synaptic junctions where the nerves almost touch (Figure 6.3.2). The transmitters generate a local flow of sodium and potassium cations that raises or lowers the electrical potential. If the potential exceeds a certain threshold, a soliton-like excitation propagates from the cell body down to the axon. This then leads to the release of transmitters at the synapses to the next nerve cell. Obviously, the nervous system may be interpreted as a large cellular automaton of identical cells but with complicated topological connections. In

[1] but not necessarily in the mathematical model.

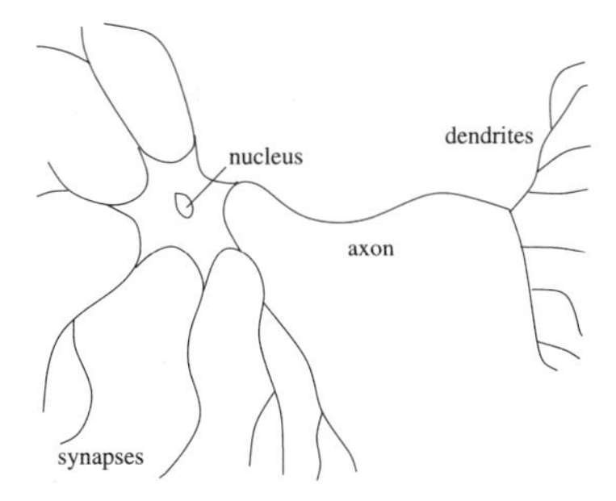

Fig. 6.1. Schematic representation of a nerve cell

particular, each cell has effectively just two states, an active one and a passive one. We adopt a spin analogy: the state of the cell α ($\alpha = 1, ..., N$) may be given by $S_\alpha = \pm 1$, where $+1$ characterizes the active state and -1 the passive state. The electrical potential may be a weighted sum of the activity of the neighboring nerve cells

$$V_\alpha = \sum_\beta J_{\alpha\beta} S_\beta . \tag{6.25}$$

The coupling parameters $J_{\alpha\beta}$ describe the influence of cell β on cell α. We remark that there is usually no symmetry (i.e., $J_{\alpha\beta} \neq J_{\beta\alpha}$). Of course, the absolute value and the sign of the parameters $J_{\alpha\beta}$ depend on the strength of the biochemically synaptic junction from cell β to cell α. The transition rule (5.16) of this cellular automaton reads

$$S_\alpha(t_{n+1}) = \operatorname{sgn}\left(V_\alpha(t_n) - \theta_\alpha\right) = \operatorname{sgn}\left(\sum_\beta J_{\alpha\beta} S_\beta(t_n) - \theta_\alpha\right), \tag{6.26}$$

where θ_α is the specific threshold of the cell [68, 235, 261]. Let us now transform this deterministic cellular automaton model into a probabilistic one. To do this, we introduce the probability that the cell α becomes active at t_{n+1},

$$p_\alpha^+(t_{n+1}) = \psi(V_\alpha(t_n) - \theta_\alpha), \tag{6.27}$$

where ψ is a sigmoidal function with the boundaries $\psi(-\infty) = 0$ and $\psi(\infty) = 1$. Equation (6.27) implies that $p_\alpha^- = 1 - p_\alpha^+$. This generalization is really observed in nervous systems. The amount of transmitter substance released at a synapse can fluctuate so that a cell remains in the passive state even though $V_\alpha(t_n)$ exceeds the threshold θ_α. For the sake of simplicity, we focus on the symmetric case $J_{\alpha\beta} = J_{\beta\alpha}$. The special choice

$$\psi(x) = \frac{1}{1 + \exp\{-2x/T\}} \tag{6.28}$$

is particularly convenient because it corresponds to an Ising model with so-called Glauber dynamics. This means that a cell changes its state independent of possible changes of other cells. For symmetric $J_{\alpha\beta}$, the system reaches, after a sufficiently long relaxation time, the thermodynamic equilibrium characterized by the stationary Gibb's distribution $\exp\{-H/T\}$ with the Hopfield Hamiltonian [8, 174, 175]

$$H = -\frac{1}{2}\sum_{\alpha\beta} J_{\alpha\beta} S_\alpha S_\beta + \sum_\alpha \theta_\alpha S_\alpha \tag{6.29}$$

and the temperature T. From here, we can reproduce (6.27) and (6.28) in a very simple manner. The cell α can undergo the transitions $+1 \to +1$, $-1 \to -1$, $-1 \to +1$, and $+1 \to -1$ with the corresponding energy differences $\Delta H_{+,+} = \Delta H_{-,-} = 0$ and $\Delta H_{-,+} = -\Delta H_{+,-} = 2(V_\alpha - \theta_\alpha)$, which follow directly from (6.29). Thus, the Gibb's measure requires the conditional probabilities

$$p_\alpha(+ \mid +) = \frac{\exp(-\Delta H_{+,+}/T)}{\exp(-\Delta H_{+,+}/T) + \exp(-\Delta H_{-,+}/T)} \tag{6.30}$$

and

$$p_\alpha(+ \mid -) = \frac{\exp(-\Delta H_{+,-}/T)}{\exp(-\Delta H_{+,-}/T) + \exp(-\Delta H_{-,-}/T)}. \tag{6.31}$$

Considering the values of the energy differences, we get $p_\alpha^+ = p_\alpha(+ \mid +) = p_\alpha(+ \mid -)$, where p_α^+ satisfies (6.27) and (6.28). Obviously, our special model of a neural network is nothing other than a spin glass (i.e., an Ising model with stochastic but symmetric interaction constants $J_{\alpha\beta}$ and the set of spin variables $S = \{S_1, ..., S_N\}$).

Now, we come back to the question of how a neural network can store items and how it can recall the items on the basis of an incomplete input. We restrict ourselves to the simple spin-glass model introduced above [124, 299, 300]. A pattern may be defined by a particular configuration $\sigma = \{\sigma_1, \sigma_2, ...\}$. Such a pattern is called a training pattern . Usually, we have to deal with more than one training pattern $\sigma^{(m)}$ with $m = 1, 2, ..., M$. Let us define the coupling constants as [8, 168, 174, 175]

$$J_{\alpha\beta} = \frac{1}{N}\sum_{m=1}^{M} \sigma_\alpha^{(m)} \sigma_\beta^{(m)}. \tag{6.32}$$

The prefactor N^{-1} is just a convenient choice for defining the scale of the couplings. Equation (6.32) is known as the Hebb rule. In the following discussion, we set $\theta_\alpha = 0$, although the theory can also work without this simplification. Thus, due to (6.32), the Hamiltonian (6.29) becomes

$$H = -\frac{N}{2}\sum_{m=1}^{M} \left(\sigma^{(m)}, S\right)^2, \tag{6.33}$$

where we have introduced the scalar product

$$(\sigma, \sigma') = \frac{1}{N} \sum_{\alpha=1}^{N} \sigma_\alpha \sigma'_\alpha. \tag{6.34}$$

In the case of only one pattern, $M = 1$, the Hamiltonian can be written as $H = -N\left(\sigma^{(1)}, S\right)^2/2$. In other words, the configurations with the lowest energy ($H = -N/2$) are given by $S = \sigma^{(1)}$ and by $S = -\sigma^{(1)}$. This means that an initially given pattern $S(0)$ approaches one of the two ground states $\sigma^{(1)}$ or $-\sigma^{(1)}$ in the course of the dynamics of the neural network for sufficiently low temperatures T. If we have a finite number $M \ll N$ of statistically independent training patterns, every one of them is a locally stable state. We remark that two patterns $\sigma^{(m)}$ and $\sigma^{(n)}$ are completely independent if the scalar product $\left(\sigma^{(m)}, \sigma^{(n)}\right)$ vanishes, $\left(\sigma^{(m)}, \sigma^{(n)}\right) = 0$. Statistical independence means that $\sigma^{(m)}$ and $\sigma^{(n)}$ represent two random series of values ± 1. Thus, we find the estimation $\left|\left(\sigma^{(m)}, \sigma^{(n)}\right)\right| \sim N^{-1/2}$. Let us set $S = \sigma^{(k)}$. Then, we obtain from (6.33)

$$\begin{aligned} H &= -\frac{N}{2} \sum_{m=1}^{M} \left(\sigma^{(m)}, \sigma^{(k)}\right)^2 \\ &= -\frac{N}{2} \left[1 + \sum_{m \neq k} \left(\sigma^{(m)}, \sigma^{(k)}\right)^2\right] \\ &\approx -\frac{N}{2} + o(M). \end{aligned} \tag{6.35}$$

It is simple to show that the training patterns $\sigma^{(m)}$ (and the dual patterns $-\sigma^{(m)}$) define the ground states of the Hamiltonian. This means that the dynamics of the neural network with a finite number of training patterns again find the stable state that most resembles the initial state $S(0)$. This is the main property of an adaptive memory . Each configuration learned by the neural network is stored in the coupling constants (6.32). A given initial configuration $S(0)$ of the network is now interpreted as a disturbed training pattern. The neural network acts to correct these errors in the input just by following its dynamics to the nearest stable state. Hence, the neural network assigns an input pattern to the nearest training pattern.

The neural network can still recall all M patterns (and the M dual patterns) as long as the temperature is sufficiently low and $M/N \to 0$ for $N \to \infty$. The critical temperature is given by $T_c = 1$ (i.e., for $T > 1$, the system reaches thermodynamic equilibrium). In other words, the neural network behaves similar to a paramagnetic lattice gas, and the equilibrium state favors no training patterns. On the other hand, for very low temperatures and a sufficiently large distance between the input pattern $S(0)$ and the training pattern, the dynamics of the system may lead the evolution $S(t)$ into spurious ghost states other than the training states. These ghost states are also minima of the free energy that occurs due to the complexity of the Hamiltonian (6.33). But it turns out that these ghost states are unstable above $T_0 = 0.46$. Hence, by choosing the temperature slightly above T_0, we can avoid these states while still keeping the training patterns stable.

Another remarkable situation occurs for $c = M/N > 0$. Here, the training states remain stable for a small enough c. But beyond a critical value $c^\star(T)$, they suddenly lose their stability, and the neural network behaves like a real spin-glass [9, 11]. Especially, the typical ultrametric structure of the spin glass states occurs in this phase. At $T = 0$, the curve $c^\star(T)$ reaches its maximum value of $c^\star(0) \approx 0.138$. For completeness, we remark that above a further curve $c_p(T)$, the spin-glass phase melts to a paramagnetic phase. However, both the spin-glass phase and the paramagnetic phase are useless for an adaptive memory. Only the phase capturing the training patterns is meaningful for the application of neural networks.

6.3.3 Topology of Neural Networks

The physical approach to neural networks discussed above is only a small contribution to the mainstream of mathematical and technical efforts concerning the development of this discipline.

Beginning in the early 1960s [334, 335, 428], the degree of scientific development of neural networks and the number of practical applications grew exponentially [67, 161, 174, 207, 341, 429]. In neural networks, computational models, or nodes, are connected through weights that are adapted during use to improve performance. The main idea is equivalent to the concept of cellular automata: High performance occurs due to interconnection of the simple computational elements. A simple node labeled by α provides a linear combination of Γ weights $J_{\alpha 1}$, $J_{\alpha 2}$,..., $J_{\alpha \Gamma}$ and Γ input values x_1, x_2,..., x_Γ and passes the result through a usually nonlinear transition or activation function ψ,

$$y_\alpha = \psi\left(\sum_{\beta=1}^{\Gamma} J_{\alpha\beta} x_\beta\right). \tag{6.36}$$

The function ψ is monotone and continuous, most commonly of a sigmoidal type. In this representation, the output of the neuron is a deterministic result. In general, the output can be formulated also on the basis of probabilistic rules (see above).

The neural network is then an interconnected set of such nodes. But in contradiction to most of the cellular automata models, the nodes or neurons of a neural network have a large number of nearest neighbors so that a dense interconnection appears. There is the theoretical experience that massively interconnected neural networks provide a greater degree of robustness than weakly interconnected networks. By robustness, we mean that small perturbations in parameters and in the input data will result in small deviations of the output data from their nominal values.

Besides their node characteristics, neural networks are characterized by their network topology . The topology can be determined by the connectivity matrix Θ with the components $\Theta_{\alpha\beta} = 1$ if a link from the node α to the node β exists and $\Theta_{\alpha\beta} = 0$ otherwise. A link from α to β means that the output of α is the input of β.

Only such weights $J_{\alpha\beta}$ can have nonzero values, which corresponds to $\Theta_{\alpha\beta} = 1$. In other words, we may write

$$J_{\alpha\beta} = \Theta_{\alpha\beta} g_{\alpha\beta}. \tag{6.37}$$

where $\Theta_{\alpha\beta}$ is fixed by the respective network architecture and remains unchanged during the learning process, while the $g_{\alpha\beta}$ should capture the training patterns.

Obviously, the connectivity matrix is not necessarily a symmetric one. We may describe this matrix symbolically by a corresponding network graph that consists of arrows and nodes. In particular, each arrow stands for an existing link, and the direction of the arrow indicates the flow of information.

The Hopfield network discussed above has the ideal connectivity $\Theta_{\alpha\beta} = 1$ for all $\alpha \neq \beta$. Thus, the topology of the Hopfield network is represented by a graph in which each node is connected to each other node by a double arrow (Figure 6.3.3). The dilution of such a topology by a random pruning procedure leads to a stochastic neural network or a so-called neural cluster. From the topological point of view, both types of neural networks distinguish not at all or only very weakly between input neurons and output neurons. The only exception is the case of a diluted network containing nodes with only outgoing arrows or only incoming arrows so that these nodes can be classified as input nodes or output nodes. Usually, these nodes are defined by the underlying program structure but not by the topology of the network.

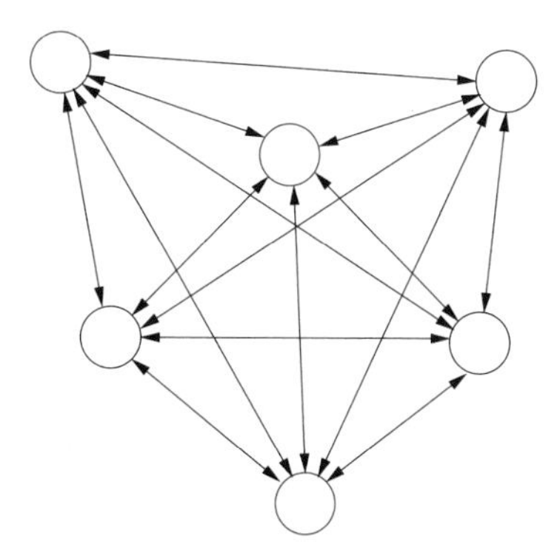

Fig. 6.2. The graph of a Hopfield network with six nodes.

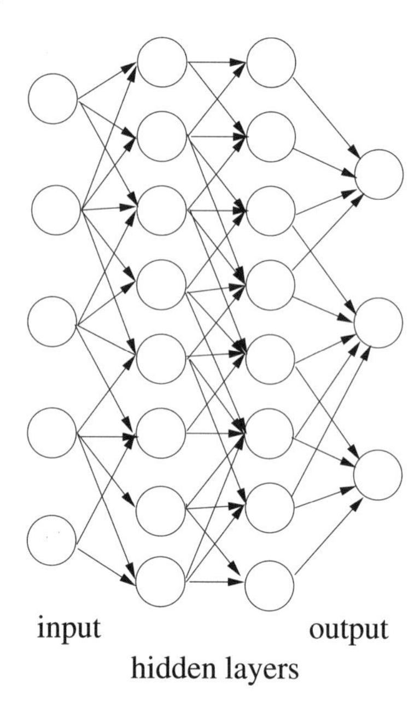

Fig. 6.3. Typical graph of a layer network.

Another version of neural networks shows a so-called layer structure , where the input nodes and output nodes can be identified on the basis of the topological structure. Formally, these networks consist of an input layer, several hidden layers, and an output layer (Figure 6.3).

Topologically, these neural networks contain no loops. Therefore, layer networks are sometimes called filters or feedforward networks. The input pattern is transformed by deterministic or, more rarely, by probabilistic rules into several intermediate patterns at the hidden layers and the final pattern at the output layer.

Modern layer networks imply several feedback mechanisms between subsequent and previous layers. Therefore, we distinguish between two categories of neural networks: feedforward networks or filters without any loops and recurrent networks, where loops occur because of feedback connections. In other words, subsequent layers have the possibility to send data to previous layers that may be used for a change in the weights or the activation functions of the previous layer in order to obtain an improved treatment of the next input. Another frequently used version consists of multiple restarts of the computation using the output of subsequent layers as a new input of previous layers. Such a technique can be used to stabilize the final output.

Between the Hopfield network and the feedforward network exist a lot of intermediate levels. The so-called Kohonen network [207] or feature map consists of a regular d-dimensional lattice and an input layer. Each node of the regular lattice is bidirectionally connected with all nodes of a neighborhood shell, and each node of the input layer is connected by directed links with all nodes of the Kohonen layer. The important property of such a network is that at the end of the computation steps the node with the largest output is set to 1 while all other nodes are defined to be zero. Thus, a Kohonen network can be used for the classification of incoming patterns.

The bidirectional associative memory [52] consists of two layers: the input and the output layers. All components of the connectivity matrix corresponding to links between both layers have the value 1, while all other coefficients vanish. Thus, the network topology of such a network is characterized by a symmetric matrix. Similar to the Hopfield model, the bidirectional associative memory approaches a stationary state after a sufficiently large number of iterative computation steps with the difference that for odd steps the data flow from the input to the output nodes while a data backflow from the output nodes to the input nodes occurs for even computation steps.

Other neural networks, such as the adaptive resonance network [161] or the learning vector quantizers [321], are further realizations of combinations of layer structures.

6.3.4 Training of Neural Networks

A neural network is characterized by its topology and its node characteristics and the training patterns captured in the values of the weights $J_{\alpha\beta}$. The remaining question is how a neural network can store the training patterns. As discussed above, the problem can be solved straightforwardly for a Hopfield network. A similar situation occurs for the bidirectional adaptive memory. But other networks with complicated loops and asymmetric connectivity matrices need a special learning procedure in order to prepare the originally nonspecified system for the subsequent working phase. The training requires a sufficiently strong adaptability of the network.

In general, adaptability may be interpreted as the ability to react to changes in the environment through a learning process [322]. In our case, the environment of a neural network is given by a real system, such as a market, the internal dynamics of which are widely unknown. In order to use a neural network for predictions, it is fed with the same input signal $x(t_n)$ (for instance, several economic indicators or a set of the last stock prices) as the real system at every discrete time step t_n. The output of the neural system may be $y(t_n)$, while $r(t_n)$ is the response of the unknown system. The error signal $e(t_n)$ is formed as the difference of both output signals, $e(t_n) = r(t_n) - y(t_n)$, and the parameters of the weights of the neural network are adjusted using this error information.

The aim of a learning procedure is to update iteratively at each time step t_n the weights $J_{\alpha\beta}(t_n)$ of an adaptive system so that a nonnegative error measure $\mathcal{E}$ is reduced at each time step t_n, $\mathcal{E}(J(t_{n+1})) \leq \mathcal{E}(J(t_n))$. This will generally ensure that, after the training process, the neural network has captured the relevant properties of the unknown system that we are trying to model. Using $\Delta J(t_n) = J(t_{n+1}) - J(t_n)$, we obtain

$$\Delta\mathcal{E}\left(J(t_n)\right) = \mathcal{E}\left(J(t_{n+1})\right) - \mathcal{E}\left(J(t_n)\right) = \sum_{\alpha\beta} \left.\frac{\partial\mathcal{E}\left(J\right)}{\partial J_{\alpha\beta}}\right|_{J=J(t_n)} \Delta J_{\alpha\beta}(t_n) \tag{6.38}$$

and therefore

$$\sum_{\alpha\beta} \left.\frac{\partial\mathcal{E}\left(J\right)}{\partial J_{\alpha\beta}}\right|_{J=J(t_n)} \Delta J_{\alpha\beta}(t_n) \leq 0. \tag{6.39}$$

This equation is always fulfilled for the special choice

$$\Delta J_{\alpha\beta}(t_n) = -\Lambda \left.\frac{\partial\mathcal{E}\left(J\right)}{\partial J_{\alpha\beta}}\right|_{J=J(t_n)}, \tag{6.40}$$

where Λ is a small positive scalar called the learning rate or the adaptation parameter [96]. A learning procedure controlled by (6.40) is also called a gradient-descent-based learning process. We remark that gradient-based algorithms inherently forget old data, which has particular importance for performance of the learning procedure.

The quasi-Newton learning algorithm is based on the second-order derivative of the error function. If we expand the error function in a Taylor series, we have

$$\Delta\mathcal{E}\left(J(t_n)\right) = \sum_{\alpha\beta} \left.\frac{\partial\mathcal{E}\left(J\right)}{\partial J_{\alpha\beta}}\right|_{J=J(t_n)} \Delta J_{\alpha\beta}(t_n) + \frac{1}{2}\sum_{\alpha\beta\gamma\delta} \left.\frac{\partial^2\mathcal{E}\left(J\right)}{\partial J_{\alpha\beta}\partial J_{\gamma\delta}}\right|_{J=J(t_n)} \Delta J_{\alpha\beta}(t_n)\Delta J_{\gamma\delta}(t_n). \tag{6.41}$$

Using the extremum condition $\partial\Delta\mathcal{E}\left(J(t_n)\right)/\partial\Delta J_{\alpha\beta}(t_n) = 0$, we get the changes

$$\Delta J_{\alpha\beta}(t_n) = -\sum_{\gamma\delta}\left[\left(\frac{\partial}{\partial J}\circ\frac{\partial}{\partial J}\mathcal{E}\left(J\right)\right)^{-1}\right]_{\alpha\beta\gamma\delta} \left.\frac{\partial\mathcal{E}\left(J\right)}{\partial J_{\gamma\delta}}\right|_{J=J(t_n)}. \tag{6.42}$$

As a simple example, let us calculate the changes $\Delta J_{\alpha\beta}(t_n)$ for a neural network with only one node and an input vector of dimension Γ. Such a simple neural network is called a perceptron. The error function may be given by

$$\mathcal{E} = e^2(t_n) = \left[r\left(t_n\right) - \psi\left(\sum_{\beta=1}^{\Gamma} J_\beta\left(t_n\right) x_\beta\left(t_n\right)\right)\right]^2 \tag{6.43}$$

with $J_\beta = J_{1\beta}$. Therefore, we obtain

$$\frac{\partial \mathcal{E}}{\partial J_\alpha(t_n)} = -2e(t_n)\psi'\left(\sum_{\beta=1}^{\Gamma} J_\beta(t_n)\, x_\beta(t_n)\right) x_\alpha(t_n)\,, \tag{6.44}$$

and the gradient-descent-based learning process is defined by the equation

$$J_\alpha(t_{n+1}) = J_\alpha(t_n) + 2\Lambda\psi'\left(\sum_{\beta=1}^{\Gamma} J_\beta(t_n)\, x_\beta(t_n)\right) e(t_n)x_\alpha(t_n)\,. \tag{6.45}$$

When deriving a learning algorithm for a general neural network, the network architecture should be taken into account. This leads, of course, to relatively complicated nonlinear equations that must be treated during the training procedure of a network.

In principle, the learning algorithms introduced above are special procedures referring to the class of adaptive learning. Roughly speaking, the idea behind this concept is to forget the past when it is no longer relevant and adapt to the changes in the environment. We remark that the term gear-shifting is sometimes used for the gradient-descent-based learning discussed above when the learning rate is changed during training. Other popular learning algorithms are deterministic and stochastic learning methods [203, 333, 393].

Finally, we mention another learning procedure, which is called constructive learning. This modern version deals with the change of architecture or topological interconnections in the network during training. Neural networks for which the topology can change in the course of the learning procedure are called ontogenic neural networks [122]. The standard procedures of constructive learning are network growing and network pruning. The growing mechanism begins with a very simple network, and if the error is too big, new subnetwork units or single network units are added to the network [173]. In contrast, network pruning starts from a large neural network, and if the error is smaller than a lower limit, the size of the network is reduced [327, 392].

6.3.5 Neural Networks for Forecasting

After the storage of a sufficiently large number of training patterns, the neural network can be used for the prediction of trends. As mentioned above, the complete training patterns contain both the input and the expected output. The treatment of both data groups may differ according to the architecture of the network and the corresponding training method. Typically, the input is a financial or economic time series, while the output may be the continuation of this series or discrete information, such as to buy or sell a certain asset. After finishing the training period, the network gets only the last input and completes this pattern as a result of its internal dynamics. The interesting part of the complete pattern is then the output, which can be understood as a prediction about the future evolution.

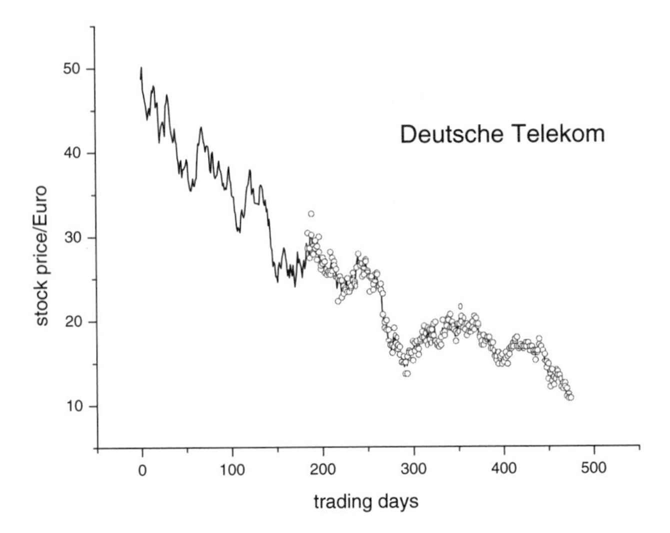

Fig. 6.4. The prediction of the future evolution (open cirles) in comparison with the real data (straight line) for Deutsche Telekom stock. The time interval for the prediction is 3 trading days. The last data are from 07/02. Source: http://www.jt-prognosis.com

In principle, there exist a large list of technical realizations of neural network models used for forecasting financial processes. Here, we will give a few examples in place of this increasing set of prognosis techniques.

A modern version of neural networks consists of coupling different forecasting neural networks [201] with only a few output nodes each and a final comparison of the results by a master network or another comparison procedure. Such a procedure may be the computation of a simple average [201] or another weighted average considering the effectiveness of the neural network models by application of a Bayesian analysis. Furthermore, old training patterns are continuously replaced by new data. This situation looks similar to the application of a moving time window. Such networks may give reasonable predictions on the basis of digital output information, such as that a stock price will increase (decrease) during a given time period.

A combination of autoregression methods and a neural network consisting of an input layer, a hidden Hopfield neural network and an output layer [177] can be used for the simultaneous prediction of the future time evolution of $M = 10^6$ stock prices up to 1 trading month (see Figures 6.4 and 6.5). The accuracy of the predictions may be estimated by the measure

$$\Omega = \frac{1}{NM} \sum_{n=1}^{N} \sum_{\alpha=1}^{M} \frac{\left(X_\alpha(t_n) - Y_\alpha(t_n)\right)^2}{\sigma_\alpha^2\left(T_{\mathrm{P}}\right)}, \tag{6.46}$$

where $Y_\alpha(t_n)$ is the prediction of the stock price α for the time t_n, while $X_\alpha(t_n)$ is the realization of this price after the end of the prediction time. Furthermore, N is the current number of predictions. The variance σ_α is here the mean square change of the stock price during the prediction time period T_P, namely

$$\sigma_\alpha^2(T_\mathrm{P}) = \frac{1}{N}\sum_{n=1}^{N}\left[X_\alpha(t_n) - X_\alpha(t_n - T_\mathrm{P})\right]^2 . \tag{6.47}$$

Obviously, if the predicted change of a stock price $Y_\alpha(t_n) - X_\alpha(t_n - T_\mathrm{P})$ is independent from the real change $X_\alpha(t_n) - X_\alpha(t_n - T_\mathrm{P})$ and both changes have the same variance, the quantity Ω approaches the value 2 for sufficiently large N. The simple prediction $Y_\alpha(t_n) = X_\alpha(t_n - T_\mathrm{P})$ leads to $\Omega = 1$ for $N \to \infty$. Thus, a real prognosis requires $\Omega < 1$. The above-mentioned combined technique [177] has a maximum accuracy of $\Omega \approx 0.60$ ($T_\mathrm{P} = 5$ trading days) for the stocks of the Dow Jones and $\Omega \approx 0.72$ ($T_\mathrm{P} = 5$ trading days) for the stocks of the German stock index DAX.

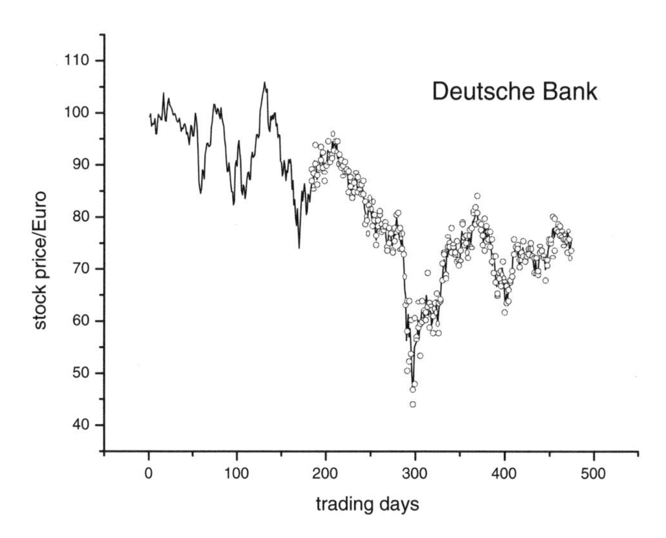

Fig. 6.5. The prediction of the future evolution (open cirles) in comparison with the real data (straight line) for Deutsche Bank stock. The time interval for the prediction is 2 trading days and the last data are from 07/02. Source: http://www.jt-prognosis.com

Finally, we remark that neural networks can be used also for other prediction problems, such as the bankruptcy of companies and banks [394, 395], the optimization of portfolios [434], marketing [176], the classification of consumers and demands [92], or project management [263].

References

1. S. Abe: Phys. Lett. A **224**, 326 (1997).
2. S. N. Afriat: *Logic of Choice and Economic Theory* (Clarendon Press, Oxford, 1987).
3. V. Akgiray: J. Bus. **62**, 55 (1989).
4. C. O. Alexander: Appl. Fin. Econ. **5**, 1 (1995).
5. C. O. Alexander: "Volatility and Correlation Forecasting." In *Handbook of Risk Management and Analysis*, ed. by Carol Alexander (John Wiley and Sons Ltd., New York, 1996) p. 233.
6. K. Amin, R. Jarrow: Math. Fin. **2**, 217 (1992).
7. D. J. Amit: *Field Theory, the Renormalization Group, and Critical Phenomena* (McGraw–Hill, New York, 1978).
8. D. J. Amit, H. Gutfreund, H. Sompolinsky: Phys. Rev. A **32**, 1007 (1985).
9. D. J. Amit, H. Gutfreund, H. Sompolinsky: Phys. Rev. Lett. **55**, 1530 (1985).
10. D. J. Amit: *Modeling Brain Function: The World of Attractor Neural Networks* (Cambridge University Press, Cambridge, 1989).
11. D. J. Amit, H. Gutfreund, H. Sompolinsky: Ann. Phys. (NY) **173**, 30 (1987).
12. L. Andersen: J. Comput. Fin. **3**, 5 (2000).
13. S. Angel, G. M. Hyman: *A Geometry of Movement for Regional Science* (Pion, London, 1976).
14. A. Arnéodo, J.-F. Muzy, D. Sornette: Eur. Phys. J. B **2**, 277 (1998).
15. V. I. Arnold: *Geometrische Methoden in der Theorie gewöhnlicher Differentialgleichungen* (Deutscher Verlag der Wissenschaften, Berlin, 1987).
16. V. I. Arnold, A. Avez: *Ergodic Problems of Classical Mechanics* (Benjamin, New York, 1968).
17. S. Asmussen, P. Glynn, J. Pitman (1995): Ann. Appl. Probab. **5**, 875 (1995).
18. C. L. Attfield, D. Demery, N. W. Duck: *Rational Expectations in Macroeconomics: An Introduction to Theory and Evidence* (Blackwell, Oxford, 1991).
19. E. Aurell, R. Baviera, O. Mammarlid, M. Serva, A. Vulpiani: Int. J. Theor. Appl. Fin. **3**, 1 (2000).
20. V. Z. Averbukh: *Pricing American Options Using Monte Carlo Simulation* (Doctoral Dissertation, Cornell University, Ithaca, 1997).
21. R. U. Ayres, A. V. Kneese: Am. Econ. Rev. **59**, 282 (1969).
22. D. F. Babbel, C. Merrill, W. Panning: "Default Risk and the Effective Duration of Bonds." Working Paper (World Bank, Financial Sector, Development Department, Washington, 1995).
23. Y. S. Badyal, R. A. Howe: J. Phys. Cond. Matter **8**, 3733 (1996).
24. J. Baez, J. Gilliam: Lett. Math. Phys. **31**, 205 (1994).
25. P. Bak: *How Nature Works: The Science of Self-Organized Criticality* (Copernicus, New York, 1996).
26. P. Bak, C. Tang, K. Wiesenfeld: Phys. Rev. Lett. **59**, 381 (1987).
27. P. Bak, C. Tang, K. Wiesenfeld: Phys. Rev. A **38**, 364 (1988).

28. J. Banks: *Handbook of Simulation – Principles, Methodology, Advances, Applications, and Practice* (Wiley, New York, 1998).
29. P. Bantay, I. M. Janosi: Phys. Rev. Lett. **68**, 2058 (1992).
30. A. L. Barabasi, H. E. Stanley: *Fractal Concepts in Surface Growth* (Cambridge University Press, Cambridge, 1995).
31. B. Drossel, F. Schwabl: Phys. Rev. Lett. **71**, 4083 (1993).
32. H. R. Barkai: Economica **26**, 240 (1959).
33. P. W. Barkley, D. W. Secker: *Economic Growth and Environmental Decay* (Harcourt Brace Jovanovich, New York, 1972).
34. J. Barraquand, D. Martineau (1995): J. Fin. Quant. Anal. **30**, 383 (1994).
35. J. W. Barret, G. Moore, P. Wilmott: Risk **5**, 82 (1992).
36. C. C. Barton, P. R. La Pointe: *Fractals in the Earth Sciences* (Plenum Press, New York, 1995).
37. E. Basar: *Chaos in Brain Function* (Springer, Berlin, 1990).
38. D. S. Bates: J. Fin. **46**, 1009 (1991).
39. W. J. Baumol: Econ. J. **68**, 665 (1958).
40. M. J. Beckmann: Econometrica **20**, 643 (1952).
41. M. J. Beckmann, T. Puu: *Spatial Economics: Potential, Density and Flow* (North-Holland, Amsterdam, 1979).
42. M. J. Beckmann, T. Puu: *Spatial Structures* (Springer, Berlin, 1990).
43. A. C. Berry: Trans. Am. Math. Soc. **49**, 122 (1941).
44. B. Biais: *Financial Mathematics* (Springer, Berlin, 1997).
45. O. Biham, A. A. Middleton, D. Levine: Phys. Rev. A **46**, R6124 (1992).
46. K. Binder: *Monte Carlo Methods in Statistical Physics* (Springer, Berlin, 1986).
47. K. Binder: *Applications of the Monte Carlo Methods in Statistical Physics* (Springer, Berlin, 1987).
48. K. Binder, D. W. Heermann: *Monte Carlo Simulations in Statistical Physics* (Springer, Berlin, 1988).
49. M. Bionducci: Z. Naturforsch. A **51**, 71 (1996).
50. F. Black, M. Scholes: J. Polit. Econ. **81**, 637 (1973).
51. M. Blaug: *Economic Theory in Retrospect*, 4th Edition (Cambridge University Press, Cambridge, 1985).
52. U. Blien, H.-G. Lindner: Jahrb. Nationalökon. Stat. **212**, 497 (1993).
53. N. N. Bogoliubov, D. V. Shirkov: *Introduction to the Theory of Quantized Fields* (Interscience, New York, 1959).
54. T. Bollerslev: J. Econ. **31**, 307 (1986).
55. T. Bollerslev, R. Y. Chou, K. F. Kroner: J. Econ. **52**, 5 (1992).
56. T. Bollerslev, R. F. Engle, D. B. Nelson: "ARCH models." In *Handbook of Econometrics, Vol. 4*, ed. by R. F. Engle, D. L. McFadden (Elsevier, North-Holland, Amsterdam, 1994).
57. L. Boltzmann: J. Math. **100**, 201 (1887).
58. J.-P. Bouchaud, R. Cont: Euro. Phys. J. B **6**, 543 (1998).
59. J.-P. Bouchard, D. Sornette: J. Phys. I France **4**, 863 (1994).
60. N. Bouleau, D. Lépingle: *Numerical Methods for Stochastic Processes* (Wiley, New York, 1994).
61. J. N. Bodurtha, Jr., G. R. Courtadon: J. Fin. Quant. Anal. **22**, 153 (1987).
62. G. E. P. Box, G. M. Jenkins: *Time Series Analysis: Forecasting and Control* (Holden–Day, New York, 1976).
63. P. Boyle, M. Broadie, P. Glasserman (1997): J. Econ. Dyn. Control **21**, 1267 (1997).
64. W. H. Branson: *Macroeconomic Theory and Policy* (Harper & Row, New York, 1979).

65. P. Bratley, B. L. Fox, L. E. Schrage: *A Guide to Simulation* (Springer, New York, 1987).
66. R. Brotherton-Ratcliffe: Risk **7**, 53 (1994).
67. D. J. Burr: "Artificial Neural Networks: A Decade of Progress." In *Artificial Neural Networks for Speech and Vision*, ed. by R. J. Mammone (Chapman and Hall, New York, 1993).
68. E. R. Caianiello: J. Theor. Biol. **1**, 204 (1961).
69. S. Carlson: *A Study on the Pure Theory of Production* (P. S. King & Son Ltd., London, 1939).
70. B. Castaing: J. Phys. II France **6**, 105 (1997).
71. E. Castillo: *Extreme Value Theory in Engineering* (Academic Press, Boston, 1988).
72. P. L. Chebyshev: Acta Math. **14**, 305 (1890).
73. S. Chen, H. Chen, D. Martinez, W. Matthaeus: Phys. Rev. Lett. **67**, 3776 (1991).
74. M. Chesney, L. O. Scott: J. Fin. Quant. Anal. **24**, 267 (1989).
75. D. Chowdbury, J. Kertezs, K. Nagel, L. Santen, A. Schadschneider: Phys. Rev. E **61**, 3270 (2000).
76. P. Cizeau, Y. Liu, M. Meyer, C.-K. Peng, H. E. Stanley: Physica A **245**, 441 (1997).
77. L. Clewlow, C. Strickland: *Implementing Derivatives Models* (Wiley, New York, 1998).
78. C. W. Cobb, P. H. Douglas: Am. Econ. Rev. **18**, 139 (1928).
79. P. Coullet, J. Tresser: J. Phys. (Paris) **C5**, 25 (1978).
80. J. C. Cox, S. A. Ross, M. Rubinstein: J. Fin. Econ. **7**, 229 (1979).
81. J. C. Cox, M. Rubinstein: *Options Markets* (Prentice–Hall, Englewood Cliffs, 1985).
82. J. S. Cramer: *Econometric Applications of Maximum Likelihood Methods* (Cambridge University Press, Cambridge, 1986).
83. M. Cremer, A. D. May: *An Extended Traffic Model for Freeway Control.* Technical Report UCB-ITS-RR-85-7 (Institute of Transportation Studies, University of California, Berkeley).
84. A. Crisanti, G. Paladin, A. Vulpiani: *Products of Random Matrices in Statistical Physics* (Springer, Berlin, 1993).
85. A. Crisanti, M. H. Jensen, A. Vulpiani, G. Paladin: Phys. Rev. A **46**, R7363 (1992).
86. E. M. F. Curado, C. Tsallis: J. Phys. A **24**, L69 (1991).
87. T. D. Crocker, A. J. Rogers: *Environmental Economics* (Dryden Press, Hinsdale, 1971).
88. M. M. Dacorogna, U. A. Müller, R. J. Nagler, R. B. Olsen, O. V. Pictet: J. Intl. Money and Fin. **12**, 413 (1993).
89. J. H. Dales: *Pollution, Property and Prices* (University of Toronto Press, Toronto, 1968).
90. H. E. Daly: *Toward a Steady State Economy* (W. H. Freeman & Co., San Francisco, 1973).
91. S. Das: *Risk Management and Financial Derivatives: A Guide to the Mathematics* (Macmillan, Basingstoke, 1998).
92. C. G. Dasgupta, G. S. Dispensa, S. Ghose: Int. J. Forecasting **10**, 235 (1994).
93. R. T. Deam, S. F. Edwards: Philos. Trans. R. Soc. London Ser. A **280**, 317 (1976).
94. H. D. Dickinson: Econ. J. **79**, 894 (1969).
95. R. M. Dillon, C. N. Manikopoulos: Electron. Lett. **27**, 824 (1991).

96. S. C. Douglas: IEEE Trans. Circuits Syst. II: Analog and Digital Signal Proc.: **44**, 209 (1997).
97. J. E. Doran, N. Gilbert: "Simulating Societies: An Introduction." In *Simulating Societies: The Computer Simulation of Social Processes*, ed. by N. Gilbert, J. E. Doran (University College London Press, London, 1994), p. 1.
98. P. B. Downing: *Air Pollution and the Social Sciences* (Praeger Publishers, New York, 1973).
99. F. C. Drost, T. E. Nijman: Econometrica **61**, 909 (1993).
100. B. Dubrulle, F. Graner, D. Sornette: *Scale Invariance and Beyond* (EDP Sciences and Springer, Berlin, 1997).
101. E. J. Dudewicz: *Modern Mathematical Statistics* (Wiley, New York, 1988).
102. D. Duffie: *Dynamic Asset Pricing Theory*, Second Edition (Princeton University Press, Princeton, 1996).
103. D. Duffie, P. Glynn: Ann. Appl. Probab. **5**, 897 (1995).
104. D. Duffie, K. J. Singleton: "Econometric Modeling of Term Structures of Defaultable Bonds" Working Paper (World Bank, Financial Sector, Development Department, Washington, 1994).
105. E. Eberlein, U. Keller: Bernoulli **1**, 281 (1995).
106. M. Edel: *Economics and Environment* (Prentice–Hall, Englewood Cliffs, 1973).
107. M. Elshamy: *Bivariate Extreme Value Distributions* (National Aeronautics and Space Administration, Office of Management, Scientific and Technical Information Program, Springfield, 1992).
108. E. J. Elton, M. J. Gruber: *Modern Portfolio Theory and Investment Analysis* (Wiley, New York, 1991).
109. P. Embrechts, C. Kluppelberg, T. Mikosch: *Modelling Extremal Events for Insurance and Finance* (Springer, New York, 1997).
110. R. F. Engle: Econometrica **50**, 987 (1982).
111. R. F. Engle, G. G. J. Lee: "Long Run Volatility Forecasting for Individual Stocks in a One Factor Model." Working Paper (Department of Economics, University of California at San Diego, La Jolla (1993).
112. R. F. Engle, G. G. J. Lee: "A Permanent and Transitory Component Model of Stock Return Volatility." Working Paper (Department of Economics, University of California at San Diego, La Jolla, (1993).
113. G. B. Ermentrout, L. Edlestein-Keshet: J. Theor. Bio. **160**, 97 (1993).
114. C. G. Esséen: Acta Math. **77**, 1 (1945).
115. F. Family, T. Vicsek: *Dynamics of Fractal Surfaces* (World Scientific, Singapore, 1991).
116. K.-T. Fang, Y. Wang: *Number-Theoretic Methods in Statistics* (Chapman and Hall, London, 1994).
117. D. E. Farrar, R. R. Glauber: Rev. Econ. Stat. **49**, 92 (1967).
118. M. J. Feigenbaum: J. Stat. Phys. **19**, 25 (1978).
119. W. Feller: *An Introduction to Probability Theory and Its Applications, Vol. 1*, Third Edition (Wiley, New York, 1968).
120. W. Feller: *An Introduction to Probability Theory and Its Applications, Vol. 2*, Second Edition (Wiley, New York, 1971).
121. B. Felderer, S. Homburg: *Makroökonomik und neue Makroökonomik* (Springer, Berlin, 1989).
122. E. Fiesler, R. Beale: *Handbook of Neural Computation* (Oxford University Press, Oxford, 1997).
123. D. E. Fischer, R. J. Jordan: *Security Analysis and Portfolio Management* (Prentice–Hall, Englewood Cliffs, 1995).

124. K. H. Fischer, J. A. Hertz: *Spin Glasses* (Cambridge University Press, Cambridge, 1991).
125. G. S. Fishman: *Discrete-Event Simulation — Modeling, Programming and Analysis* (Springer, New York, 2001).
126. A. W. Flux, Econ. J. **4**, 308 (1894).
127. J. C. Francis, Cheng-Few Lee, and Donald E. Farrar: *Readings in Investments* (McGraw–Hill Book Co., New York, 1980).
128. S. Franz, G. Parisi, M. A. Virasoro: Europhys. Lett. **17**, 5 (1992).
129. S. Franz, G. Parisi, M. A. Virasoro: J. Phys. I France **2**, 1869 (1992).
130. A. M. Freeman, R. H. Haveman, A. V. Kneese: *The Economics of Environmental Policy* (Wiley, New York, 1973).
131. K. R. French: J. Fin. Econ. **8**, 55 (1980).
132. R. Frisch: *Theory of Production* (D. Reidel, Dordrecht, 1965).
133. U. Frisch: *Turbulence, The Legacy of A. N. Kolmogorov* (Cambridge University Press, Cambridge, 1995).
134. U. Frisch, B. Hasslacher, Y. Pomeau: Phys. Rev. Lett. **56**, 1505 (1986).
135. U. Frisch, D. d'Humières, B. Hasslacher, P. Lallemand, Y. Pomeau, J.-P. Rivet: Complex Syst. **1**, 649 (1987).
136. U. Frisch, D. Sornette: J. Phys. I France **7**, 1155 (1997).
137. H. Funke: *Eine allgemeine Theorie der Polypol- und Oligopolpreisbildung* (Springer, Berlin, 1985).
138. J. Galambos: *The Asymptotic Theory of Extreme Order Statistics*, Second Edition (R.E. Krieger, Malabar, 1987).
139. J. Galambos, J. Lechner, E. Simiu: *Extreme Value Theory and Applications: Proceedings of the Conference on Extreme Value Theory and Applications, Gaithersberg, Maryland, 1993* (Kluwer Academic, Boston, 1994).
140. D. Gale: *The Theory of Linear Economic Models* (McGraw–Hill, New York, 1960).
141. A. Galli, M. Armstrong, B. Jehl: J. Petrol.Technology **51**, 44 (1999).
142. G. Gandolfo: *International Economics, Vol.2* (Springer, Berlin, 1995).
143. C. W. Gardiner: *Handbook of Stochastic Methods* (Springer, Berlin, 1997).
144. M. S. Geisel: "Comparing and Choosing among Parametric Statistical Models: A Bayesian Analysis with Macroeconomic Applications." (Ph.D. dissertation, University of Chicago, 1970).
145. M. S. Geisel: "Bayesian Comparisons of Simple Macroeconomic Models." In *Studies in Bayesian Econometrics and Statistics*, ed. by S. Feinberg, A. Zellner (North-Holland, Amsterdam, 1974).
146. P.-G. de Gennes: *Scaling Concepts in Polymer Physics* (Cornell University Press, Ithaca, 1979).
147. C. R. Gent, C. P. Sheppard: Comput. Control Eng. J., **3**, 109 (1992).
148. J. E. Gentle: *Random Number Generation and Monte Carlo Methods* (Springer, New York, 1998).
149. N. Georgescu-Roegen: *The Entropy Law and the Economic Process* (Harvard University Press, Cambridge, 1971).
150. N. Georgescu-Roegen: "The Entropy Law and the Economic Problem." In *Toward a Steady State Economy*, ed. by H. E. Daly (W. H. Freeman & Co., San Francisco, 1973), p. 33.
151. N. A. Gershenfeld, A. S. Weigend: "The Future of Time Series: Learning and Understanding." In *Time Series Prediction: Forecasting the Future and Understanding the Past*, ed. by A. S. Weigend, N. A. Gershenfeld (Addison–Wesley, Reading, 1993).
152. B. V. Gnedenko, A. N. Kolmogorov: *Limit Distribution for Sum of Independent Random Variables* (Addison–Wesley, Reading, 1954).

153. J. Goldstone, A. Salam, S. Weinberg: Phys. Rev. **127**, 965 (1962).
154. P. Gopikrishnan, M. Meyer, L. A. N. Amaral, H. E. Stanley: Eur. Phys. J. B **3**, 139 (1998).
155. P. Gopikrishnan, V. Plerou, L. A. N. Amaral, M. Meyer, H. E. Stanley: Phys. Rev. E **60**, 1390 (1999).
156. D. Grant, G. Vora, D. E. Weeks: J. Fin. Eng. **5**, 221 (1996).
157. D. Grant, G. Vora, D. E. Weeks: Manage. Sci. **43**, 1589 (1997).
158. Z. Griliches: *Handbook in Economics* (North-Holland, Amsterdam, 1992).
159. P. Groeneboom, J. A. Wellner: *Information Bounds and Nonparametric Maximum Likelihood Estimation* (Birkhäuser, Basel, 1992).
160. S. Grossmann, S. Thomae: Z. Naturforsch. **32** A, 1353 (1977).
161. S. Grossberg: Prog. Theor. Biol. **3**, 51 (1974).
162. B. Grosz: AI Mag. **17**, 67 (1996).
163. T. Guhr, A. Müller-Groeling, H. A. Weidenmüller: Phys. Rep. **299**, 189 (1998).
164. E. J. Gumbel: *Statistical Theory of Extreme Values and Some Practical Applications: A Series of Lectures* (U.S. Government Printing Office, Washington, 1954).
165. E. J. Gumbel: *Statistics of Extremes* (Columbia University Press, New York, 1958).
166. H. Haken: *Information and Self-Organization* (Springer, Berlin, 2000).
167. J. M. Harrison, D. Kreps: J. Econ. Theory **20**, 381 (1979).
168. D. O. Hebb: *The Organization of Behaviour* (Wiley, New York, 1949).
169. R. Hegselmann: "Cellular Automata in the Social Science." In *Modelling and Simulation in the Social Sciences from the Philosophy of Science Point of View*, ed. by R. Hegselmann (Kluwer Academic Publishers, Dordrecht, 1996), p. 209.
170. P. Hellekalek, G. Larcher: *Random and Quasi-Random Point Sets* (Springer, New York, 1998).
171. S. Heston: Rev. Fin. Stud. **6**, 327 (1987).
172. J. R. Hicks: *The Theory of Wages* (Macmillan, New York, 1932).
173. M. Hochfeld, S. E. Fahlman: IEEE Trans. Neural Networks **3**, 603 (1992).
174. J. J. Hopfield: Proc. Nat. Acad. Sci. USA **79**, 2554 (1982).
175. J. J. Hopfield: Proc. Nat. Acad. Sci. USA **81**, 3088 (1984).
176. H. Hruschka: Marketing ZFP **4**, 217 (1991).
177. http://www.JT-Prognosis.com.
178. C. Huang, R. H. Litzenberger: *Foundations for Financial Economics* (Prentice–Hall, Englewood Cliffs, 1988).
179. J. C. Hull: *Options, Futures and Other Derivatives*, Third Edition (Prentice–Hall, Upper Saddle River, 1997).
180. J. C. Hull, A. White: J. Fin. **42**, 281 (1987).
181. S. P. Huntington: *Political Order in Changing Societies* (Yale University Press, New Haven, 1968).
182. J. F. Ingersoll: *Theory of Financial Decision Making* (Rowman and Littlefield, Savage, 1987).
183. C. Itzykson, J.-M. Drouffe: *Statistical Field Theory, Vol. 1* (Cambridge University Press, Cambridge, 1989).
184. C. Itzykson, J.-M. Drouffe: *Statistical Field Theory, Vol. 2* (Cambridge University Press, Cambridge, 1989).
185. P. Jäckel, *Monte Carlo Methods in Finance* (Wiley, New York, 2002).
186. R. A. Jarrow, D. Lando, S. M. Turnbull: Rev. Fin. Stud. **10**, 481 (1997).
187. R. A. Jarrow, S. M. Turnbull: J. Fin. **50**, 53 (1995).
188. E. T. Jaynes: Phys. Rev. **106**, 4 (1957).
189. E. T. Jaynes: Phys. Rev. **108**, 171 (1957).

190. E. T. Jaynes: Am. J. Phys. **33**, 391 (1965).
191. G. Jona-Lasinio: Nuovo Cimento **34**, 1790 (1964).
192. C. Joy, P. P. Boyle: Manage. Sci. **42**, 926 (1996).
193. L. Kadanoff, G. McNamara, G. Zanetti: Phys. Rev. A **40**, 4527 (1989).
194. E. Kamke: *Differentialgleichungen, Lösungsmethoden und Lösungen* (Teubner, Stuttgart, 1983).
195. E. R. Kandel, J. H. Schwartz: *Principles of Neural Science* (Elsevier, Amsterdam, 1985).
196. T. Kariya: *Quantitative Methods for Portfolio Analysis: MTV model Approach* (Kluwer, Dordrecht, 1993).
197. K. Kawasaki: Phys. Rev. **150**, 291 (1966).
198. D. A. Keen, R. L. McGreevy: Nature (London) **344**, 423 (1990).
199. W. T. Lord Kelvin: Br. Assoc. Adv. Sci. Rep. **51**, 526 (1882).
200. A. Ya. Khintchine, P. Lévy: C. R. Acad. Sci. Paris **202**, 374 (1936).
201. T. Kimoto, K. Asakawa, M. Yoda, M. Takeoka: "Stock Market Prediction System with Modular Neural Networks." In *Proceedings of the International Joint Conference on Neural Networks, Vol. 1* (IEEE Network Council, San Diego, 1990), p. 1.
202. R. R. Kinnison: *Applied Extreme Value Statistics* (Battelle Press, Columbus, 1985, distributed by Macmillan, New York).
203. S. Kirkpatrick, C. D. Gelatt, Jr., M. P. Vecchi: Science **220**, 671 (1983).
204. N. Koblitz: The Math. Intell. **10** (1), 4 (1988).
205. N. Koblitz: The Math. Intell. **10** (1), 14 (1988).
206. N. Koblitz: The Math. Intell. **10** (2), 11 (1988).
207. T. Kohonen: Biol. Cybernet. **43**, 59 (1982).
208. A. N. Kolmogorov: Dokl. Akad. Nauk. SSSR **30**, 9 (1941).
209. A. N. Kolmogorov: Dokl. Akad. Nauk. SSSR **31**, 9538 (1941).
210. A. N. Kolmogorov: Dokl. Akad. Nauk. SSSR **32**, 16 (1941).
211. I. Koponen: Phys. Rev. E **52**, 1197 (1995).
212. D. H. Krantz, R. D. Luce, P. Suppes, A. Tversky: *Foundations of Measurement* (Academic Press, New York, 1971).
213. U. Krause: "Impossible Models." In *Modelling and Simulation in the Social Sciences from the Philosophy of Science Point of View*, ed. by R. Hegselmann (Kluwer Academic Publishers, Dordrecht, 1996).
214. D. M. Kreps: *A Course in Microeconomic Theory* (Prentice–Hall, New York, 1999).
215. R. Kubo: J. Phys. Soc. Jpn. **17**, 1100 (1962).
216. R. Kubo, K. Matsuo, K. Kitahara: J. Stat. Phys. **9**, 51 (1973).
217. S. W. Kuffler, J. G. Nichols, A. R. Martin: *From Neuron to Brain* (Sinauer Associates, Sunderland, 1984).
218. S. Kulback: Ann. Math. Stat. **22**, 79 (1951).
219. S. Kulback: *Information Theory and Statistics* (Wiley, New York, 1951).
220. L. Laloux, P. Cizeau, J.-P. Bouchaud: Phys. Rev. Lett. **83**, 1467 (1999).
221. R. Lämmel: *Sozialphysik: Naturkraft, Mensch und Wirtschaft* (Franckh'sche Verlagshandlung, Stuttgart, 1925).
222. S. Lang: *The File* (Springer, New York, 1981).
223. L. D. Landau: *Statistische Physik*, Third Edition (Akademische Verlagsgesellschaft, Berlin, 1971).
224. D. Lando: "A Continuous-Time Markov Model of the Term Structure of Credit Spreads." Working Paper (Graduate Schoal of Management, Cornell University, Ithaca, 1994).

225. O. E. Lanford: "Entropy and Equilibrium States in Classical Mechanics." In *Statistical Mechanics and Mathematical Problems*, ed. by A. Lenard Lecture Notes in Physics, Vol. 20 (Springer, Berlin, 1973), p. 1.
226. H. A. Latane, D. L. Tuttle, C. P. Jones: *Security Analysis and Portfolio Management*, Second Edition (Ronald, New York, 1975).
227. W. Launhardt: *Mathematische Begründung der Volkswirtschaftslehre* (B. G. Teubner, Leipzig, 1885).
228. D. A. Lavis, G. M. Bell: *Statistical Mechanics of Lattice Systems, Vol. 2* (Springer, Berlin, 1999).
229. C. F. Lee, J. E. Finnerty, D. H. Wort: *Security Analysis and Portfolio Management* (Scott, Foresman/Little Brown Higher Education, Glenview, 1990).
230. H. E. Leland, K. B. Toft: "Optimal Capital Structure, Endogenous Bankruptcy, and the Term Structure of Credit Spreads." Working Paper (University of California, Berkeley, 1995).
231. V. R. Lesser: ACM Comput. Surv. **27**, 340 (1995).
232. P. Lévy: *Caleul des probabilités* (Gauthier-Villars, Paris, 1925).
233. J. X. Li: Rev. An. Econ. **15**, 111 (2000).
234. C. W. Li, W. K. Li: J. Appl. Econ. **11**, 253 (1996).
235. W. A. Little: Math. Biosci. **109**, 101 (1974).
236. J. S. Liu: *Monte Carlo Strategies in Scientific Computing* (Springer, New York, 2001).
237. Y. Liu, P. Cizeau, M. Meyer, C.-K. Peng, H. E. Stanley: Physica A **245**, 437 (1997).
238. Y. Liu, P. Gopikrishnan, P. Cizeau, M. Meyer, C.-K. Peng, H. E. Stanley: Phys. Rev. E **59**, 1390 (1999).
239. L. Ljung, T. Soderstrom: IEEE Trans. Neural Networks **5**, 803 (1983).
240. P. J. Lloyd: J. Pol. Econ. **77**, 21 (1969).
241. F. Longstaff, E. Schwartz: J. Fin. **47**, 1259 (1992).
242. E. Lukacs: *Probability and Mathematical Statistics* (Academic Press, New York, 1972).
243. T. Lux: J. Econ. Behav. Organ. **33**, 143 (1998).
244. D. B. Madan, H. Unal: "Pricing of the Risk Default." Working Paper (College of Business, University of Maryland, College Park, 1994).
245. J. Makhoul: Proc. IEEE **63**, 561 (1995).
246. B. B. Mandelbrot: J. Bus. **36**, 394 (1963).
247. B. B. Mandelbrot: *The Fractal Geometry of Nature* (W. H. Freeman, San Francisco, 1982).
248. B. B. Mandelbrot: Science **279**, 783 (1998).
249. R. N. Mantegna, H. E. Stanley: Phys. Rev. Lett **73**, 2946 (1994).
250. R. N. Mantegna, H. E. Stanley: Nature (London) **376**, 46 (1995).
251. R. N. Mantegna, H. E. Stanley: Nature (London) **383**, 587 (1996).
252. R. N. Mantegna, H. E. Stanley: "Physics Investigations of Financial Markets." In *Proceedings of the International School of Physics Enrico Fermi', Course CXXXIV*, ed. by F. Mallamace, H. E. Stanley (IOS Press, Amsterdam, 1997).
253. R. N. Mantegna, H. E. Stanley: *Introduction to Econophysics* (Cambridge University Press, Cambridge, 1999).
254. J. Marcienkiewicz: Math. Z. **44**, 612 (1939).
255. H. M. Markowitz: J. Fin. **1**, 77 (1952).
256. H. M. Markowitz: *Portfolio Selection: Efficient Diversification of Investments* (Blackwell, Cambridge, 1993).
257. M. Markus, B. Hess: Nature (London) **347**, 56 (1990).
258. A. Mas-Colell, M. D. Whinston, J. R. Green: *Microeconomic theory* (Oxford University Press, New York, 1995).

259. A. Mataez: "Financial Modeling on Option Theory with Truncated Lévy Process." Working Paper School of Mathematics and Statistics, Report 97-28 (University of Sidney, Sidney, 1997).
260. R. M. May: Nature (London) **261**, 459 (1976).
261. W. S. McCullough, W. Pitts: Bull. Math. Biophys. **5**, 115 (1943).
262. R. L. McGreevy, L. Pusztai: Mol. Simulation **1**, 359 (1988).
263. R. A. McKim: Project Manage. J. **24**, 28 (1993) .
264. D. Meadows, J. Richardson, G. Bruckmann: *Groping in the Dark* (Wiley, New York, 1982).
265. R. Merton: Bell J. Econ. Manage. Sci. **4**, 141 (1973).
266. R. C. Merton: J. Fin. Econ. **3**, 125 (1976).
267. R. C. Merton: J. Fin. Econ. **5**, 241 (1977).
268. R. C. Merton: *Continuous-Time Finance* (Blackwell, Cambridge, 1990).
269. M. Metha: *Random Matrices* (Academic Press, New York, 1995).
270. A. Metha, S. F. Edwards: Physica A **157**, 1091 (1991).
271. N. Metropolis, S. Ulam: J. Am. Stat. Assoc. **44**, 335 (1949).
272. N. Metropolis, A. W. Rosenbluth, M. N. Rosenbluth, A. H. Teller, E. Teller: J. Chem. Phys. **21**, 1087 (1953).
273. E. A. Mitscherlich: Landw. Jahrb. **38**, 537 (1909).
274. T. Mommsen: Corpus Inscriptionum Latinarum **III**, 801-841, 1055-1058, 1909-1953 and 2208-2211, (de Gruyter, Berlin, 1873).
275. W. J. Morokoff: SIAM Rev. **40**, 765 (1998).
276. M. Musiela, M. Rutkowski: *Martingale Methods in Financial Modeling* (Springer, Berlin, 1997).
277. K. Nagel, M. Schreckenberg: J. Phys. I France **2**, 2221 (1992).
278. N. J. Nagelkerke: *Maximum Likelihood Estimation of Functional Relationships*, Lecture Notes in Statistics (Springer, Berlin, 1992).
279. L. Narens: Theory and Decision **13**, 1 (1971).
280. D. B. Nelson: Econometrica **59**, 347 (1991).
281. H. Niederreiter: SIAM CBMS **63**, 241 (1992).
282. H. Niederreiter, P. Hellekalek, G. Larcher, P. Zinterhof: *Monte Carlo and Quasi-Monte Carlo Methods 1996*, Lectures Notes in Statistics (Springer, New York, 1998).
283. S. S. Nielsen, E. I. Ronn: Adv. Futures Options Res. **9**, 175 (1997).
284. S. P. Nishenko, C. C. Barton: Geol. Soc. Am. Abstr. Programs **25**, 412 (1993).
285. A. Noguchi: "General Equilibrium Models." In *Economic and Financial Modelling with Mathematica*, ed. by H. R. Varian (Springer-Verlag, Berlin, 1993).
286. A. Nowak, M. Lewenstein: "Modeling Social Change with Cellular Automata." In *Modelling and Simulation in the Social Sciences from the Philosophy of Science Point of View*, ed. by R. Hegselmann (Kluwer Academic Publishers, Dordrecht, 1996), p. 249.
287. M. A. Nowak, R. M. May: Nature (London) **359**, 826 (1992).
288. E. P. Odum: *Fundamentals of Ecology* (Saunders, Philadelphia, 1971).
289. K. Okuguchi: *The Theory of Oligopoly with Multi-product Firms* (Springer, Berlin, 1990).
290. A. M. Okun: Am. Stat. Assoc. Proc. Bus. Econ. Stat. Sec., 98 (1962).
291. Z. H. Olami, J. S. Feder, K. Christensen: Phys. Rev. Lett. **68**, 1244 (1992).
292. C.-A. Olsson: Econ. Hist. **14**, 64 (1971).
293. D. F. Owen: *What Is Ecology?* (Oxford University Press, London, 1974).
294. T. F. Palander: *Beiträge zur Staatstheorie* (Almquist & Wiksells, Uppsala, 1935).
295. T. I. Palley: Int. Rev. Appl. Econ. **7**, 144 (1963).
296. A. Pagan: J. Empirical Fin. **3**, 15 (1996).

297. S. V. Panyukov, Y. Rabin: Phys. Rep. **269**, 1 (1996).
298. C. H. Papadimitriou, K. Steigitz: *Combinatorial Optimization* (Prentice–Hall, Englewood Cliffs, 1996).
299. G. Parisi: Phys. Rev. Lett. **43**, 1754 (1979).
300. G. Parisi: J. Phys. A **13**, 1101 (1980).
301. G. Parisi: J. Stat. Phys. **72**, 857 (1993).
302. G. Parisi, N. Sourlas: Phys. Rev. Lett. **43**, 744 (1979).
303. W. Paul, J. Baschnagel: *Stochastic Processes. From Physics to Finance* (Springer, Berlin, 2000).
304. H. O. Peitgen, P. H. Richter: *The Beauty of Fractals* (Springer, New York, 1986).
305. G. Peng, H. J. Heermann: Phys. Rev. E **49**, 1796 (1994).
306. I. Peterson: Sci. News **158** (November 11, 2000).
307. A. W. H. Phillips: Economica **25**, 283 (1958).
308. J. A. Picazo: "American Option Pricing: A Classification-Monte Carlo (CMC) Approach." In *Monte Carlo and Quasi-Monte Carlo Methods 2000*, ed. by K.-T. Fang, F. J. Hickernell, H. Niederreiter (Springer, Berlin, 2002), pp. 422–433.
309. V. F. Pisarenko, Hydrol. Proc. **12**, 461 (1998).
310. P. A. Samuelson: J. Econ. Lit. **15**, 24 (1977).
311. V. Plerou, P. Gopikrishnan, B. Rosenow, L. A. N. Amaral, H. E. Stanley: Phys. Rev. Lett. **83**, 1471 (1999).
312. Y. Pormeau: J. Phys. **43**, 859 (1982).
313. M. Postan: *The Cambridge Economic History of Europe, Vols. 1–3* (1966–1977).
314. R. B. Potts: Proc. Cambridge Philos. Soc. **48**, 106 (1952).
315. A. A. Powell, C. W. Murphy: *Inside a Modern Macroeconometric Model: A Guide to the Murphy model* (Springer, Berlin, 1995).
316. T. Puu: Reg. Sci. Urban Econ. **8**, 225 (1978).
317. T. Puu: *The Allocation of Road Capital in Two-Dimensional Space: A Continuous Approach* (North-Holland, Amsterdam, 1979).
318. T. Puu: Reg. Sci. and Urban Econ. **11**, 317 (1981).
319. T. Puu: Chaos, Solitons and Fractals **5**, 35 (1995).
320. T. Puu: Chaos, Solitons and Fractals **3**, 99 (1993).
321. M. Pytlik: *Diskriminierungsanalyse und künstliche Neuronale Netze zur Klassifizierung von Jahresabschlüssen* (Peter Lang GmbH, Frankfurt, 1995).
322. S. Haykin: IEEE Signal Processing Mag. **15**, 66 (1999).
323. M. Raberto, E. Scalas, G. Gumberti, M. Riani: Physica A **269**, 148 (1999).
324. R. Rammal, G. Toulouse, M.A. Virasoro: Rev. Mod. Phys. **58**, 765 (1986).
325. J. B. Ramsey, P. Rothman: J. Money Credit Banking **28**, 1 (1996).
326. S.B. Raymar, M.J. Zwecher (1997): J. Derivatives **5**, 7 (1997).
327. R. Reed: IEEE Trans. Neural Networks **4**, 740 (1993).
328. D. Ricardo: *On the Principles of Political Economy and Taxation* (Murray, London, 1817).
329. B.D. Ripley: *Stochastic Simulation* (Wiley, New York, 1987).
330. H. Risken: *The Fokker–Planck Equation* (Springer, Berlin, 1996).
331. F. Robert: *Microeconomics and Behavior* (McGraw–Hill, New York, 1994).
332. J. Robinson: *Economics of Imperfect Competition* (Macmillian, London, 1933).
333. K. Rose: Proc. IEEE **86**, 2210 (1998).
334. F. Rosenblatt: Psychol. Rev. **65**, 386 (1958).
335. F. Rosenblatt: *Principles of Neurodynamics* (Spartan, Washington, 1962).
336. A. Rosenthal: Ann. Phys. **42**, 796 (1913).
337. S. Ross: J. Econ. Theory **13**, 341 (1976).

338. S.M. Ross: *Simulation*, Second Edition (Academic Press, New York, 1997).
339. R. Y. Rubinstein: *Simulation and the Monte Carlo Method* (Wiley, New York, 1981).
340. M. Rudolf: *Algorithms for Portfolio Optimization and Portfolio Insurance* (Haupt, Bern, 1994).
341. D. E. Rumelhart, G. E. Hinton, R. Williams: Nature (London) **323**, 533 (1986).
342. J. B. Rundle, W. Klein, S. Gross, D. L. Turcotte: Phys. Rev. Lett. **75**, 1658 (1995).
343. J. B. Rundle, W. Klein, S. Gross, D. L. Turcotte: Phys. Rev. Lett. **78**, 3798 (1997).
344. R. R. Russel: *Microeconomics* (Wiley, New York, 1979).
345. G. Samorodnitsky, M. S. Taqqu: *Stable Non-Gaussian Random Processes: Stochastic Models with Infinite Variance* (Chapman and Hall, New York, 1994).
346. P. A. Samuelson: J. Pol. Econ. **87**, 923 (1979).
347. B. Sandelin: Econ. and Hist. **19**, 117 (1976).
348. S. B. Savage: Adv. Appl. Mech. **24**, 289 (1994).
349. E. Schneider: *Pricing and Equilibrium* (Allen and Unwin Ltd., Crows Nest, 1962).
350. J. A. Schumpeter: *History of Economic Analysis* (Allen and Unwin, London, 1954).
351. B. M. Schulz, M. Schulz, S. Trimper: Phys. Rev. E **58**, 3368 (1998).
352. B. M. Schulz, M. Schulz, S. Trimper: Phys. Lett. A **291**, 87 (2001).
353. B. M. Schulz, S. Trimper: Phys. Lett. A **256**, 266 (1999).
354. B. M. Schulz, S. Trimper, M. Schulz: Eur. Phys. J. B **15**, 499 (2000).
355. B. M. Schulz, S. Trimper, M. Schulz: J. Chem. Phys. **114**, 10402 (2001).
356. M. Schulz: J. Chem. Phys. **133**, 10793 (2000).
357. M. Schulz, B. M. Schulz: Phys. Rev. B **58**, 8178 (1998).
358. M. Schulz, B. M. Schulz, S. Trimper: Phys. Rev. E **64**, 026104 (2001).
359. M. Schulz, P. Reineker: Phys. Rev. B **48**, 9369 (1993).
360. M. Schulz, P. Reineker: Phys. Rev. B **52**, 4131 (1995).
361. M. Schulz, P. Reineker: Chem. Phys. **284**, 331 (2002)
362. M. Schulz, S. Stepanow: Phys. Rev. B **59**, 13528 (1999).
363. M. Schulz, S. Trimper: Phys. Rev. B **96**, 8421 (1996).
364. M. Schulz, S. Trimper: Phys. Rev. B **58**, 8178 (1998).
365. M. Schulz, S. Trimper: J. Phys. A: Math. Gen. **33**, 7289 (2000).
366. M. Schulz, S. Trimper: Phys. Rev. B **64**, 233101 (2001).
367. H. G. Schuster: *Deterministic Chaos: An Introduction*, Second Edition (VCH Verlagsgesellschaft, Weinheim, 1988).
368. L. O. Scott: J. Fin. Quant. Anal. **22**, 419 (1987).
369. J. J. Seneca, H. K. Taussig: *Environmental Economics* (Prentice–Hall, Englewood Cliffs, 1974).
370. A. C. Séror: "Simulation of Complex Organizational Processes: A Review of Methods and Their Epistomological Foundations." In *Simulating Societies: The Computer Simulation of Social Processes*, ed. by N. Gilbert, J. E. Doran (University College London Press, London, 1994), p. 19.
371. C. E. Shannon: Bell Syst. Tech. J. **27**, 379 (1948).
372. R. J. Shiller: *Macro Markets: Creating Institutions for Managing Society's Largest Economic Risks* (Clarendon Press, Oxford, 1993).
373. M. Shubik, R. Levitan: *Market Structure and Behavior* (Harvard University Press, Cambridge, 1980).
374. C. L. Siegel: Ann. Math. **46**, 423 (1945).

375. H. A. Simon: Math. Intell. **10** (1), 11 (1988).
376. H. A. Simon: Math. Intell. **10** (2), 10 (1988).
377. H. A. Simon: Math. Intell. **10** (2), 12 (1988).
378. I. Sinai: Russ. Math. Surv. **25**, 137 (1970).
379. M. Sipper, E. Ruppin: Physica D **99**, 428 (1997).
380. W. C. Snyder: Math. Comput. Simulation **54**, 131 (2000).
381. A. K. Soper: Chem. Phys. **202**, 295 (1996).
382. W. J. Spillman: *The Law of Diminishing Returns* (World Book Co., Yonkers-on-Hudson, 1924).
383. D. Sornette: *Critical Phenomena in Natural Sciences* (Springer, Berlin, 2000).
384. D. Sornette, P. Miltenberger, C. Vanneste: Pure Appl. Geophys. **142**, 491 (1994).
385. D. Sornette, C. Vanneste, L. Knopoff: Phys. Rev. A **45**, 8351 (1992).
386. D. Sornette, A. Sornette: Bull. Seismol. Soc. Am. **89**, 1121 (1999).
387. H. v. Stackelberg: Jahrb. Nationalökon. Stat. **148**, 680 (1938).
388. P. Stalder: *Regime Transitions, Spillover and Buffer Stocks* (Springer, Berlin, 1991).
389. D. Stauffer: J. Phys. A **24**, 909 (1991).
390. G. Straubeta: Phys. Rev. E **53**, 3505 (1996).
391. S. H. Strogatz: *Nonlinear Dynamics and Chaos* (Addison–Wesley, Reading, 1994).
392. J. Sum, C. S. Leung, G. H. Young, W. K. Kan: IEEE Trans. Neural Networks **10**, 161 (1999).
393. H. Szu, R. Harley: Proc. IEEE **75**, 1538 (1987).
394. K. Tam, M. Kiang: Appl. Artif. Intell. **4**, 429 (1990).
395. K. Tam, M. Kiang: Manage. Sci. **38**, 926 (1992).
396. S. Tezuka: "Financial Applications of Monte Carlo and Quasi-Monte Carlo Methods". In *Random and Quasi-Random Point Sets*, ed. by P. Hellekalek, G. Larcher (Springer, New York, 1998), p. 303.
397. J. H. v. Thünen: *Der isolierte Staat in Beziehung auf Nationaleinkommen und Landwirtschaft*, reprint of the 1826 edition (Gustav Fischer, Stuttgart, 1966).
398. B. Townshend: Signal-Processing ICASSP **91**, 429 (1991).
399. W. M. Troburn: Mind **23**, 297 (1915).
400. W. M. Troburn: Mind **26**, 345 (1918).
401. G. W. Trivoli: *Personal Portfolio Management: Fundamentals and Strategies* (Prentice–Hall, Upper Saddle River, 1999).
402. C. Tsallis: J. Stat. Phys. **52**, 479 (1988).
403. C. Tsallis: Chaos, Solitons and Fractals **6**, 539 (1995)
404. C. Tsallis: Braz. J. Phys. **29**, 1 (1999).
405. R. S. Tsay: J. Am. Stat. Assoc. **81**, 590 (1987).
406. A. R. J. Turgot: "Observations sur le Memoire de M. de Saint Péravy." In *Oeuvres de Turgot*, ed. by E. Daire (Guillaumin, Paris, 1844).
407. M. A. Usábel: Insur. Math. Econ. **23** 71 (1998).
408. S. Uvell: Maximum Likelihood Estimates Based on Incomplete Information. (Thesis, University of Umea, Umea, 1975).
409. K. Velupillai: Econ. Hist. **16**, 111 (1973).
410. A. Vercelli: *Methodological Foundations of Macroeconomics* (Cambridge University Press, Cambridge, 1991).
411. P. A. Victor: *Pollution: Economy and Environment* (Allen and Unwin, London, 1972).
412. V. K. Vijay: *An Introduction to Probability Theory and Mathematical Statistics* (Wiley, New York, 1976).

413. T. Vicsek: *Fractal Growth Phenomena*, Second Edition (World Scientific, Singapore, 1992).
414. L. Wagner: Phys. Rev. E **49**, 2115 (1994).
415. L. Walras: *Note on Mr. Wicksteed's Refutation of the English Theory of Rent, Appendix III of Elements deconomie politique pure*, Third Edition (F. Rouge, Lausanne, 1874).
416. J. K. Walters: Phys. Rev. B **53**, 2405 (1996).
417. T. J. Watsham: *Options and Futures in International Portfolio Management* (Chapman and Hall, London, 1992).
418. A. Weber: *Über der Standort der Industrien* (J. C. B. Mohr, Tübingen, 1909).
419. W. Weidlich: Brit. J. Math. Stat. Psychol. **24**, 251 (1971).
420. W. Weidlich: Collect. Phenom. **1**, 51 (1972).
421. A. A. Weiss: Econ. Theory **2**, 107 (1986).
422. D. B. West: *Introduction to Graph Theory* (Prentice–Hall, Englewood Cliffs, 1982).
423. R. E. Whaley: J. Fin. **41**, 127 (1982).
424. M. L. Whicker, L. Sigelman: *Computer Simulation Applications: An Introduction* (Sage, Newbury Park, 1991).
425. J. K. Whitaker: *The Early Economic Writings of Alfred Marshall 1867–1890, Vol. 2* (The Free Press, New York, 1975).
426. K. Wicksell: *Selected Papers on Economic Theory* (Allen and Unwin, London, 1958).
427. P. H. Wicksteed: *An Essay on the Co-ordination of the Laws of Distribution* (Macmillan and Co., London, 1894).
428. B. Widrow, M. E. Hoff: Proc. WESCON Convention **4**, 96 (1960).
429. B. Widrow, M. E. Hoff: Proc. IEEE **78**, 1415 (1990).
430. J. B. Wiggins: J. Fin. Econ. **19**, 351 (1987).
431. S. Wolfram: Nature (London) **311**, 419 (1984).
432. S. Wolfram: *Theory and Application of Cellular Automata* (World Scientific, Singapore, 1986).
433. S. Wolfram: *Cellular Automata and Complexity* (Addison–Wesley, Reading, 1994).
434. F. S. Wong: Neurocomputing **2**, 147 (1990).
435. F. Y. Wu: Rev. Mod. Phys. **54**, 235 (1982).
436. G. U. Yule: Philos. Trans. R. Soc. London A **226**, 267 (1927).
437. D. Zajdenweber: *Hasard et Prévision* (Economica, Paris, 1976).
438. D. Zajdenweber: Fractals **3**, 601 (1995).
439. D. Zajdenweber: Risk Insur. **63**, 95 (1996).
440. D. Zajdenweber: "Scale Invariance in Economics and Finance." In *Scale Invariance and Beyond*, ed. by B. Dubrulle, F. Graner, D. Sornette (EDP Sciences and Springer, Berlin, 1997).
441. J.-M. Zakoian: J. Econ. Dyn. Control **18**, 931 (1994).
442. A. Zellner: *An Introduction to Bayesian Inference in Econometrics* (Wiley, New York, 1971).
443. A. Zellner: *Basic Issues in Econometrics* (University of Chicago Press, Chicago, 1984).
444. J. Zinn-Justin: *Quantum Field Theory and Critical Phenomena* (Clarendon Press, Oxford, 1989).
445. G. K. Zipf: *Human Behavior and the Principle of Least Effort* (Addison–Wesley, Cambridge, 1949).
446. G. O. Zumbach, M. M. Dacorogna, J. L. Olsen, R. B. Olsen: *Introducing a Scale of Market Shocks* (Internal document GOZ.1998-10-01, Olsen & Associates, Seefeldstrasse 233, 8008 Zürich, Switzerland, 1998).

447. G. O. Zumbach, M. M. Dacorogna, J. L. Olsen, R. B. Olsen: *Measuring Shock in Financial Markets* (Internal document GOZ.1999-03-18, Olsen & Associates, Seefeldstrasse 233, 8008 Zürich, Switzerland, 1999).

Index

Springer Tracts in Modern Physics

146 **Low-Energy Ion Irradiation of Solid Surfaces**
By H. Gnaser 1999. 93 figs. VIII, 293 pages

147 **Dispersion, Complex Analysis and Optical Spectroscopy**
By K.-E. Peiponen, E.M. Vartiainen, and T. Asakura 1999. 46 figs. VIII, 130 pages

148 **X-Ray Scattering from Soft-Matter Thin Films**
Materials Science and Basic Research
By M. Tolan 1999. 98 figs. IX, 197 pages

149 **High-Resolution X-Ray Scattering from Thin Films and Multilayers**
By V. Holý, U. Pietsch, and T. Baumbach 1999. 148 figs. XI, 256 pages

150 **QCD at HERA**
The Hadronic Final State in Deep Inelastic Scattering
By M. Kuhlen 1999. 99 figs. X, 172 pages

151 **Atomic Simulation of Electrooptic and Magnetooptic Oxide Materials**
By H. Donnerberg 1999. 45 figs. VIII, 205 pages

152 **Thermocapillary Convection in Models of Crystal Growth**
By H. Kuhlmann 1999. 101 figs. XVIII, 224 pages

153 **Neutral Kaons**
By R. Beluševi 1999. 67 figs. XII, 183 pages

154 **Applied RHEED**
Reflection High-Energy Electron Diffraction During Crystal Growth
By W. Braun 1999. 150 figs. IX, 222 pages

155 **High-Temperature-Superconductor Thin Films at Microwave Frequencies**
By M. Hein 1999. 134 figs. XIV, 395 pages

156 **Growth Processes and Surface Phase Equilibria in Molecular Beam Epitaxy**
By N.N. Ledentsov 1999. 17 figs. VIII, 84 pages

157 **Deposition of Diamond-Like Superhard Materials**
By W. Kulisch 1999. 60 figs. X, 191 pages

158 **Nonlinear Optics of Random Media**
Fractal Composites and Metal-Dielectric Films
By V.M. Shalaev 2000. 51 figs. XII, 158 pages

159 **Magnetic Dichroism in Core-Level Photoemission**
By K. Starke 2000. 64 figs. X, 136 pages

160 **Physics with Tau Leptons**
By A. Stahl 2000. 236 figs. VIII, 315 pages

161 **Semiclassical Theory of Mesoscopic Quantum Systems**
By K. Richter 2000. 50 figs. IX, 221 pages

162 **Electroweak Precision Tests at LEP**
By W. Hollik and G. Duckeck 2000. 60 figs. VIII, 161 pages

163 **Symmetries in Intermediate and High Energy Physics**
Ed. by A. Faessler, T.S. Kosmas, and G.K. Leontaris 2000. 96 figs. XVI, 316 pages

164 **Pattern Formation in Granular Materials**
By G.H. Ristow 2000. 83 figs. XIII, 161 pages

165 **Path Integral Quantization and Stochastic Quantization**
By M. Masujima 2000. 0 figs. XII, 282 pages

166 **Probing the Quantum Vacuum**
Pertubative Effective Action Approach in Quantum Electrodynamics and its Application
By W. Dittrich and H. Gies 2000. 16 figs. XI, 241 pages

167 **Photoelectric Properties and Applications of Low-Mobility Semiconductors**
By R. Könenkamp 2000. 57 figs. VIII, 100 pages

Springer Tracts in Modern Physics

168 **Deep Inelastic Positron-Proton Scattering in the High-Momentum-Transfer Regime of HERA**
By U.F. Katz 2000. 96 figs. VIII, 237 pages

169 **Semiconductor Cavity Quantum Electrodynamics**
By Y. Yamamoto, T. Tassone, H. Cao 2000. 67 figs. VIII, 154 pages

170 **d–d Excitations in Transition-Metal Oxides**
A Spin-Polarized Electron Energy-Loss Spectroscopy (SPEELS) Study
By B. Fromme 2001. 53 figs. XII, 143 pages

171 **High-T_c Superconductors for Magnet and Energy Technology**
By B. R. Lehndorff 2001. 139 figs. XII, 209 pages

172 **Dissipative Quantum Chaos and Decoherence**
By D. Braun 2001. 22 figs. XI, 132 pages

173 **Quantum Information**
An Introduction to Basic Theoretical Concepts and Experiments
By G. Alber, T. Beth, M. Horodecki, P. Horodecki, R. Horodecki, M. Rötteler, H. Weinfurter, R. Werner, and A. Zeilinger 2001. 60 figs. XI, 216 pages

174 **Superconductor/Semiconductor Junctions**
By Thomas Schäpers 2001. 91 figs. IX, 145 pages

175 **Ion-Induced Electron Emission from Crystalline Solids**
By Hiroshi Kudo 2002. 85 figs. IX, 161 pages

176 **Infrared Spectroscopy of Molecular Clusters**
An Introduction to Intermolecular Forces
By Martina Havenith 2002. 33 figs. VIII, 120 pages

177 **Applied Asymptotic Expansions in Momenta and Masses**
By Vladimir A. Smirnov 2002. 52 figs. IX, 263 pages

178 **Capillary Surfaces**
Shape – Stability – Dynamics, in Particular Under Weightlessnes
By Dieter Langbein 2002. 182 figs. XVIII, 364 pages

179 **Anomalous X-ray Scattering for Materials Characterization**
Atomic-Scale Structure Determination
By Yoshio Waseda 2002. 132 figs. XIV, 214 pages

180 **Coverings of Discrete Quasiperiodic Sets**
Theory and Applications to Quasicrystals
Edited by P. Kramer and Z. Papadopolos 2002. 128 figs., XIV, 274 pages

181 **Emulsion Science**
Basic Principles. An Overview
By J. Bibette, F. Leal-Calderon, V. Schmitt, and P. Poulin 2002. 50 figs., IX, 140 pages

182 **Transmission Electron Microscopy of Semiconductor Nanostructures**
An Analysis of Composition and Strain State
By A. Rosenauer 2003. 136 figs., XII, 238 pages

183 **Transverse Patterns in Nonlinear Optical Resonators**
By K. Staliunas, V. J. Sánchez Morcillo 2003. 132 figs., XII, 226 pages

184 **Statistical Physics and Economics**
Concepts, Tools and Applications
By M. Schulz 2003. 45 figs., XII, 250 pages

185 **Electronic Defect States in Alkali Halides**
Effects of Interaction with Molecular Ions
By V. Dierolf 2003. 79 figs., XII, 196 pages

186 **Electron-Beam Interactions with Solids**
Application of the Monte Carlo Method to Electron Scattering Problems
By M. Dapor 2003. 27 figs., XII, 116 pages